Encyclopaedia of Mathematical Sciences

Volume 69

Editor-in-Chief: R. V. Gamkrelidze

W. Barth R. Narasimhan (Eds.)

Several Complex Variables VI

Complex Manifolds

Springer-Verlag
Berlin Heidelberg New York
London Paris Tokyo
Hong Kong Barcelona

Title of the Russian edition:
Itogi nauki i tekhniki, Sovremennye problemy matematiki,
Fundamental'nye napravleniya, Vol. 69
Kompleksnyj analiz - mnogie peremennye 6
Publisher VINITI, Moscow

Mathematics Subject Classification (1980):
32 Axx, 32 Cxx, 32 Hxx, 32 Lxx, 32 Mxx, 14 Jxx

With 4 Figures

ISBN 3-540-52788-5 Springer-Verlag Berlin Heidelberg New York
ISBN 0-387-52788-5 Springer-Verlag New York Berlin Heidelberg

Printing: Druckhaus Beltz, Hemsbach/Bergstr.; binding: Schäffer, Grünstadt
2141/3140-543210 – Printed on acid-free paper

A Volume in Honour
of
Reinhold Remmert

List of Editors and Contributors

Editor-in-Chief

R. V. Gamkrelidze, Academy of Sciences of the USSR, Steklov Mathematical Institute, ul. Vavilova 42, 117966 Moscow, Institute for Scientific Information (VINITI), ul. Usievicha 20a, 125219 Moscow, USSR

Consulting Editors

W. Barth, Universität Erlangen-Nürnberg, Mathematisches Institut, Bismarckstraße 1½, D-8520 Erlangen, FRG

R. Narasimhan, The University of Chicago, Department of Mathematics, 5734 University Avenue, Chicago, IL 60637, USA

Contributors

S. R. Bell, Purdue University, Department of Mathematics, Mathematical Sciences Building, West Lafayette, IN 47907, USA

J.-L. Brylinski, The Pennsylvania State University, Department of Mathematics, 305 McAllister Building, University Park, PA 16802, USA

A. T. Huckleberry, Fakultät und Institut für Mathematik, Ruhr-Universität Bochum, Universitätsstraße 150, D-4630 Bochum 1, FRG

R. Narasimhan, The University of Chicago, Department of Mathematics, 5734 University Avenue, Chicago, IL 60637, USA

C. Okonek, Mathematisches Institut der Universität Bonn, Wegelerstraße 10, D-5300 Bonn 1, FRG

G. Schumacher, Fakultät und Institut für Mathematik, Ruhr-Universität Bochum, Universitätsstraße 150, D-4630 Bochum 1, FRG

A. Van de Ven, Mathematisch Institut, Rijksuniversiteit Leiden, Wassenaarseweg 80, NL-2300 RA Leiden, The Netherlands

S. Zucker, The Johns Hopkins University, Department of Mathematics, Baltimore, MD 21218, USA

Preface

This volume was originally intended to appear in June 1990 as an independent book. During the final stages of production we felt that it would be a valuable addition to this series, and we would reach a wider audience with its publication in English (Springer-Verlag), as well as in Russian (VINITI).

I am very glad that the authors of the articles and the editors of the volume unanimously agreed to the suggestion about its inclusion in the Encyclopaedia. We hope the volume will thus be of use to a larger section of the mathematical community.

August 1990 *R. Gamkrelidze*

Foreword

This volume is to appear as Reinhold Remmert starts out on his seventh decade. There are five contributions, each dealing with a different aspect of Complex Analysis. These essays are not meant to present new results; rather, they are meant to facilitate access to some, at least, of the many branches of the subject by collecting together those topics which, in the author's view, seemed the most useful as an introduction to the respective fields. Forty years ago, the fundamental problems of Complex Analysis still awaited solution, and it was essential to find and sharpen the appropriate tools. Today, it is difficult to discern even the main directions of the subject and their mutual relations. Hence the essays in this volume are addressed to mathematicians who are not specialists in these aspects of complex analysis.

If complex analysis now appears as a well organized modern branch of mathematics, Reinhold Remmert was among the few who made it so. His discoveries, his high standards and his books and editorial work have influenced not only complex analysis, but many other fields of mathematics as well.

Remmert's work touches, and unites, the most diverse aspects of complex analysis. The subject is of the greatest importance because of its long and distinguished history and because of its connections with so many branches of science, both inside and outside mathematics. Thus, it is not surprising if there is a tendency for different parts of the subject to diverge in different directions. Keeping these tendencies in balance was always one of Remmert's main intentions.

We hope that the present volume will reflect at least some of the richness and balance of Complex Analyis. The volume is dedicated to Reinhold Remmert by his many friends, students, and colleagues as a token of their gratitude and admiration.

June 1990

Wolf Barth
Raghavan Narasimhan

Table of Contents

Proper Holomorphic Mappings of Complex Spaces

Steven R. Bell* and Raghavan Narasimhan

Contents

* Research supported by NSF Grant DMS 8420754 and an AMS Fellowship

Introduction

In this article, we attempt to describe some of the most important results concerning proper holomorphic mappings between complex spaces. The first two sections deal respectively with the Remmert Proper Mapping Theorem and the Grauert Direct Image Theorem. These are certainly the most important results of what may be called the general theory. Sect. 3 deals with embeddings in $\mathbb{C}^N$. Finally, Sect. 4 treats proper maps between bounded domains, which is, perhaps, the domain of greatest recent and current research. We have not dealt with this in as great detail as we might have because of the availability of the excellent survey articles of Bedford [Bed] and Forstnerič [FF 1].

We shall be using the terminology of several complex variables without defining the standard notions. For complex spaces, coherent sheaves and Stein and holomorphically convex spaces, one may consult the books of Grauert-Remmert [GR 1, GR 2]. For pseudoconvexity, strict (= strong) pseudoconvexity and plurisubharmonic functions, one may consult the notes of Fornaess-Stensønes [FS].

The study of proper holomorphic maps between complex spaces is of great importance for several reasons. [We shall often call them, simply, proper maps.] In the first place, the existence of a non-trivial proper map from one complex space X to another, Y, often places strong restrictions on these spaces. Examples of this are:

1. If X, Y are compact Riemann surfaces and if there is a non-constant holomorphic map $\pi : X \to Y$, then, the genus of Y is $\leq$ that of X.
2. If X is a Stein space and $\pi : X \to Y$ a proper map, then $\pi(X)$ is again a Stein space. [If Y is Stein, then X need only be holomorph-convex in general.]

Further, even when proper maps exist, they tend to be "rigid" and rather few in number. For instance, if $n \geq 2$, a proper map of the unit ball in $\mathbb{C}^n$ into itself is an analytic automorphism.

On the other hand, proper holomorphic mappings do occur in important contexts. We now mention some of these contexts.

1. Finite Mappings. Let X, Y be complex spaces. A holomorphic map $\pi : X \to Y$ is said to be *finite* if π is proper and $\pi^{-1}(y)$ is finite $\forall y \in Y$.

If $\pi : X \to Y$ is finite, then $\pi(X)$ is an analytic set in Y (see Sect. 1 below) and has the following property: if $x \in X$ and $y = \pi(x)$, then the local ring $\mathcal{O}_{X,x}$ of X at x is a *finite module* over the local ring $\mathcal{O}_{Y,y}$ of Y at y. A finite map $\pi : X \to Y$ is called a *(finite) ramified covering* if X, Y are reduced, if $\pi(X) = Y$ and if $\dim_x X = \dim_y Y$ for all $x \in X$, $y \in Y$. In this case, there exists a nowhere dense analytic set $B \subset Y$ such that $\pi | X - \pi^{-1}(B) \to Y - B$ is a finite *unramified* covering and $\pi^{-1}(B)$ is nowhere dense in X. The smallest such B is called the branch locus of π; further, if $x \in X - \pi^{-1}(B)$ and $y = \pi(x)$, then the induced map $\pi^* : \mathcal{O}_{Y,y} \to \mathcal{O}_{X,x}$ is an isomorphism (i.e. π is a local isomorphism at x). If $\pi : X \to Y$ is a ramified covering and Y is irreducible, then the cardinality of $\pi^{-1}(y)$ is independent of $y \in Y - B$ and is called the *degree* of π (other names being the *number of sheets* or *multiplicity*).

Examples are, of course, provided by finite unramified coverings. Moreover, any irreducible projective variety of dimension n can be realized as a ramified covering of the (complex) projective space $\mathbb{P}^n$, while any irreducible affine variety of dimension n can be realized as a ramified covering of $\mathbb{C}^n$. Also, any pure dimensional complex space (of dimension n) can *locally* be realized as a ramified covering of an open set in $\mathbb{C}^n$. (These are versions of the so-called Noether normalization lemma.)

In this connection, we mention a deep theorem of Grauert and Remmert [GR 3] which implies that a ramified covering $\pi : X \to \mathbb{C}^n$ is an affine variety if the closure in $\mathbb{P}^n$ of the branch locus of π is an analytic set in $\mathbb{P}^n$. Their theorem can be formulated as follows:

Let S be a connected normal complex space, $T \subset S$ a proper analytic set. Let $Y = S - T$, let X be a normal complex space and $\pi : X \to Y$, a ramified covering. Then π extends to a ramified covering $\tilde{\pi} : \tilde{X} \to S$ (in an obvious sense) if and only if the closure $\overline{B}$ in S of the branch locus $B \subset Y$ of π is an analytic set in Y.

It is not hard to find a topological extension $\tilde{\pi} : \tilde{X} \to S$; this is done in just the same way that one adds algebraic branch points to the "Riemann domain" of a holomorphic function. There is also a sheaf $\mathcal{O}_{\tilde{X}}$ of $\mathbb{C}$-algebras of continuous functions on $\tilde{X}$ forming the natural candidate for the analytic extension: it consists of continuous functions on open sets in $\tilde{X}$ which satisfy monic polynomial equations whose coefficients are holomorphic functions on open sets in S. The proof given by Grauert-Remmert that $(\tilde{X}, \mathcal{O}_{\tilde{X}})$ is an analytic space is difficult. The central point is to find, given any $t_0 \in T$, sections of $\mathcal{O}_{\tilde{X}}$ in a neighbourhood of $\tilde{\pi}^{-1}(t_0)$ separating the points of $\tilde{\pi}^{-1}(y) = \pi^{-1}(y)$ for a general point $y \in S - T$ near t_0.

It may be worth remarking that this can be done rather simply using Skoda's results [Sk] on L^2 methods for the $\bar{\partial}$-operator. Choose a small connected Stein neighbourhood U of t_0 and a holomorphic function f on U vanishing on $(T \cup \overline{B}) \cap U$ as well as the set of singular points of U; we may assume that $f \not\equiv 0$. Let $A = \{s \in U | f(s) = 0\}$. Skoda's results imply easily that there exist holomorphic functions F on $\pi^{-1}(U - A)$ separating points of $\pi^{-1}(y_0)$ for a given $y_0 \in U - A$, and growing at most like a fixed negative power of the distance to $\tilde{\pi}^{-1}(A)$ at points of $\tilde{\pi}^{-1}(A)$. Multiplying F by a large power of f gives us the function being sought.

2. Embeddings. A holomorphic map $f : X \to Y$ is called (a closed or proper) embedding if f is proper, injective and has the following property: $\forall x \in X$ and $h \in \mathcal{O}_{X,x}$, there exists $\tilde{h} \in \mathcal{O}_{Y,f(x)}$ such that $h = \tilde{h} \circ f$. When X and Y are manifolds, the latter condition means simply that the rank of the tangent map of f at x equals $\dim_x X$.

The complex spaces which can be embedded in $\mathbb{P}^N$ are exactly the projective algebraic varieties; those which can be embedded in $\mathbb{C}^N$ for some N are exactly the Stein spaces whose (Zariski) tangent spaces have bounded dimension (see Sect. 3 below for a precise statement). They include Stein *manifolds.*

3. Birational Morphisms and Modifications. Let X, Y be complex spaces and $\pi : X \to Y$, a *proper* holomorphic map. π is called a *birational morphism* if there is a nowhere dense analytic set $A \subset Y$ such that $\pi^{-1}(A)$ is nowhere dense in X and $\pi|X - \pi^{-1}(A) \to Y - A$ is an analytic isomorphism (= biholomorphic map). If, in addition, $R^0\pi_*(\mathcal{O}_X) = \mathcal{O}_Y$ (see Sect. 2 for the definition of the direct image $R^0\pi_*(\mathcal{O}_X)$), π is called a *modification.* If Y is normal, this last condition is automatic and the fibres of π are then *connected* ("Zariski's main theorem" as it is called in algebraic geometry).

Examples are provided by blowing up a point (or, more generally, an analytic subset of codimension ≥ 2) in Y.

4. Deformations. Let X_0 be a compact complex space, S a complex space and let $s_0 \in S$ be a base point. Consider a proper map $\pi : X \to S$ together with an analytic isomorphism of $\pi^{-1}(s_0)$ onto X_0. Then $\pi : X \to S$ is called a *deformation* of X_0 if the map π is flat, i.e. if $\forall x \in X$, the local ring $\mathcal{O}_{X,x}$ is a flat module over the local ring $\mathcal{O}_{S,\pi(x)}$. This is the case if X, S are smooth and π is a submersion, in which case one speaks of smooth deformations.

Although we shall mention in Sect. 2 some results of importance in deformation theory, we shall not pursue the subject in this article. The literature is very extensive. For the deformation theory of manifolds, see the book of Kodaira [Kod] and the references given there (work of Kodaira-Spencer, Kuranishi ...). The general case is a transposition to the analytic category of Grothendieck's framework in algebraic geometry, although the analytic theory often involves difficult convergence proofs. See e.g. Sém. Cartan [C 3], the seminar of Douady-Verdier [DV], the paper of Forster [Fo 2] and the references given there.

5. Proper Holomorphic Mappings Between Bounded Domains in $\mathbb{C}^n$. We shall discuss these in some detail later (Sect. 4); see also the survey articles of Bedford [Bed] and Forstnerič [FF 1]. While they impose strong restrictions on the domains in question, it has turned out recently that strictly pseudoconvex domains admit many more proper maps into balls or polydiscs than was previously suspected.

1. Remmert's Proper Mapping Theorem

The first basic theorem is due to R. Remmert [R 1] and may be thought of as a far reaching generalization of classical elimination theory.

Theorem 1.1. *Let X, Y be complex spaces and $\pi : X \to Y$ a proper holomorphic map. Then $\pi(X)$ is an analytic subset of Y.*

We shall sketch a proof of this theorem. A slightly stronger version is proved in Whitney [Wh]. The theorem is also an immediate consequence of Grauert's direct image theorem (see Sect. 2). Further, some of the basic ideas are explained in Chapters 14, 15 of Rudin [Ru] in the case of mappings between domains in $\mathbb{C}^N$.

The main tool is the following extension theorem of Remmert-Stein [RS 1].

Theorem 1.2. *Let X be a complex space, p an integer ≥ 1 and A an analytic subset of X such that $\dim_x A \leq p-1$ $\forall x \in A$. Let Y be an analytic subset of $X - A$. Then, if $\dim_x Y \geq p$ $\forall x \in Y$, the topological closure $\overline{Y}$ of Y in X is an analytic subset of X.*

Actually, Remmert and Stein proved a stronger theorem generalizing an earlier theorem of P. Thullen. It asserts that if A is analytic in X and Y is analytic in $X - A$, and if $\dim_y Y = \dim_a A$ $\forall a \in A, y \in Y$, then $\overline{Y}$ is analytic in X if and only if $\overline{Y}$ does not contain any irreducible component of A (equivalently, if and only if $\overline{Y}$ is analytic in the neighbourhood of at least one point on each irreducible component of A). The case when X is a manifold and A has codimension 1 at each of its points is the one treated by Thullen.

Sketch of a Proof of Theorem 1.2. We may assume that X is an open set in $\mathbb{C}^n$. An easy induction on p shows that it is enough to consider the case when $A = X \cap \{z_p = \ldots = z_n = 0\}$ ($(z_1, \ldots, z_n)$ being coordinate functions on $\mathbb{C}^n$) and $\dim_y Y = p$ $\forall y \in Y$; in fact, we have only to remark that the singular locus of A has dimension $< p-1$.

Let $\lambda : \mathbb{C}^n \to \mathbb{C}^p$ be a generic (linear) projection. Then $\lambda^{-1}(0) \cap (A \cup Y)$ is countable, and there is a countable dense set $\{y_\nu\} \subset Y$ so that y_ν is isolated in $\lambda^{-1}\lambda(y_\nu)$. Choose coordinates $w_1, \ldots, w_p, w_{p+1}, \ldots, w_n$ in $\mathbb{C}^n$, the first p of the w_j being coordinates for $\mathbb{C}^p$, and numbers $\varepsilon > 0$, $\delta > 0$ such that

$$(A \cup Y) \cap \{|w_1| \leq \varepsilon, \ldots, |w_p| \leq \varepsilon,\ |w_{p+1}|^2 + \ldots + |w_n|^2 = \delta\} = \emptyset$$

(possible since we can choose $\delta > 0$ such that the intersection is empty when $\varepsilon = 0$ because $\lambda^{-1}(0) \cap (A \cup Y)$ is countable). Let $U' = \{w \in \mathbb{C}^p \mid |w_j| < \varepsilon, j = 1, \ldots, p\}$, $U'' = \{(w_{p+1}, \ldots, w_n) \in \mathbb{C}^{n-p} \mid \Sigma |w_k|^2 < \delta\}$, and $U = U' \times U''$. Then $\lambda | (A \cup Y) \to U'$ is proper and $\lambda(A)$ is a hyperplane in U' (of dimension $p-1$). It follows that $\lambda | Y - \lambda^{-1}\lambda(A) \to U' - \lambda(A)$ is a finite ramified covering. Let d be the degree of this map, and let f be holomorphic on Y. Then, there are functions $a_1, \ldots, a_d$ holomorphic on $U' - \lambda(A)$ with the following property: if we set

$$P_f(w, W) = W^d + a_1(w_1, \ldots, w_p) W^{d-1} + \ldots + a_d(w_1, \ldots, w_p),$$

then $P_f(w, f(w)) \equiv 0$ on $Y - \lambda^{-1}\lambda(A)$. In fact, we may take $a_\nu(w_1, \ldots, w_p) = \nu^{\text{th}}$ elementary symmetric function in $f(x_1), \ldots, f(x_d)$, where $\{x_1, \ldots, x_d\}$ are the points of $Y - \lambda^{-1}\lambda(A)$ with $\lambda(x_j) = (w_1, \ldots, w_p)$ when $(w_1, \ldots, w_p)$ is outside the branch locus. One now checks, using the Riemann extension theorem and the continuous dependence of the roots of a polynomial on its coefficients, the following two facts: (i) if f is the restriction to Y of a function holomorphic on U, then the a_ν are holomorphic on U', so that $\widetilde{f}(w) = P_f(w, f(w))$ defines a function holomorphic on U; (ii) $U \cap \overline{Y}$ is the set of common zeros of the $\widetilde{f}$ when f runs over all holomorphic functions on U.

This, of course, implies Theorem 1.2.

Sketch of a Proof of Theorem 1.1. We start with the following remark. If X is a pure dimensional complex space and $\pi : X \to Y$ is a holomorphic map, define the *rank of* π at x by

$$\operatorname{rk}_x(\pi) = \dim_x X - \dim_x(\pi^{-1}\pi(x)), \qquad x \in X$$

and the *rank of* π by $\operatorname{rk}(\pi) = \sup_{x \in X} \operatorname{rk}_x(\pi)$. Then, if π maps X *onto* Y, we have $\dim(Y) = \operatorname{rk}(\pi)$. [In fact, Theorem 1.2 implies easily that the $\sup \operatorname{rk}_x(\pi)$ is attained on any open dense set of X, in particular at a smooth point. One can then use the implicit function theorem.]

To prove Remmert's theorem, we may assume that Y is an open set in $\mathbb{C}^n$ and that X is irreducible. We use induction on $\dim X$. Let S be the singular locus of X and $T = \pi(S)$; T is analytic by induction hypothesis. If $\pi^{-1}(T) = X$, the theorem is proved. If $\pi^{-1}(T) \neq X$, then $\operatorname{rk}(\pi|\pi^{-1}(T)) < \operatorname{rk}(\pi)$, so that $\dim T < \operatorname{rk}(\pi)$. Thus, because of the Remmert-Stein theorem, it suffices to show that $\pi(X - \pi^{-1}(T))$ is analytic in $Y - T$ (of dimension $\operatorname{rk}(\pi|X - \pi^{-1}(T)) = \operatorname{rk}(\pi)$). If $B \subset X - \pi^{-1}(T)$ is the set of points where the jacobian matrix of π has rank $< \operatorname{rk}(\pi)$, then B is analytic and $\dim B < \dim X$. By induction, $\pi(B)$ is analytic in $Y - T$. Repeating the argument given above with S replaced by B, we may assume that $\pi^{-1}\pi(B) \neq X - \pi^{-1}(T)$ and must show that $\pi(X - \pi^{-1}(T) - \pi^{-1}(B))$ is analytic of dimension $\operatorname{rk}(\pi)$ in $Y - T - \pi(B)$. This follows at once from the rank theorem.

Theorem 1.2 has several other applications. First, if $\pi : X \to Y$ is a holomorphic map, then, for any integer $k \geq 0$, the set $\{x \in X \mid \dim_x \pi^{-1}\pi(x) \geq k\}$ is closed in X (this was used in the proof of the remark at the beginning of our sketch of Theorem 1.1). This set is, in fact, an analytic subset of X.

Secondly, H. Cartan pointed out that Theorem 1.2 implies Chow's theorem that *any analytic subset* V *of* $\mathbb{P}^n$ *is algebraic*. In fact, Theorem 1.2 implies that the cone $C(V)$ over V in $\mathbb{C}^{n+1}$ is analytic at 0. If $\{f_1 = \ldots = f_N = 0\}$ are local holomorphic equations of $C(V)$ at 0, we expand each f_j in a series of homogeneous polynomials $f_j = \sum_{k=0}^{\infty} f_{jk}$, f_{jk} being a homogeneous polynomial of degree k. V is then defined by $\{z \in \mathbb{P}^N \mid f_{jk}(z) = 0 \ \forall j, k\}$.

2. Grauert's Direct Image Theorem

Let X, Y be complex spaces and let $\pi : X \to Y$ be a holomorphic map. If $\mathscr{F}$ is a sheaf of abelian groups on X and q is an integer, we denote by $R^q\pi_*(\mathscr{F})$ the sheaf on Y associated to the presheaf $U \mapsto H^q(\pi^{-1}(U), \mathscr{F})$, U open in Y. [Of course $R^q\pi_*(\mathscr{F}) = 0$ for $q < 0$.] If $\mathscr{F}$ is a sheaf of $\mathcal{O}_X$-modules, then $R^q\pi_*(\mathscr{F})$ is, in a natural way, a sheaf of $\mathcal{O}_Y$-modules on Y. $R^q\pi_*(\mathscr{F})$ is called the q-th direct image of $\mathscr{F}$ under π. When $q = 0$, we sometimes denote $R^0\pi_*(\mathscr{F})$ by $\pi_*(\mathscr{F})$ and call it simply the direct image of $\mathscr{F}$. If U is open in Y, we have $\Gamma(U, \pi_*(\mathscr{F})) = \Gamma(\pi^{-1}(U), \mathscr{F})$; the corresponding statement is false in general when $q > 0$.

It was shown by Grothendieck that if π is a proper algebraic morphism between algebraic varieties X, Y, then, for any $\mathcal{O}_X$-coherent sheaf $\mathscr{F}$, all the direct images $R^q\pi_*(\mathscr{F})$ are $\mathcal{O}_Y$-coherent. He pointed out the importance of this result and raised the question of whether a corresponding result holds for complex spaces. This was proved by Grauert [G 3] in 1960.

Theorem 2.1. *Let $\pi : X \to Y$ be a proper holomorphic map between the complex spaces X and Y. Let $\mathscr{F}$ be a coherent analytic sheaf on X. Then, for any $q(\geq 0)$, the q-th direct image $R^q\pi_*(\mathscr{F})$ is a coherent analytic sheaf on Y.*

Grauert's proof of this theorem is difficult. A variant was published by K. Knorr [Kn] in 1971. Simpler proofs have since been found (Kiehl-Verdier [KV], Forster-Knorr [FK]; see also the book by Grauert-Remmert [GR 2]).

We shall briefly indicate the formalism used by Forster-Knorr [FK] and state the technical results to which they reduce the proof.

Let Ω be an open set in $\mathbb{C}^n$ and $\pi : X \to \Omega$ a proper holomorphic map. Let $\mathscr{F}$ be a coherent analytic sheaf on X. By restricting Ω, we can find a finite *atlas*, consisting of triples $(U_k, j_k, D_k(r) \times \Omega)$, $(k = 1, \ldots, N)$, where U_k is open in X, $D_k(r) = \{z \in \mathbb{C}^{N_k} \mid |z_j| < r\}$, $j_k : U_k \to D_k(r) \times \Omega$ is an embedding such that $\pi|U_k = pr_2 \circ j_k$ (pr_2 being projection on Ω) and $X = \cup_k U_k$. If $V \subset \Omega$, we set $X(V) = \pi^{-1}(V)$ and, for $0 < \rho \leq r$, we set $U_k(\rho, V) = j_k^{-1}(D_k(\rho) \times V)$. If $C^\bullet(\mathscr{U}(\rho, V), \mathscr{F})$ is the Čech (alternating) cochain complex relative to the covering $\mathscr{U}(\rho, V) = \{U_k(\rho, V)\}$ of $X(V)$ (ρ close to r), and V is Stein, then $H^q(C^\bullet(\mathscr{U}(\rho, V), \mathscr{F}) \simeq H^q(X(V), \mathscr{F})$. Fix $0 < r_0 < r$ such that this holds for $r_0 \leq \rho \leq r$.

We now consider what Forster and Knorr call *linked systems of sheaves* (relative to this atlas).

Let A_q denote the class of increasing sequences $k_0 < k_1 < \ldots < k_q$ (of $q+1$ elements) where $1 \leq k_i \leq N$. If $\alpha \in A_q$, we set $U_\alpha(\rho, V) = \cap_{i=0}^q U_{k_i}(\rho, V)(\alpha = (k_0, \ldots, k_q))$, $D_\alpha(\rho) = \prod_{i=0}^q D_{k_i}(\rho)$. The product of the j_{k_i} induces an embedding $j_\alpha : U_\alpha(\rho, V) \to D_\alpha(\rho) \times V$. If $\alpha \in A_q, \beta \in A_{q'}$, and $\beta \subset \alpha$, let $\pi_{\beta\alpha} : D_\alpha(\rho) \times V \to D_\beta(\rho) \times V$ be the projection onto the factors $D_{k_i}(\rho)$ occurring in β. A linked system of sheaves is now a family of sheaves $\{\mathscr{G}_\alpha\}$ (indexed by $\cup_{q \geq 0} A_q$), where $\mathscr{G}_\alpha$ is a coherent sheaf on $D_\alpha(r) \times \Omega$, together with a family of morphisms

$\{\varphi_{\alpha\beta}\}_{\beta\subset\alpha}$:

$$\varphi_{\alpha\beta} : \mathscr{G}_\beta \to (\pi_{\beta\alpha})_*\mathscr{G}_\alpha$$

such that $\varphi_{\alpha\alpha} = id$ and, if $\gamma \subset \beta \subset \alpha$, $\varphi_{\alpha\gamma} = ((\pi_{\gamma\beta})_*\varphi_{\alpha\beta}) \circ \varphi_{\beta\gamma}$.

A coherent sheaf $\mathscr{F}$ on X induces a linked system $j_*(\mathscr{F}) : j_*(\mathscr{F})_\alpha = (j_\alpha)_*(\mathscr{F})$.

Morphisms between linked systems are defined in the obvious way. We define the chain complex $C^\bullet(\rho, V, \mathscr{G})$ with $C^q = \prod_{\alpha\in A_q} \Gamma(D_\alpha(\rho) \times V, \mathscr{G}_\alpha)$ and the "usual" coboundary map. We say that a linked system $\{\mathscr{G}_\alpha\}$ is free (of finite rank) if each $\mathscr{G}_\alpha$ is.

It is not hard to show, using Theorem B for Stein spaces, that if $V \Subset \Omega$ is Stein and $r_0 \le \rho_0 < r$, then, relative to $\{D_\alpha(\rho_0) \times V\}$, there exists a resolution

$$\cdots \mathscr{L}^{p+1} \to \mathscr{L}^p \to \ldots \to \mathscr{L}^0 \to j_*(\mathscr{F}) \to 0$$

such that each $\mathscr{L}^p$ is a linked system which is free of finite rank.

Let $\rho_0 > r_0$ and let $C^\bullet(\rho, V)$ $(r_0 \le \rho \le \rho_0)$ be the total complex associated to the double complex $\{C^q(\rho, V, \mathscr{L}^p)\}$, i.e., $C^n(\rho, V) = \prod_{q-p=n} C^q(\rho, V, \mathscr{L}^p)$]. Projection on $C^\bullet(\rho, V, \mathscr{L}^0)$ followed by composition with the map $\mathscr{L}^0 \to j_*(\mathscr{F})$ defines a morphism of complexes $C^\bullet(\rho, V) \to C^\bullet(\rho, V, j_*(\mathscr{F}))$. The spectral sequence of the double complex $\{C^q(\rho, V, \mathscr{L}^p)\}$ degenerates (because of Theorem B for Stein spaces), so that, for $r_0 \le \rho \le \rho_0$, the morphism $C^\bullet(\rho, V) \to C^\bullet(\rho, V, j_*(\mathscr{F}))$ induces an isomorphism of the q-th cohomology groups of the two complexes for all q.

Note: A morphism of chain complexes inducing isomorphisms on the cohomology groups is called a *quasi-isomorphism.*

The main part of the proof consists in the following two propositions.

Proposition $I(n)$. *Let K be a compact subset of V. Then, there exist: a Stein open set V_n with $K \subset V_n \subset V$, a number ρ_n with $r_0 < \rho_n < \rho_0$, a complex $\mathscr{E}^\bullet$ of finitely generated free $\mathcal{O}_{V_n}$-modules of the form*

$$0 \to \mathscr{E}^n \to \mathscr{E}^{n+1} \to \ldots \to \mathscr{E}^N \to 0$$

and a morphism $\mathscr{E}^\bullet(V_n) \to C^\bullet(\rho_n, V_n)$ which induces isomorphisms on the q-th cohomology groups for all $q > n$.

Let $K^\bullet(\rho)$ be the cone of the morphism $u : \mathscr{E}^\bullet \to C^\bullet(\rho)$ given by $I(n)$ for $r_0 < \rho \le \rho_n$ [i.e. $K^n(\rho) = C^p(\rho) \oplus \mathscr{E}^{p+1}$ with differential given by $(d_c^p - u^{p+1}, d_{\mathscr{E}}^{p+1})$ (with the obvious notation)]. Let $Z^p(\rho) \subset K^p(\rho)$ be the space of p-cocycles of $K^\bullet(\rho)$.

Proposition $II(n-1)$. *If $W \Subset V_n$ is a Stein open set and $r_0 < \rho' < \rho \le \rho_n$, there exists a morphism $\psi : K^{n-1}(\rho) \to Z^{n-1}(\rho')$ of $\mathcal{O}_W$-modules such that $\psi | Z^{n-1}(\rho)$ is the restriction map from ρ to ρ'.*

These are proved by descending induction on n. One shows that $I(n)$ and $II(n) \Rightarrow II(n-1)$ and that $I(n)$ and $II(n-1) \Rightarrow I(n-1)$; for the details, see Forster-Knorr [FK]. Grauert's theorem follows from Proposition $I(-1)$.

The above scheme actually suggests a stronger form of Grauert's theorem. Grothendieck had obtained such a stronger version in the algebraic context. The following theorem can be proved by combining the argument sketched above with Grothendieck's techniques. However, the argument given by Mumford [M] enables one to obtain the theorem as a consequence of Grauert's theorem if one notes that because of a theorem of Frisch [Fris], any Stein space can be exhausted by compact sets K which have fundamental systems of open Stein neighbourhoods and are such that the ring $\mathcal{O}(K)$ of functions holomorphic in a neighborhood of K (depending on the function) is actually noetherian.

Theorem 2.2. *Let X, Y be complex spaces, $\pi : X \to Y$ a proper holomorphic map. Suppose that Y is Stein, and let $\mathcal{F}$ be a coherent analytic sheaf on X. Let Y_0 be a relatively compact open set in Y. Then, there is an integer $N > 0$ such that the following hold.*

I. There exists a complex

$$\mathcal{E}^\bullet : \ldots \to \mathcal{E}^{-1} \to \mathcal{E}^0 \to \ldots \to \mathcal{E}^N \to 0$$

of finitely generated locally free $\mathcal{O}_{Y_0}$-modules on Y_0 [i.e. each $\mathcal{E}^p$ is the sheaf of holomorphic sections of a holomorphic vector bundle on Y_0] such that for any Stein open set $W \subset Y_0$, we have

$$H^q(\Gamma(W,\mathcal{E}^\bullet)) \simeq \Gamma(W, R^q\pi_*(\mathcal{F})) \simeq H^q(\pi^{-1}(W),\mathcal{F}) \qquad \forall q \in \mathbb{Z}.$$

II. (Base Change Theorem). Assume, in addition, that $\mathcal{F}$ is π-flat [i.e. $\forall x \in X$, the stalk $\mathcal{F}_x$ is flat as a module over $\mathcal{O}_{Y,\pi(x)}$]. Then, there exists a complex

$$\mathcal{E}^\bullet : 0 \to \mathcal{E}^0 \to \mathcal{E}^1 \to \ldots \to \mathcal{E}^N \to 0$$

of finitely generated locally free $\mathcal{O}_{Y_0}$-sheaves $\mathcal{E}^p$ with the following property:

Let S be a Stein space and $f : S \to Y$ a holomorphic map. Let $X' = X \times_Y S$ and $f' : X' \to X$ and $\pi' : X' \to S$ be the two projections. Then, if T is an open Stein subset of $f^{-1}(Y_0)$, we have, for all $q \in \mathbb{Z}$

$$H^q(\Gamma(T, f^*(\mathcal{E}^\bullet)) \simeq \Gamma(T, R^q\pi'_*(\mathcal{F}')) \simeq H^q(\pi'^{-1}(T),\mathcal{F}')$$

where $\mathcal{F}' = (f')^*(\mathcal{F})$.

In these results, the complex $\mathcal{E}^\bullet$ is, of course, not unique; however, the cohomology sheaves of $\mathcal{E}^\bullet$ are (by construction). This suggests considering a new category in which quasi-isomorphic complexes of sheaves become isomorphic. There is, in fact, a category $\mathcal{D}$ (the so-called derived category of the category of sheaves) with the right universal properties which achieves this. The basic idea is to "invert" a class of morphisms in an abelian category, much as one "inverts" a set of elements in a ring to define a quotient field or a localization. There are, however, some technical problems in applying this to the category of complexes of sheaves and inverting quasi-isomorphisms, and one has to pass through the category whose objects are complexes of sheaves, but whose morphisms are homotopy classes of usual morphisms of complexes. For the details of this construction, see the article of Grivel in the book of Borel [Bo].

One can then define the *total* direct image $\mathbb{R}\pi_*(\mathscr{F})$ of a coherent sheaf $\mathscr{F}$ on X under an *arbitrary* holomorphic map $\pi : X \to Y$. It is an object in the derived category of the category of sheaves on Y and is represented by a complex $\mathscr{C}^\bullet$ of sheaves such that $H^q(\mathscr{C}^\bullet) \simeq R^q\pi_*(\mathscr{F})$. [There are several such complexes; one is given by a sheaf version of the Čech complex of $\mathscr{F}$.] Theorem 2.2 simply asserts that if π is proper, then $\mathbb{R}\pi_*(\mathscr{F})$ is represented, on Y_0, by a complex, bounded above, of finitely generated locally free sheaves on Y_0 (with a considerable sharpening when $\mathscr{F}$ is π-flat).

The introduction of the direct image $\mathbb{R}\pi_*(\mathscr{F})$ is of the greatest importance in several situations. It is central in the duality theory of coherent sheaves, a vast generalization of the Serre duality theorem initiated by Grothendieck (see Hartshorne [Har], Ramis-Ruget [RR], Ramis-Ruget-Verdier [RRV]). It is fundamental in the Grothendieck-Riemann-Roch theorem (see Chapters 15, 18 in Fulton's book [Fu]; also Fulton-Lang [FuL]). It should be noted that there is a version of the Grothendieck-Riemann-Roch theorem for proper maps between complex *manifolds* (O'Brian-Toledo-Tong [OTT]), but the general case of proper maps between complex spaces remains open.

Theorems 2.1 and 2.2 have many other applications. We turn now to some of them.

First, Theorem 2.1 implies Remmert's proper mapping theorem; in fact, if $\pi : X \to Y$ is proper, then $\pi(X) = \operatorname{supp}(R^0\pi_*(\mathscr{O}_X))$ and the support of any coherent sheaf is an analytic set.

The following theorem is very useful especially in deformation theory.

Let X, Y be complex spaces, $\pi : X \to Y$ a proper map. Let $\mathscr{F}$ be a π-flat coherent sheaf on X. For $y \in Y$, denote by $\mathfrak{m}_y$ the $\mathscr{O}_Y$-sheaf of germs of holomorphic functions "vanishing at y": the stalk of $\mathfrak{m}_y$ at y is the maximal ideal of $\mathscr{O}_{Y,y}$; that at $t \neq y$ is $\mathscr{O}_{Y,t}$. We set $\mathscr{F}(y) =$ analytic restriction of $\mathscr{F}$ to $\pi^{-1}(y) = \mathscr{F} \otimes_{\mathscr{O}_Y} (\mathscr{O}_Y/\mathfrak{m}_y)$.

Theorem 2.3. *With the above notation and hypotheses we have:*

(a) If $k \geq 0$ is an integer, then, for any $q \geq 0$, the set

$$\{y \in Y \mid \dim_{\mathbb{C}} H^q(\pi^{-1}(y), \mathscr{F}(y)) \geq k\}$$

is a closed analytic set in Y.

(b) If Y is reduced, let $q \geq 0$ be given. Then $R^q\pi_*(\mathscr{F})$ is locally free on Y if the function $y \mapsto \dim_{\mathbb{C}} H^q(\pi^{-1}(y), \mathscr{F}(y))$ is locally constant on Y.

(c) The function $y \mapsto \chi(\mathscr{F}(y)) = \sum_{q=0}^{\infty}(-1)^q \dim_{\mathbb{C}} H^q(\pi^{-1}(y), \mathscr{F}(y))$ is locally constant on Y.

Proof. Since the theorem is local on Y, we may assume, in view of Theorem 2.2, part II, that there is a complex

$$\mathscr{E}^\bullet : 0 \to \mathcal{O}_Y^{p_0} \xrightarrow{\alpha_0} \mathcal{O}_Y^{p_1} \xrightarrow{\alpha_1} \cdots \xrightarrow{\alpha_{N-1}} \mathcal{O}_Y^{p_N} \to 0$$

with the base change property in Theorem 2.2, part II. In particular, if $y \in Y$, we have

$$H^q(\pi^{-1}(y), \mathscr{F}(y)) \simeq H^q(\mathscr{E}^\bullet \otimes (\mathcal{O}_Y/\mathfrak{m}_y)).$$

Now, the morphisms α_ν are given by matrices of holomorphic functions and we have

$$\begin{aligned} d_q(y) &= \dim_{\mathbb{C}} H^q(\pi^{-1}(y), \mathscr{F}(y)) = \dim_{\mathbb{C}}(\ker \alpha_q(y)/\operatorname{im}\alpha_{q-1}(y)) \\ &= p_q - \operatorname{rank}(\alpha_q(y)) - \operatorname{rank}(\alpha_{q-1}(y)). \end{aligned}$$

Hence the set $\{y \in Y | d_q(y) \geq k\}$ is the (finite) union of the sets $Y_{a,b} = \{y \in Y | \operatorname{rank}\alpha_q(y) \leq a\}, \operatorname{rank}\alpha_{q-1}(y) \leq b\}$ with $a + b = k$, $a, b \geq 0$; since each $Y_{a,b}$ is clearly a closed analytic set, this gives (a). Moreover, the above formula shows that $d_q(y)$ is locally constant if and only if $\operatorname{rank}\alpha_q(y)$ (and hence $\dim \ker \alpha_q(y)$) and $\operatorname{rank}\alpha_{q-1}(y)$ are *both* locally constant. This gives (b). As for (c), we have $\chi(\mathscr{F}(y)) = \sum_{q=0}^N (-1)^q p_q$.

For a more detailed description of the circle of ideas around the above theorem, see Chapter III, Sect. 3 in the book of Bănică-Stănăşilă [BS].

We next describe the application made by H. Cartan [C 2] of Grauert's theorem in the study of quotients of complex spaces. This subject was initiated by K. Stein [S]; there has also been some more recent work by H. Grauert [G 5].

We shall suppose, for simplicity of exposition, that our complex spaces are reduced although the results are valid in general.

Let then X be a reduced complex space with structure sheaf $\mathcal{O}_X$. Let R be an equivalence relation on X, and $Y = X/R$ the topological quotient space (with the quotient topology) and let $p : X \to Y$ be the natural projection. We define a sheaf $\mathcal{O}_Y$ of continuous functions on Y by: $\mathcal{O}_Y(U) =$ set of continuous functions f on U such that $f \circ p$ is holomorphic on $p^{-1}(U)$ ($U \subset Y$ is any open set). We also denote this sheaf by $\mathcal{O}_X/R$.

Recall that R is said to be *proper* if, for any compact set $K \subset X$, the set of all $y \in K$ related by R to some point of K (= saturation of K) is again compact. Equivalently, $Y = X/R$ is Hausdorff and $p : X \to Y$ is proper.

Cartan's main theorem is as follows.

Theorem 2.4. *Let R be a proper equivalence relation on the reduced complex space X. Then $(Y, \mathcal{O}_Y)$ is a complex space if, and only if, the following condition is fulfilled.*

Every point $y_0 \in Y$ has a neighborhood U such that the class of continuous maps $\varphi : U \to S$ of U into a reduced complex space S (depending on φ) such that $\varphi \circ p$ is holomorphic on $p^{-1}(U)$ separates the points of U.

Proof of Theorem 2.4. Let U satisfy the conditions above and let $\Omega = p^{-1}(U)$. If S is a reduced complex space, consider the class $H_S = H_S(U)$ of holomorphic maps $\Phi : \Omega \to S$ of the form $\Phi = \varphi \circ p$ where φ is continuous on U. Let $\Delta_S = \{(s,s) \in S \times S\}$ be the diagonal of S and let $\Delta_\Phi = (\Phi \times \Phi)^{-1}(\Delta_S) \subset \Omega \times \Omega$ ($\Phi \in H_S$). Let $V \Subset U$ and $W = p^{-1}(V) \times p^{-1}(V) \subset \Omega \times \Omega$. The hypothesis of the theorem means that

$$\bigcap_S \bigcap_{S \in H_S} \Delta_\Phi = \Delta_\Omega$$

(where Δ_Ω is the diagonal of Ω). Since W is relatively compact in Ω, there is a finite subset $\{\Phi_\alpha\}_{\alpha=1,\dots,N}$ of $\cup_S H_S$ such that

$$W \cap \bigcap_{S,H_S} \Delta_\Phi = W \cap \bigcap_{\alpha=1}^{N} \Delta_{\Phi_\alpha} = W \cap \Delta_\Omega = \Delta_{p^{-1}(V)}.$$

Thus, if we replace X by $p^{-1}(V)$ and consider the map $F = \Phi_1 \times \ldots \times \Phi_N : p^{-1}(V) \to S_1 \times \ldots \times S_N$, we may assume the following: there is a holomorphic map $F : X \to S$ such that two points x, y are equivalent under R if and only if $F(x) = F(y)$. Further, *we may assume that F is proper*. To see this, let $F = \varphi \circ p$, $\varphi : Y \to S$. The above assumption means that φ is injective. Let U_0 be a relatively compact open set in Y and $L = \partial U_0$; L is compact, so that $S' = S - \varphi(L)$ is open in S. But $\varphi | U_0$, considered as a map into S' is proper: if $C \subset S'$ is compact, $\varphi^{-1}(C) \cap U_0$ is disjoint from an open set in Y containing ∂U_0 (e.g. $\varphi^{-1}(S - C)$) and so is closed in Y; since $U_0 \Subset Y, \varphi^{-1}(C)$ is compact. We have only to replace X by $p^{-1}(U_0)$ and S by S'.

Thus, it remains to prove the following:

Proposition 2.5. *Let $F : X \to S$ be a proper holomorphic map between reduced complex spaces and let R be the equivalence relation defined by F : $xRy \Leftrightarrow F(x) = F(y)$. Then $(X/R, \mathcal{O}_X/R)$ is a complex space.*

Proof of 2.5. Replacing S by $F(X)$ (which is analytic by Remmert's theorem), we may assume that F is onto S. Hence F induces a homeomorphism of X/R onto S, and we may consider $\widetilde{\mathcal{O}} = \mathcal{O}_X/R$ as a sheaf on S. We shall show below that $\widetilde{\mathcal{O}}$ is $\mathcal{O}_S$-coherent. This implies the result: we may assum shrinking S that $\widetilde{\mathcal{O}}$ is generated (everywhere on S) by finitely many sections $f_1, \dots, f_k \in \Gamma(S, \widetilde{\mathcal{O}})$. Let $g_j = f_j \circ p$ and A the image of X in $S \times \mathbb{C}^k$ by the holomorphic map $x \mapsto (F(x), g_1(x), \dots, g_k(x))$. A is analytic by Remmert's theorem, and we see at once that $(S, \widetilde{\mathcal{O}})$ is isomorphic to $(A, \mathcal{O}_A)$.

To show that $\widetilde{\mathcal{O}}$ is coherent, consider $D = (F \times F)^{-1}(\Delta_S) \subset X \times X$ (Δ_S is the diagonal of S). Let $h : D \to S$ be the map $(x, y) \mapsto F(x) (= F(y))$. If U is open in X and f is holomorphic on U, we define two holomorphic functions f_1, f_2 on $(U \times U) \cap D$ by: $f_1(x, y) = f(x), f_2(x, y) = f(y)$. This defines two $\mathcal{O}_S$-morphisms α_1, α_2 from $F_*(\mathcal{O}_X)$ to $h_*(\mathcal{O}_D)$. Moreover, to say that f is constant on fibres of F means simply that $f(x) = f(y)$ if $(x, y) \in D$, i.e. that

$f_1 = f_2$. Hence $\widetilde{\mathcal{O}} = \ker(\alpha_1 - \alpha_2)$. Since $F_*(\mathcal{O}_X)$ and $h_*(\mathcal{O}_D)$ are $\mathcal{O}_S$-coherent, so also is $\widetilde{\mathcal{O}}$.

Cartan's theorem implies the existence of the quotient complex space under the action of a properly discontinuous group; this is due again to Cartan. In fact he had proved this earlier without the use of Grauert's direct image theorem.

A group Γ of homeomorphisms acting on a locally compact space X is said to be properly discontinuous if, for any compact set $K \subset X$, the set $\{\gamma \in \Gamma \,|\, \gamma(K) \cap K \neq \emptyset\}$ is finite.

Theorem 2.6. *Let X be a reduced complex space and Γ a properly discontinuous group of analytic automorphisms of X. Then X/Γ carries a natural structure of complex space.*

If $x \in X$, the isotropy group $\Gamma_x = \{\gamma \in \Gamma \,|\, \gamma(x) = x\}$ is finite. It is easy to find a neighborhood U of x such that $\Gamma_x(U) = U$ and such that if $\gamma \in \Gamma$ and $\gamma(U) \cap U \neq \emptyset$, then $\gamma \in \Gamma_x$. We have thus to consider U/Γ_x; here the conditions of Theorem 2.5 are easily verified. In fact Γ_x-invariant holomorphic functions on U separate points of U/Γ_x if U is small enough as is seen by averaging holomorphic functions on U: consider $z \mapsto \sum_{\gamma \in \Gamma_x} f(\gamma z)$ where f is holomorphic on U (and $z \in U$).

Another application of Cartan's theorem is the proof of a result of Remmert on holomorphically convex spaces (which again was first proved, in a slightly weaker form, before the direct image theorem).

Recall that a complex space X is called *holomorphically convex* if for any non-compact set $E \subset X$, there is a function f holomorphic on X such that $f(E)$ is unbounded (in $\mathbb{C}$). Equivalently, for any compact set $K \subset X$, the set $\hat{K}_X = \{x \in X \,|\, |f(x)| \leq \sup_{y \in K} |f(y)| \ \forall f \text{ holomorphic on } X\}$ is again compact.

Consider the following equivalence relation R on a holomorphically convex space X : xRy if $f(x) = f(y)$ for all f holomorphic on X. This relation is *proper*: in fact, the saturation of a compact set K is contained in $\hat{K}_X$. Moreover, holomorphic functions on X separate distinct equivalence classes under R. Thus, by Cartan's theorem, $(X/R, \mathcal{O}_X/R) = (Y, \mathcal{O}_Y)$ is a complex space. One sees at once that Y is holomorphically convex and that holomorphic functions on Y separate points. The following theorem of Grauert [G 1] now implies that Y is Stein:

Theorem 2.7. *A holomorphically convex complex space is Stein if and only if it contains no compact analytic sets of positive dimension.*

Thus, we obtain the *Remmert Reduction Theorem:*

Theorem 2.8. *Let X be a holomorphically convex complex space. There exists a Stein space Y and a proper holomorphic map $\pi : X \to Y$ which induces an* *isomorphism on the spaces of global holomorphic functions. Moreover, the fibres of* π *are connected and* π *has the following universal property: if* $\varphi : X \to S$ *is a holomorphic map into a Stein space* S*, there is a (unique) holomorphic map* $\tilde{\varphi} : Y \to S$ *with* $\varphi = \tilde{\varphi} \circ \pi$*. Finally,* Y *and* π *are uniquely determined (up to unique isomorphism).*

The connectedness of the fibres follows from the coherence of $\pi_*(\mathcal{O}_X)$ and the fact that global sections of $\pi_*(\mathcal{O}_X)$ (= holomorphic functions on X) generate its stalks at every point since Y is Stein (Theorem A). The other properties follow easily from the construction.

We mention also a very important theorem of Grauert [G 4] which uses this circle of ideas.

Theorem 2.9. *Let* M *be a connected compact complex submanifold of a complex manifold* X*. Suppose that the normal bundle* N *of* M *in* X *is weakly negative in the sense that the zero section of* N *has a strictly pseudoconvex neighborhood. Then* M *can be blown down to a point, i.e., there is a normal complex space* Y *and a proper map* $\pi : X \to Y$ *such that* $\pi(M) = \{y_0\}$ *is a single point and* $\pi|X - M \to Y - \{y_0\}$ *is an analytic isomorphism.*

Grauert proves this theorem by showing that M has a strictly pseudoconvex neighbourhood U in X in which all compact analytic sets of positive dimension lie in M. Given this, one can simply apply the Remmert reduction theorem.

Finally, the direct image theorem can be used to prove the Stein factorization theorem (see Stein [S]) which is very useful in both analytic and algebraic geometry.

Theorem 2.10. *Let* X, Y *be complex spaces,* $\pi : X \to Y$*, a proper holomorphic map. Then,* π *can be factored, uniquely up to unique isomorphism, as follows*

$$X \xrightarrow{f} S \xrightarrow{g} Y, \qquad \pi = g \circ f$$

where (a) g *is a finite map and (b)* f *is a proper map such that the natural map (given by composition with* f*) of* $\mathcal{O}_S$ *into* $f_*(\mathcal{O}_X)$ *is an isomorphism.*

Condition (b) implies that f is surjective and has connected fibres. Thus, the space S is obtained from X by collapsing connected components of fibres of π to points.

We sketch the proof when X is reduced; the general case is similar. (See also the books of Grauert-Remmert [GR 2] and Bănică-Stănăşilă [BS].)

Since $\pi_*(\mathcal{O}_X)$ is $\mathcal{O}_Y$-coherent, any $y_0 \in Y$ has a neighbourhood V such that finitely many holomorphic functions $f_1, \ldots, f_k$ on $\pi^{-1}(U)$ generate $\pi_*(\mathcal{O}_X)$ on U. The map $f_U : \pi^{-1}(U) \to U \times \mathbb{C}^k$, $x \mapsto (\pi(x), f_1(x), \ldots, f_k(x))$ is proper and so its image is an analytic set Z_U. The f_j are constant on connected components of π (by the maximum principle). Further, if g_U is the restriction to Z_U of the projection of $U \times \mathbb{C}^k$ to U, g_U is a finite map. In fact, it is clearly proper, and $\mathcal{O}_{Z_U,z_0}(z_0 \in Z_U)$ is generated, over $\mathcal{O}_{U,g_U(z_0)}$, by the coordinate functions of $\mathbb{C}^k$. This gives the model of S locally over Y. The (global) theorem follows from the uniqueness.

3. Embeddings in $\mathbb{C}^n$

One of the basic theorems in differential topology is Whitney's theorem that any smooth n-dimensional manifold can be embedded properly in $\mathbb{R}^{2n+1}$. There are many refinements and related results. In this section, we shall describe holomorphic analogues of some of these problems.

Closed subspaces of Stein spaces being again Stein, only Stein spaces can be embedded in $\mathbb{C}^N$. For singular spaces, this is not enough for an embedding in $\mathbb{C}^N$ to be possible.

Let $(X,0)$ be the germ of a complex space embedded in $(\mathbb{C}^N,0)$. Let $\mathfrak{m}$ be the maximal ideal of $\mathcal{O}_{X,0}$ and set $T_0^*(X) = \mathfrak{m}/\mathfrak{m}^2$. Let $T_0(X)$ be the dual vector space of $T_0^*(X)$. $T_0(X)$ is called the (Zariski) tangent space to X at 0, and $T_0^*(X)$ is called the cotangent space. Since clearly the restriction map $T_0^*(\mathbb{C}^N) \to T_0^*(X)$ is surjective, the map $T_0(X) \to T_0(\mathbb{C}^N)$ is injective; clearly we must have $\dim_{\mathbb{C}} T_0(X) \leq N$. This has a converse: if $\dim_{\mathbb{C}} T_0(X) = d$, the germ $(X,0)$ can be realized as a subspace of $(\mathbb{C}^d,0)$. In particular, X is smooth at 0 if and only if $\dim_0 X = \dim_{\mathbb{C}} T_0(X)$.

Thus, if a complex space X can be embedded in $\mathbb{C}^N$, we have $\dim_{\mathbb{C}} T_x(X) \leq N\ \forall x \in X$ (and X must be Stein).

In 1956, Remmert announced that any Stein space of dimension n admits a proper injective holomorphic map in some $\mathbb{C}^N$. A more refined theorem was proved in Bishop [Bi] and Narasimhan [N 1]. The results can be summarized as follows.

Theorem 3.1. *Let X be a Stein space of dimension n.*

1. There exists a proper injective holomorphic map of X into $\mathbb{C}^{2n+1}$ whose tangent map is injective at every smooth point. In particular, any Stein manifold of dimension n can be embedded properly in $\mathbb{C}^{2n+1}$.

2. If there is an integer m such that $\dim_{\mathbb{C}} T_a(X) \leq m\ \forall a \in X$ and if $m > n$, then X can be embedded properly in $\mathbb{C}^{n+m}$.

In the case of C^∞ manifolds, any such manifold admits a proper smooth map into $\mathbb{R}$; in the complex case, if $\pi : X \to Y$ is a proper holomorphic map into a complex space Y and X is Stein, then $\dim X \leq \dim Y$. Also one sees easily that there is no proper map $\pi : \Omega \to \mathbb{C}^n$ if Ω is a bounded domain in $\mathbb{C}^n$. [π must be a ramified covering, say of degree k. If f is a bounded holomorphic function on Ω, then there are *bounded* holomorphic functions $a_1, \ldots, a_k$ on $\mathbb{C}^n$ with $f(x)^k + a_1(\pi(x))f(x)^{k-1} + \ldots + a_k(\pi(x)) \equiv 0$ (see Sect. 1, sketch of Theorem 1.2). Since non-constant holomorphic functions on $\mathbb{C}^n$ cannot be bounded, this implies that f is constant.] Hence, the following theorem proved by Bishop in the course of his proof of Theorem 3.1 is best possible.

Theorem 3.2. *Let X be a Stein space of dimension n.*

1. There exists a holomorphic map $f : X \to \mathbb{C}^n$ such that for any compact $K \subset \mathbb{C}^n$, every connected component of $f^{-1}(K)$ is compact.

2. There exists a proper holomorphic map of X into $\mathbb{C}^{n+1}$.

The main step in Bishop's proof of this theorem is the following. Let X be a Stein space of dimension n and let Ω be a relatively compact open subset. Suppose that $f_1,\ldots,f_k$ are holomorphic functions on X such that

$$P_0 = \{x \in \Omega \mid |f_j(x)| < 1, j = 1,\ldots,k\} \Subset \Omega.$$

Then, given a compact set $K \subset P_0$, there exist n holomorphic functions $g_1,\ldots,g_n$ on X (in fact, polynomials in $f_1,\ldots,f_k$ with complex coefficients) such that

$$K \subset \{x \in P_0 \mid |g_1(x)| < 1,\ldots,|g_n(x)| < 1\} \Subset P_0.$$

Bishop's idea is explained very well in Hörmander's book [Hö] where, however, only a weaker result is proved (with n replaced by $2n$).

In the case of C^∞ manifolds, Whitney himself had improved his embedding theorem; this was completed at one point by M. Hirsch to give the following theorem: a smooth manifold of dimension n can be properly embedded in $\mathbb{R}^{2n}$. Again, it is a basic principle that if a smooth manifold of dimension n behaves homotopically like a "simple" manifold (like S^n) or one of smaller dimension, then one can expect to embed it in $\mathbb{R}^k$ with $k < 2n$ (work of A. Haefliger, S. Smale, M. Hirsch, A. Phillips). Now Stein spaces do behave homotopically like a space of real dimension $< 2n$.

Theorem 3.3. *A Stein space of dimension n has the homotopy type of a CW-complex of real dimension n.*

This theorem was proved by Andreotti-Frankel [AF] in the case of Stein manifolds; the proof uses Morse theory and the formulation above is that of Milnor [Mi]. In the singular case, this was proved for complex affine varieties by Karchyauskas [Ka] and in the general case by Hamm (see [Ham]). The theorem is also a very special case of the results Goresky-Macpherson obtain in their study of Morse theory on singular spaces; see [GM].

On the other hand, the so-called *Oka principle* would state that on Stein spaces, holomorphic problems admitting continuous solutions must admit holomorphic solutions. This received great impetus from Grauert [G 2] who proved this principle for the classification of principal bundles with a complex Lie group as structure group (the continuous and the holomorphic classifications are the same). If one believed that the Oka principle holds for embeddings and immersions, one would be led to the following conjecture on the basis of the remarks made above.

Conjecture 3.4. *Any Stein manifold of dimension n can be holomorphically immersed in* $\mathbb{C}^N$ *and properly (holomorphically) embedded in* $\mathbb{C}^{N+1}$ *with* $N = n + \left[\frac{n}{2}\right]$.

A word of caution. There are certainly limits to the Oka principle. Thus the complex 7-sphere $\sum_1^8 z_i^2 = 1$ in $\mathbb{C}^8$ is diffeomorphic to an open set in $\mathbb{C}^7$, but is *not* analytically ismorphic to an open set in $\mathbb{C}^7$ (Stout-Zame [SZ]).

In 1970, O. Forster [Fo 1] took the first major step in the direction of Conjecture 3.4. We shall now outline his method.

Let X be a Stein manifold of dimension n. Let $m > n$ and let $f_0 : X \to \mathbb{C}^m$ be a proper holomorphic map (f_0 exists by Bishop's theorem). If $g_1, \ldots, g_p$ are holomorphic functions on X which generate all holomorphic functions on X as an *algebra* over the ring of polynomials in the components of f_0, then the map $f_0 \times (g_1, \ldots, g_p) : X \to \mathbb{C}^{m+p}$ is an embedding. Thus, it is sufficient (but not necessary) to bound the number of generators of the sheaf $f_{0*}(\mathcal{O}_X)/(1)$ as a sheaf of $\mathcal{O}_{\mathbb{C}^m}$-modules ((1) being the subsheaf generated by the constants).

In a series of papers, Forster and Ramspott ([FR] and the references given there) had generalized Grauert's Oka principle to coherent sheaves: their study gives the following estimate for the number of generators of a coherent sheaf on a Stein space.

Let S be an n-dimensional Stein space and $\mathcal{F}$ a coherent sheaf on S. If $s \in S$, set $L_s(\mathcal{F}) = \mathcal{F}_s/\mathfrak{m}_s\mathcal{F}_s$, where $\mathcal{F}_s$ is the stalk of $\mathcal{F}$ at s and $\mathfrak{m}_s$ is the maximal ideal in $\mathcal{O}_{S,s}$. The set $S_p(\mathcal{F}) = \{s \in S \mid \dim_{\mathbb{C}} L_s(\mathcal{F}) \geq p\}$ is a closed analytic set in S; let $d_p(\mathcal{F})$ be its dimension. Then, we have

Theorem 3.5 (Forster-Ramspott [FR]). *$\mathcal{F}$ is generated over $\mathcal{O}_S$ by $d + \left[\frac{n}{2}\right]$ global sections, where $d = \sup_p d_p(\mathcal{F})$.*

(*Note:* Forster-Ramspott state a slightly weaker result, but remark that this is because they only had a weaker version of Theorem 3.3 at their disposal.) Thus, to study embeddings of X, it suffices to find a good upper bound for the following number $b = b(f_0)$.

Let $f_0 : X \to \mathbb{C}^m$ be proper and let $\mathcal{G} = f_{0*}(\mathcal{O}_X)/(1)$; set $b = \sup_{p>0}\{p + \left[\frac{1}{2}d_p\right]\}$ where $d_p = d_p(\mathcal{G})$.

To obtain bounds for the numbers d_p, Forster proceeds as follows. Let $J^{k,m}(X)$ be the vector bundle of k-jets of germs of holomorphic maps of open sets in X into $\mathbb{C}^m$. For $x \in X$, we denote by $J_x^{k,m}$ the fibre (not the stalk) of $J^{k,m}(X)$ at x; it is naturally identified with $\oplus_{m\,\text{copies}}(\mathcal{O}_{X,x}/\mathfrak{m}_x^{k+1})$, $\mathfrak{m}_x$ being, as usual, the maximal ideal in $\mathcal{O}_{X,x}$. If $\tau = (\tau_1, \ldots, \tau_m) \in J_x^{k,m}$, the value $\tau_j(x)$ of τ_j at x is well-defined (with the above identification). Denote by $c(\tau)$ the codimension in $\mathcal{O}_{X,x}/\mathfrak{m}_x^{k+1}$ of the ideal generated by $(\tau_1 - \tau_1(x), \ldots, \tau_m - \tau_m(x))$. If $p \geq 1$, the set $M_p = \{\tau \in J^{k,m}(X) | c(\tau) \geq p\}$ is an analytic set. The central point is that M_p *has large codimension*, in particular, at least $(p-1)(m-n+1)$ if $p \leq 4$ (although we do not have an exact formula for this codimension). Let $r \geq 1$, X^r the product of X with itself r times and $X_0^r = \{(x_1, \ldots, x_r) \in X^r \mid x_i \neq x_j \text{ for } i \neq j\}$. For integers $p_1, \ldots, p_r \geq 1$, set $M(p_1, \ldots, p_r) = \{(\tau_1, \ldots, \tau_r) \in \oplus_{r\,\text{times}} J^{k,m}(X) | c(\tau_j) \geq p_j\}$ [$\oplus$ is the direct sum of the vector bundle $J^{k,m}(X)$ with itself r times]. We can think of $M(p_1, \ldots, p_r)$ as an analytic set in the Cartesian product $(J^{k,m}(X))^r$. Forster's main technical lemma is that if we choose f_0 so that the induced map $j_r^k(f_0) : X_0^r \to (J^{k,m}(X))^r$ is transversal to $M(p_1, \ldots, p_r)$ (for all r and p_j), then $d_p \leq n - (p-1)(m-n)$ (if $m > n$ is not too small). Further, part 2 of Bishop's Theorem 3.2 can easily be strengthened to show that a dense set of holomorphic maps of X into $\mathbb{C}^m$ consists of proper maps if $m > n$ (in the topology of uniform convergence on compact sets), so that this transversality condition can be fulfilled. By suitably choosing $m > n$, Forster obtains the following

Theorem 3.6. *Let X be a Stein manifold of dimension n. If $n \geq 2$, X can be properly embedded in $\mathbb{C}^{2n}$. If $n \geq 6$, it can be embedded in $\mathbb{C}^{2n-k}$ with $k = \left[\frac{n-2}{3}\right]$.*

It should be remarked that by a more careful analysis of the codimension of M_p than was undertaken by Forster, U. Schafft [Sch] has shown that if $\ell > 0$ is an integer, then there is $n(\ell) > 0$ such that Stein manifolds of dimension $n > n(\ell)$ embed in $\mathbb{C}^N$ with $N = n + 2 + \left[\frac{1}{2}n + \frac{1}{\ell}n\right]$.

In 1971, Gromov and Eliashberg [GE] announced much stronger results. These form a small part of the program in Gromov's book [Gr]. In the special case of the problem of (not necessarily proper) immersions of Stein manifolds, the results may be described as follows.

Let X be an n-dimensional Stein manifold and let $q > n$. In the jet bundle $J^{1,q}(X)$, consider the open set U_0 of elements $(\tau_1, \ldots, \tau_q) \in J_x^{1,q}$, $x \in X$, such that the covectors $\tau_j - \tau_j(x) \in \mathfrak{m}_x/\mathfrak{m}_x^2 = T_x^*(X)$ generate $T_x^*(X)$. Then, *any continuous section $s : X \to U_0$ is homotopic to a section of the form $j^1(f) : X \to U_0$, where f is a holomorphic map of X into $\mathbb{C}^q$* ($j^1(f)(x)$ is, of course, the 1-jet of f at x).

This implies that a Stein manifold of dimension n can be holomorphically immersed in $\mathbb{C}^q$ with $q > n$ if and only if the cotangent bundle $T^*(X)$ is generated by q global sections (continuous or holomorphic; the conditions are equivalent by Grauert's Oka principle). Moreover, this statement has an analogue for affine varieties: if X is a smooth affine algebraic variety and $T^*(X)$ is generated by $q > n$ algebraic 1-forms, X can be algebraically immersed in $\mathbb{C}^q$. The case $q = n$ is one of the major unsolved problems in the subject.

Question. Can a parallelizable Stein manifold of dimension n be holomorphically immersed in $\mathbb{C}^n$?

The only case known is when $n = 1$: every open Riemann surface carries a holomorphic function f with $(df)(x) \neq 0$ $\forall x \in X$. (See

Gunning-Narasimhan [GN].) When $n > 1$, there is certainly a difference between the algebraic and the holomorphic categories. The complex spheres $S^n_{\mathbb{C}} = \{(z_0, \dots, z_n) \in \mathbb{C}^{n+1} \mid z_0^2 + \dots + z_n^2 = 1\}$ and the complement of a nonsingular cubic in $\mathbb{P}^2$ are algebraically parallelizable but cannot be *algebraically* immersed in $\mathbb{C}^n$ (resp. $\mathbb{C}^2$). However, as J.J. Loeb has remarked, $S^n_{\mathbb{C}}$ can be holomorphically immersed in $\mathbb{C}^n$. More generally, one has the following result:

Let $x \in \mathbb{C}$, $y^{(1)} = (y_1^{(1)}, \dots, y_{n_1}^{(1)}), \dots, y^{(k)} = (y_1^{(k)}, \dots, y_{n_k}^{(k)})$, $n = n_1 + \dots + n_k$, be independent variables, let $d \geq 1$ and let $P_j(y^{(j)})$ be a homogeneous polynomial of degree $d_j \geq 1$ in $y^{(j)}$. Then, the hypersurface $X = \{(x, y^{(1)}, \dots, y^{(k)}) \mid x^d + P_1(y^{(1)}) + \dots + P_k(y^{(k)}) = 1\}$ can be holomorphically immersed in $\mathbb{C}^n$.

Proof. Let φ be an entire function of x, and $F : \mathbb{C}^{n+1} \to \mathbb{C}^n$ the map $(x, y^{(1)}, \dots, y^{(k)}) \mapsto (e^{\alpha_1 \varphi(x)} y^{(1)}, \cdots, e^{\alpha_k \varphi(x)})$, where the α_j are constants. The tangent map of F sends $(u, v^{(1)}, \dots, v^{(k)}) \in \mathbb{C}^{n+1}$ to $\left((\alpha_1, \varphi'(x) y^{(1)} u + v^{(1)}) e^{\alpha_1, \varphi(x)}, \dots\right)$; its kernel at $(x, y^{(1)}, \dots, y^{(k)}) \in X$ is given by

$$(*) \qquad v^{(j)} = -\alpha_j \varphi'(x) u y^{(j)} , \qquad j = 1, \dots, k .$$

Restricted to the tangent space $dx^{d-1}u + \sum_j \sum_\nu \frac{\partial P_j}{\partial y_\nu^{(j)}} v_\nu^{(j)} = 0$ of X, this gives

$$\left(dx^{d-1} - \varphi'(x) \sum_j \alpha_j \sum_\nu \frac{\partial P_j}{\partial y_\nu^{(j)}} y_\nu^{(j)} \right) u = 0 .$$

Since P_j is homogeneous of degree j, if we take $\alpha_j = \frac{1}{d_j}$, this gives

$$(**) \qquad \left(dx^{d-1} - \varphi'(x)(1 - x^d) \right) u = 0$$

(since $\sum_\nu \frac{\partial P_j}{\partial y_\nu^{(j)}} y_\nu^{(j)} = d_j P_j(y^{(j)})$ by Euler's equation). We choose an entire function ψ so that

$$dx^{d-1} - \varphi'(x)(1 - x^d) = e^{\psi(x)} , \qquad x \in \mathbb{C}$$

(possible since we have only to find a polynomial ψ with $e^{\psi(\zeta)} = d\zeta^{d-1}$ for all $\zeta \in \mathbb{C}$ with $\zeta^d = 1$); the map $F|X$ is an immersion since $(**)$ implies that $u = 0$ and $(*)$ then shows that $v^{(j)} = 0 \ \forall j$.

It seems to us that the question of a holomorphic (necessarily non-algebraic) immersion in $\mathbb{C}^2$ of the complement of a smooth cubic in $\mathbb{P}^2$ is a good test for the general question.

Gromov states in his book that there is a homotopy principle for proper embeddings similar to the one for immersions: if $f_0 : X \to \mathbb{C}^{q_0}$ is a proper generic holomorphic map, then the existence of $g : X \to \mathbb{C}^{q-q_0}$ so that $f_0 \times g$ is an embedding is equivalent to finding a global section over X with values in a suitable open subset of a jet bundle $J^{k,q-q_0}(X)$ (with $k = k(n)$ sufficiently large). Gromov indicates in his book how this can be done if f_0 is subject to certain additional requirements.

It has to be confessed that at least one of the authors of this article has been unable to carry out the proof of this theorem. This theorem would imply that *any Stein manifold of dimension n can be properly embedded in* $\mathbb{C}^q$ *if* $q \geq \frac{1}{2}(3n+3)$. This is, of course, very close to the optimal result conjectured in 3.4

None of these methods is applicable to the case $n = 1$, when conjecture 3.4 takes the form:

Conjecture 3.7. *Any open Riemann surface can be properly embedded in* $\mathbb{C}^2$.

Little is known about this problem. Perhaps the best result is the following theorem of K. Kasahara and T. Nishino (1969). The proof was published in Japanese; we are not sure if they published an English version. However, their argument has become well known, and we shall sketch it below. A different construction was given by Alexander [Al 2].

Theorem 3.8. *The open unit disc* $D = \{z \in \mathbb{C} \mid |z| < 1\}$ *can be properly embedded in* $\mathbb{C}^2$.

The proof is based on a construction going back to Fatou and Bieberbach.

Proposition 3.9. *There is a unique function* φ *holomorphic on* $\mathbb{C}^2$ *such that*

$$\varphi(z,w) = z + w + \sum_{m+n\geq 2} a_{mn} z^m w^n$$

near $0 \in \mathbb{C}^2$, *and satisfying*

$$\varphi(4z,4w) = 4\varphi(z,w) - 5\varphi(2z,-2w)^2 + 2\varphi(2z,-2w)^5.$$

The proof consists in showing that the functional equation gives a unique solution as a formal power series; convergence is checked by the majorant method. For the details, see Bochner-Martin [BM].

Let $T : \mathbb{C}^2 \to \mathbb{C}^2$ be the transformation $(z,w) \mapsto (2z,-2w)$. If $F = (f_1, f_2) : \mathbb{C}^2 \to \mathbb{C}^2$ is the map given by $f_1 = \varphi, f_2 = \varphi \circ T$, we have $f_1 \circ T = f_2, f_2 \circ T = 4f_1 - 5f_2^2 + 2f_2^5$. One checks easily that F has a constant jacobian determinant $(= -4)$. It follows easily that F is injective: if $F(\zeta_1) = F(\zeta_2)$, then, the functional equation for $F \circ T$ shows that $F(T^{-n}\zeta_1) = F(T^{-n}\zeta_2)$ $\forall n \in \mathbb{N}$; but F is locally injective near 0, so that $T^{-n}\zeta_1 = T^{-n}\zeta_2$ for large n.

We claim that there is a neighborhood U of $(1,1)$ such that $F(\mathbb{C}^2) \cap U = \emptyset$. In fact if $f_1(\zeta) = 1 + \varepsilon_1, f_2(\zeta) = 1 + \varepsilon_2$ and $T^{-1}(\zeta) = \zeta'$, then, the functional equation shows that $f_1(\zeta') = 1 + \frac{1}{4}\varepsilon_2 + O(\varepsilon_1^2), f_2(\zeta') = 1 + \varepsilon_1$. Hence, if $\delta > 0$ is small and $|\varepsilon_1|, |\varepsilon_2| < \delta$, we have $|f_1(\zeta') - 1| < \delta$, $|f_2(\zeta') - 1| < \delta$. Iterating this, we see that $|f_1(T^{-n}\zeta) - 1|$ and $|f_2(T^{-n}\zeta) - 1|$ are both $< \delta$ for all $n \geq 0$. But $f_1(0,0) = 0$, so that this is impossible.

Thus $\Omega = F(\mathbb{C}^2)$ is biholomorphic to $\mathbb{C}^2$ and $\Omega \cap U = \emptyset$, U being a neighborhood of $(1,1)$.

The main remark of Kasahara-Nishino is that Ω is a Runge domain in $\mathbb{C}^2$. To see this, consider the polynomial map $P : (z,w) \mapsto (w, 4z+2w^5-5w^2)$ of $\mathbb{C}^2$ into itself; P is an isomorphism of $\mathbb{C}^2$ onto itself and we have $F \circ T = P \circ F$.

Let $\Delta(r)$ be the bidisc $\{(z,w) \in \mathbb{C}^2 | \, |z| < r, |w| < r\}$. Choose $\rho > 0$ so small that $F(\Delta(\rho))$ is a Runge domain in $\mathbb{C}^2$ (this is possible: if $\rho < \rho_1 < \rho_0$ are so chosen that $F(\Delta(\rho)) \subset \Delta(\rho_1) \subset F(\Delta(\rho_0))$, then $F(\Delta(\rho))$ is Runge in $F(\Delta(\rho_0))$ and hence also in $\Delta(\rho_1)$, but then it is Runge also in $\mathbb{C}^2$). Since $F \circ T = P \circ F$, we have $F(\Delta(2\rho)) = P(F(\Delta(\rho)))$, and this is Runge in $\mathbb{C}^2$ since $F(\Delta(\rho))$ is and P is an isomorphism. Iterating, we find that $F(\Delta(2^k\rho))$ is Runge in $\mathbb{C}^2$ for any $k \geq 0, k \in \mathbb{N}$, and therefore so is $\Omega = \cup_{k\geq 0} F(\Delta(2^k\rho))$.

Consider now a complex line $L = \{(1,1) + \lambda\zeta_0 | \lambda \in \mathbb{C}\}$ through $(1,1)$ so that $L \cap \Omega \neq \emptyset$. If D_0 is a connected component of $L \cap \Omega$, then D_0 is Runge in $L \simeq \mathbb{C}$ and so is simply connected. Since clearly $D_0 \neq L$, the Riemann mapping theorem implies that D_0 is biholomorphic to the unit disc. Since clearly D_0 is a closed submanifold of $\Omega \simeq \mathbb{C}^2$, the theorem follows.

By intersecting Ω with other curves embedded in $\mathbb{C}^2$ (e.g. $\mathbb{C} - \{k\,\text{points}\}$), one can show that there are plane domains of arbitrary (finite) connectivity bounded by circles which embed in $\mathbb{C}^2$. Also, it is known that all annuli embed in $\mathbb{C}^2$ (see Laufer [L] and Alexander [Al 2]). But the general case of an arbitrary plane domain (even one of finite connectivity) remains open.

The Fatou-Bieberbach example of a biholomorphic map $F : \mathbb{C}^2 \hookrightarrow \mathbb{C}^2$ with interior $(\mathbb{C}^2 - F(\mathbb{C}^2)) \neq \emptyset$ contrasts sharply with what one expects in the algebraic case. Here one has the famous:

Jacobian Conjecture. *Any polynomial map $P : \mathbb{C}^n \to \mathbb{C}^n$ which is a local (analytic) isomorphism is necessarily a (global) algebraic isomorphism.*

In this case, the jacobian determinant of P, being a nowhere zero polynomial, is a constant, as was the case with the Fatou-Bieberbach example. For work on this conjecture, see [BCW] and the references given there.

We shall conclude this section with some problems of uniqueness of embeddings.

Let X be a Stein manifold and let $f_0, f_1 : X \to \mathbb{C}^N$ be two proper embeddings. We say that f_0, f_1 are (holomorphically) isotopic if there exists an open set U in $\mathbb{C}$, $U \supset [0,1]$, and a holomorphic map $F : U \times X \to \mathbb{C}^N$ such that $F(j,x) = f_j(x)$ $\forall x \in X, j = 0,1$ and such that, for each $u \in U$, the map $x \mapsto F(u,x)$ of X into $\mathbb{C}^N$ is a proper embedding.

One has the following theorem of Abhyankar and Moh [AM].

Theorem 3.10. *Let $F : \mathbb{C} \to \mathbb{C}^2$ be a polynomial embedding. Then, there is a finite sequence $\sigma_1, \ldots, \sigma_N$ of algebraic automorphisms of $\mathbb{C}^2$, where each σ_ν either belongs to* GL(2, $\mathbb{C}$) *or is of the form*

$$z \mapsto z + P(w), (P \text{ a polynomial of } w \text{ alone}), w \mapsto w,$$

such that $\sigma_N \circ \ldots \circ \sigma_1 \circ F$ *is the linear embedding* $z \mapsto (z, 0)$.

This theorem can be used to prove a classical theorem going back to Jung which asserts that the *group of algebraic automorphisms of* $\mathbb{C}^2$ *consists precisely of the transformations* $\sigma_N \circ \ldots \circ \sigma_1$ with the σ_ν as above. Further, all algebraic embeddings of $\mathbb{C}$ into $\mathbb{C}^N$ with $N \geq 4$ are isotopic to the linear embedding. This is a special case of the following (unpublished) theorem of M. V. Nori.

Theorem 3.11. *Let* X *be a smooth affine variety of dimension* n *and let* $f, g : X \to \mathbb{C}^N$ *be two (proper) algebraic embeddings of* X *into* $\mathbb{C}^N$ *with* $N \geq 2n+2$. *Then* f, g *are isotopic via proper algebraic embeddings.*

The main step is to prove the following: if A, B are generic linear transformations of $\mathbb{C}^N$ and we set $A \circ f = (F_1, \ldots, F_N)$, $B \circ g = (G_1, \ldots, G_N)$, then, for any k, $1 \leq k \leq N$, the map $(F_1, \ldots, F_k, G_{k+1}, \ldots, G_N)$ is again a proper embedding.

As for Stein manifolds, one has the following relative embedding theorem (see Narasimhan [N 2] and Acquistapace-Broglia-Tognoli [ABT]) which implies that any two embeddings of a Stein manifold of dimension n into $\mathbb{C}^N$ with $N \geq 2n+3$ are isotopic; a slight modification of the argument can be used to prove just the isotopy statement for $N \geq 2n+2$ (i.e. Theorem 3.11 has a holomorphic analogue).

Theorem 3.12. *Let* X *be a Stein manifold of dimension* n *and let* Y *be a closed complex submanifold. Let* $f_0 : Y \to \mathbb{C}^N$ *be a proper embedding with* $N \geq 2n+1$. *Then, there is a proper embedding* $f : X \to \mathbb{C}^N$ *with* $f|Y = f_0$.

The extent to which these dimensions are best possible remains open. One may conjecture (in analogy with the real case) that $\mathbb{C}$ admits "knotted" embeddings in $\mathbb{C}^3$.

Conjecture 3.13. *There exists an algebraic embedding of* $\mathbb{C}$ *into* $\mathbb{C}^3$ *which is not isotopic (via proper holomorphic embeddings) to the linear embedding.*

There are many obvious questions in this connection which await study. For instance, it would be interesting to know if the Abhyankar-Moh theorem has an analogue for *holomorphic* embeddings of $\mathbb{C}$ into $\mathbb{C}^2$ [one would, of course, have to allow also transformations of $\mathbb{C}^2$ of the form $z \mapsto ze^{h(w)}, w \mapsto w$ where h is holomorphic on $\mathbb{C}$]. Further, it would be of interest to study the analogue of this theorem for embeddings of $\mathbb{C}$ into $\mathbb{C}^N (N \geq 3)$ both algebraically and analytically. There is also the problem of embeddings of $\mathbb{C}^n$ into $\mathbb{C}^N$ *with small codimension*. The study of algebraic and holomorphic embeddings is certainly not as well developed as that of C^∞ embeddings.

4. Proper Holomorphic Maps into Bounded Domains

We shall begin with a theorem of Remmert and Stein [RS 2].

Theorem 4.1. *Let X, Y be connected non-compact (reduced) complex spaces. Let D be a bounded strictly pseudoconvex domain in $\mathbb{C}^N$. Then, there is no proper holomorphic map of $X \times Y$ into D.*

Proof. Suppose $f = (f_1, \ldots, f_N) : X \times Y \to D$ is proper. Let $\{y_\nu\}_{\nu=1,2\ldots}$ be a sequence of points in Y without limit points. By Montel's theorem, there is a subsequence $\{\nu_r\} \subset \{\nu\}$ such that $\{f_j(x, y_{\nu_r})\}$ converges as $r \to \infty$ to a holomorphic function $\varphi_j(x)$ on X (uniformly on compact subsets of X). Since, for $x \in X$, $\varphi(x) = (\varphi_1(x), \ldots, \varphi_N(x))$ is a limit point of $f(x, y_{\nu_r})$ and f is proper, we must have: $\varphi(X) \subset \partial D$. Since D is strictly pseudoconvex (and X is connected), φ is a constant. [The boundary of a strictly pseudoconvex domain cannot contain a positive dimensional germ of analytic set.] Let a be a smooth point of X and $(x_1, \ldots, x_n)$ a system of coordinates on an open set U containing a. By Weierstrass' theorem, we have

$$\frac{\partial f_j}{\partial x_k}(x, y_{\nu_r}) \to \frac{\partial \varphi_j}{\partial x_k}(x) = 0 \text{ as } r \to \infty.$$

Thus, for any sequence $\{y_\nu\}$ without limit points, there is a subsequence $\{\nu_r\}$ so that $\frac{\partial f_j}{\partial x_k}(x, y_{\nu_r}) \to 0 (x \in U)$. Hence, for $x \in U$, $\frac{\partial f_j}{\partial x_k}(x, y_\nu) \to 0$ as $\nu \to \infty$ for any such $\{y_\nu\}$. By the maximum principle, $\frac{\partial f_j}{\partial x_k} \equiv 0$ on $U \times Y$. Since a is any regular point of X, f *is independent of* $x \in X$, and so cannot be proper.

This very elegant argument of Remmert and Stein has been extended to correspondences by Rischel [Ri]. This extension implies the following theorem.

Theorem 4.2. *Let X, Y be irreducible complex spaces of dimension > 0. Let D be a bounded strictly pseudoconvex domain in $\mathbb{C}^n$. Assume that $n = \dim(X \times Y)$. Then, there is no proper holomorphic map of D into $X \times Y$.*

Of course, either of these theorems implies that a product of two complex spaces is never analytically isomorphic to a bounded strictly pseudoconvex domain; in particular the ball and the polydisc in $\mathbb{C}^n$ are analytically distinct if $n > 1$.

Before we proceed to consider proper maps between bounded domains in more detail, we mention two rigidity theorems, due again to Remmert-Stein [RS 2].

Theorem 4.3. *Let P_1, P_2 be the interiors of Euclidean polyhedra in $\mathbb{C}^2$. Assume that neither of them is complex-affine equivalent to a product of convex polygons in $\mathbb{C}$. Then, any proper holomorphic map of P_1 to P_2 is the restriction to P_1 of a complex affine transformation; in particular, it is biholomorphic.*

Theorem 4.4. *Let D be a bounded domain of holomorphy in $\mathbb{C}^n, n > 1$. Let K be a compact subset of D which is not contained in any proper analytic subset of D (e.g. $\mathring{K} \neq \emptyset$). Then any proper holomorphic map of $D - K$ into itself is the restriction to $D - K$ of an analytic automorphism of D (in particular, is itself biholomorphic).*

The proof of Theorem 4.4 is based on a fundamental theorem of H. Cartan [C 1] concerning iterates of a holomorphic self-map of a bounded domain.

In the rest of this section, we shall describe some of the results about the behavior of proper holomorphic maps from a bounded domain in $\mathbb{C}^n$ to one in $\mathbb{C}^m$, $m \geq n$. We begin with the equidimensional case. Here the results rest firmly on the foundations laid by Remmert [R 1]; they may be thought of as a synthesis of the theory of ramified coverings, the theory of conformal mappings in $\mathbb{C}$, and the modern theory of linear partial differential equations (the $\bar{\partial}$-Neumann problem). We begin with the C^∞ boundary behavior of biholomorphic and proper maps between domains in $\mathbb{C}^n$ with C^∞ boundaries. This has no real analogue in one variable, largely because of the Riemann mapping theorem and Carathéodory's theorem about continuous extensions of conformal maps to the boundary. Indeed, there is no local obstruction to mapping a boundary curve onto a segment on the real line by the boundary values of a biholomorphic map. Questions concerning C^∞ regularity, while they are interesting, do not play as important a role as they do in several variables where both the Riemann mapping theorem and Carthéodory's theorem fail spectacularly.

On the basis of analogy with the plane, one would expect that biholomorphic maps between smooth bounded pseudoconvex domains in $\mathbb{C}^n$ extend smoothly to the boundary. If this is the case, certain differential geometric properties of the boundaries must be preserved. The most basic such property is *strict* pseudoconvexity of a boundary hypersurface defined by a C^2 function. To check that strict pseudoconvexity is preserved requires the map to have at least a C^2 extension to the boundary. There is an entire hierarchy of more sophisticated differential geometric invariants, the Chern-Moser [CM] invariants, which depend on the higher order derivatives of the defining function. Thus, if one is to attempt a biholomorphic classification of at least domains with smooth boundaries in $\mathbb{C}^n$, it is important to know if biholomorphic maps between such domains extend smoothly to the boundary. [In the remaind this article, *smooth* will be understood to mean C^∞.]

In one complex variable, this smooth extension is an easy consequence of the classical theory of the Dirichlet problem for the laplacian. To illustrate this, let f be a biholomorphic map of a bounded domain $\Omega \subset \mathbb{C}$ with smooth boundary onto the unit disc. Let $f(a) = 0$ $(a \in \Omega)$ and let $\varphi \in C^\infty(\overline{\Omega})$ be 1 near $\partial\Omega$ and 0 near a. Since f extends continuously to $\partial\Omega$ (by Carathéodory's theorem), the function $u = (f - \frac{1}{f})\varphi$ is continuous on $\overline{\Omega}$ and Δu is C^∞ on $\overline{\Omega}$.

It follows that u is itself C^∞ on $\overline{\Omega}$. We can now simply differentiate u with respect to z or $\bar{z}$ to see that f extends smoothly to $\overline{\Omega}$.

This argument yields smoothness for the Riemann mapping function. In general, if f is a biholomorphic map between multiply connected domains, we can repeatedly use the Riemann mapping theorem to reduce the problem to the case of a map between domains with real analytic boundaries. In this situation, one can apply the Schwarz reflection principle and since the Riemann mappings are known to extend smoothly, it follows that f itself is smooth up to the boundary. It is interesting to note that Painlevé [Pa] proved the smooth extendibility of conformal maps between smooth domains in 1887, long before Carathéodory's work on continuous extension (see [BK] for the history of this problem).

This method does not lend itself to generalization to several variables since harmonic functions have little to do with holomorphic functions in several variables. There is an alternative approach, pioneered by Stefan Bergman [Berg], which does have potential for generalization. Bergman found that the Riemann mapping function f of a simply connected domain is related to the *Bergman kernel function* $K(z,w)$ via the formula $f'(z) = cK(z,a)$ where c is a constant and $a = f^{-1}(0)$. Thus, if it can be shown that K is smooth up to the boundary, it would follow that f is too. The Bergman kernel is related to a $\bar{\partial}$-problem and is a more useful device in a several variable setting. Indeed, C. Fefferman [F] used the Bergman kernel and the associated metric to prove the first, and still the most important, theorem about boundary regularity of mappings in several variables.

Theorem 4.5. *A biholomorphic map between smooth bounded strictly pseudoconvex domains in* $\mathbb{C}^n$ *extends smoothly to the boundary.*

We shall now outline a proof of this theorem different from Fefferman's original argument. It evolved from Fefferman's proof via simplifying ideas introduced by Webster [W 2], Ligocka [Li] and Bell [Be 1, Be 2, BL]; see [DF 2] for an account of the history of this proof.

Proof of Fefferman's Theorem. Let Ω_1, Ω_2 be smooth bounded strictly pseudoconvex domains in $\mathbb{C}^n$ and $f : \Omega_1 \to \Omega_2$ a biholomorphic map. Let P_1, P_2 denote the Bergman projections associated to Ω_1, Ω_2 respectively. (The Bergman projection associated to a domain D is the orthogonal projection of $L^2(D)$ onto the subspace of holomorphic functions in L^2.) Let $u = \det(f')$ be the (holomorphic) jacobian determinant of f. A simple Hilbert space argument (using change of variables in a multiple integral and the fact that $|u|^2$ is the real jacobian of f viewed as a mapping on $\mathbb{R}^{2n}$ with orientation given by the complex structure of $\mathbb{C}^n$) shows that these Bergman projections transform under f according to the formula $u \cdot ((P_2\varphi) \circ f) = P_1(u \cdot (\varphi \circ f))$.

We use the transformation formula in conjunction with the following fact. Let h be holomorphic on Ω_2, $h \in C^\infty(\overline{\Omega}_2)$. Then, there is a function $\varphi \in C^\infty(\overline{\Omega}_2)$

which vanishes to infinite order on $\partial\Omega_2$ and such that $P_2\varphi = h$. To construct φ, the C^∞ version of the Cauchy-Kowalewski theorem implies the existence of $\psi \in C^\infty(\overline{\Omega}_2)$ such that $h - \Delta\psi$ vanishes to infinite order on $\partial\Omega_2$, while ψ and $\nabla\psi$ both vanish on $\partial\Omega_2$. Integration by parts shows that $\Delta\psi$ is orthogonal to holomorphic functions, and we may take $\varphi = h - \Delta\psi$.

Given a holomorphic function h smooth up to the boundary of Ω_2, let φ be a C^∞ function with the properties given above (in particular $P_2\varphi = h$). The transformation formula gives $u \cdot (h \circ f) = u \cdot ((P_2\varphi) \circ f) = P_1(u \cdot (\varphi \circ f))$. We must now show that the fact that φ vanishes to infinite order on $\partial\Omega_2$ implies that $u \cdot (\varphi \circ f) \in C^\infty(\overline{\Omega}_1)$. Once this is done, we proceed as follows: since the Bergman projection associated to a smooth strictly pseudoconvex domain preserves the space of functions smooth up to the boundary (see below), we have $u \cdot (h \circ f) \in C^\infty(\overline{\Omega}_1)$. Taking $h \equiv 1$, it follows that $u \in C^\infty(\overline{\Omega}_1)$. Applying this to f^{-1}, we see that $\det((f^{-1})') = (u \circ f^{-1})^{-1}$ is smooth up to the boundary of Ω_2; in particular, u has no zeros on $\overline{\Omega}_1$. Hence $u \cdot (h \circ f) \in C^\infty(\overline{\Omega}_1)$ if and only if $h \circ f \in C^\infty(\overline{\Omega}_1)$. Taking now $h = z_j$ (the coordinate functions), it follows that f extends smoothly to $\overline{\Omega}_1$.

Thus, it remains to show that $u \cdot (\varphi \circ f) \in C^\infty(\overline{\Omega}_2)$. Since the components of f are bounded holomorphic functions, we have $|\partial^\alpha f_j / \partial z^\alpha| \leq c_\alpha d_1(z)^{-|\alpha|}$ (by the Cauchy inequalities) where the constant c_α is independent of $z \in \Omega_1$ and $d_j(z)$ denotes the distance of z from the boundary of $\Omega_j (j = 1, 2)$. Since φ vanishes to infinite order on $\partial\Omega_2$, to show that the derivatives of $u \cdot (\varphi \circ f)$ are bounded, it suffices to check that $d_2(f(z)) \leq C d_1(z)$ (with the constant C independent of $z \in \Omega_1$). To do this, let ρ be a strictly plurisubharmonic function defining $\partial\Omega_1$. Then $\rho \circ f^{-1}$ is subharmonic and < 0 on Ω_2, so that, by the Hopf lemma (see next paragraph), $\rho \circ f^{-1}(z) \leq -C d_2(z)$. Since $|\rho(w)| \leq C d_1(w)$, the required inequality follows and completes the proof.

The Hopf lemma asserts the following. If Ω is a bounded domain with C^2 boundary in $\mathbb{R}^n$, $a \in \partial\Omega$ and $\nu(a)$ is the inward normal to $\partial\Omega$ at a, then, for any subharmonic function u on Ω with $u < 0$ on Ω, there is a constant $c > 0$ such that

$$\overline{\lim} \frac{u(x)}{|x - a|} \leq -c \qquad \forall a \in \partial\Omega$$

where the limit superior is as $x \to a$ along $\nu(a)$. The proof is as follows. Let $B_R(x_0) = \{x \in \mathbb{R}^n \mid |x - x_0| < R\}$ be a ball of radius R about $x_0 \in \Omega$ tangent to $\partial\Omega$ at $a, B_R(r_0) \subset \Omega$. Then, if $0 < r < R$, the function $v(x) = e^{-\lambda|x-x_0|^2} - e^{-\lambda R^2}$ is subharmonic on $B_R(x_0) - \overline{B}_r(x_0)$ if λ is large (compared to r) and $v = 0$ for $|x - x_0| = R$. We apply the maximum principle to $u + \varepsilon v$ on $B_R(x_0) - \overline{B}_r(x_0)$ with small ε (depending only on $\sup\{u : |x - x_0| \leq r\}$ to conclude that $u(x) \leq -\varepsilon v(x)$ for $r < |x - x_0| < R$. When Ω has C^2 boundary, x_0 can be chosen in a fixed compact subset of Ω and R bounded below by $R_0 > 0$. The lemma follows easily. Of course, this inequality is only of interest when $\underline{\lim} u(x) = 0$, $x \to a$ along $\nu(a)$. E. Hopf [Hop] used this argument to prove a strong maximum principle for general second order linear elliptic operators.

The proof sketched above of Fefferman's theorem can be generalized to several other situations, notably to the case of proper maps between certain pseudoconvex domains. This was done by Bell and Catlin [BC] and by Diederich and Fornaess [DF 2]. Before stating this theorem, we give a very simple example to illustrate the behavior. Let $p, q > 0$ be integers, $f(z_1, z_2) = (z_1^p, z_2^q)$. f maps the smooth (weakly) pseudoconvex domain $E = \{z \in \mathbb{C}^2 | \, |z_1|^{2p} + |z_2|^{2q} < 1\}$ onto the unit ball $B = \{z \in \mathbb{C}^2\} | \, |z_1|^2 + |z_2|^2 < 1\}$. The (complex) jacobian $u(z_1, z_2) = pqz_1^{p-1}z_2^{q-1}$ has zeros inside E if p or $q > 1$, and, unlike the situation in one variable, Hartogs' theorem implies that branching must occur *on* ∂E; f is an unramified covering map (of degree pq) of $E - V$ onto $B - V$, where $V = \{z_1 z_2 = 0\}$.

Theorem 4.6. *Let Ω_1, Ω_2 be smooth (weakly) pseudoconvex domains in $\mathbb{C}^n$ and $f : \Omega_1 \to \Omega_2$, a proper holomorphic map. If the Bergman projection associated to Ω_1 maps the space of functions smooth up to $\partial\Omega_1$ into itself, then f extends smoothly to $\overline{\Omega}_1$.*

We shall now outline the modifications of the proof we gave above of Fefferman's theorem needed to prove this more general result.

First, f is a ramified covering, say of degree m. Let V_2 be the branch locus of f in Ω_2 and $V_1 = f^{-1}(V_2)$; f is an unramified covering of m sheets of $\Omega_1 - V_1$ onto $\Omega_2 - V_2$. (Also, $f^{-1}(w)$ contains less than m points if $w \in V_2$.) The first step is to show that the Bergman projections transform under proper maps exactly as they do under biholomorphic maps. Next, one must show that $u \cdot (\varphi \circ f) \in C^\infty(\overline{\Omega}_1)$ whenever $\varphi \in C^\infty(\overline{\Omega}_2)$ and vanishes to infinite order on $\partial\Omega_2$. The problem reduces to showing that there is an integer $N > 0$ and a constant $C > 0$ so that $d_2(f(z))^N \leq C d_1(z)$. If Ω_1 is strictly pseudoconvex, the argument given earlier shows that we may take $N = 1$. In general, we use the following theorem of Diederich-Fornaess [DF 1] (see Range [Ra] for this argument): there is a C^∞ boundary defining function p for Ω_1 and an integer $N > 0$ such that the function $\rho = -(-p)^{\frac{1}{N}}$ is continuous on $\overline{\Omega}_1$, and plurisubharmonic on Ω_1. (It is, of course, < 0 on Ω, and $= 0$ on $\partial\Omega_1$.) We consider the function $\lambda(w) = \sup\{\rho(z); z \in f^{-1}(w)\}$; it is clearly a continuous plurisubharmonic function < 0 on $\Omega_2 - V_2$ and so extends to a plurisubharmonic function on Ω_2. Since f is proper, we see that λ is an exhaustion function on Ω_2 which extends continuously to $\partial\Omega_2$ and $= 0$ on $\partial\Omega_2$. We may apply the Hopf lemma to conclude that $\lambda(w) \leq -Cd_2(w)$ and the desired inequality follows easily.

Thus, we have shown that $u \cdot (h \circ f)$ is smooth on $\overline{\Omega}_1$ if h is holomorphic on Ω_2 and smooth on $\overline{\Omega}_2$. We can no longer conclude that $u \neq 0$ on $\overline{\Omega}_1$ and the last step in the proof is a division theorem. One shows that the order of vanishing of u on $\partial\Omega_1$ is bounded, and that this, together with the fact that $u \cdot f_j^k$ is smooth on $\overline{\Omega}_1$ for *all* integers $k \geq 0$ (and all $j = 1, \ldots, n$) implies that u divides uf_j in the ring of holomorphic functions which are smooth on $\overline{\Omega}_1$.

Thus f extends smoothly to $\overline{\Omega}_1$. Details of this argument will be found in [BC] and [DF 2].

This immediately raises the question of when the Bergman projection preserves the space of functions smooth up to the boundary. In one variable, this holds for any smooth bounded domain as follows from Spencer's formula

$$P\varphi = \varphi - 4\frac{\partial}{\partial z}G\frac{\partial}{\partial \bar{z}}\varphi$$

where G is the classical Green's operator for the Dirichlet problem, i.e. G satisfies $\Delta Gu = u$ and $Gu = 0$ on the boundary. Since G is known to preserve smoothness up to the boundary, the same is true of P. Spencer's formula might give the impression that the Bergman projection is also too closely tied to harmonic functions to be of use in several variables. But, as mentioned earlier, it is tied rather to a $\bar{\partial}$-problem. If Ω is a smooth plane domain, the $\bar{\partial}$-problem can be described as follows. Given $\alpha \in C^\infty(\overline{\Omega})$, the Cauchy integral furnishes a solution of $\frac{\partial u}{\partial \bar{z}} = \alpha$ with $u \in C^\infty(\Omega)$. Since u is only determined up to addition of a holomorphic function, it is natural to consider that solution of $\frac{\partial u}{\partial \bar{z}} = \alpha$ which is $L^2(\Omega)$-orthogonal to holomorphic functions in L^2. The $\bar{\partial}$-problem *is to prove that this particular solution is in* $C^\infty(\overline{\Omega})$ *when* α *is.* The solution operator Λ in this case is given in terms of Green's operator: $\Lambda = 4\frac{\partial}{\partial z}G$. In several variables, the situation is more complicated.

Let Ω be a bounded domain in $\mathbb{C}^n$ with smooth boundary. As in one variable, if $v \in C^\infty(\overline{\Omega})$, the Bergman projection P can be written in terms of the solution operator Λ of a $\bar{\partial}$-problem: $Pv = v - \Lambda\bar{\partial}v$, where the $\bar{\partial}$-problem is now the following:

$\bar{\partial}$-Problem. *Let* $\alpha = \sum_{j=1}^n \alpha_j d\bar{z}_j$ *be a* $(0,1)$*-form on* Ω *with coefficients in* $C^\infty(\overline{\Omega})$. *Suppose that* $\bar{\partial}\alpha = 0$ *(i.e.* $\frac{\partial \alpha_j}{\partial \bar{z}_k} = \frac{\partial \alpha_k}{\partial \bar{z}_j}\ \forall j,k$*). Then* $\Lambda\alpha$ *is the function* u *in* $C^\infty(\Omega)$ *(if it exists) for which* $\bar{\partial}u = \alpha$ *(i.e.* $\frac{\partial u}{\partial \bar{z}_j} = \alpha_j\ \forall j$*) and which is orthogonal to holomorphic functions on* Ω. *Does* $\Lambda\alpha$ *lie in* $C^\infty(\overline{\Omega})$ *if the coefficients of* α *do?*

It seems a very difficult question to decide when Λ has this property. When Ω is not pseudoconvex, Λ may not even exist and Barrett [Ba] has given examples of bounded *non-pseudoconvex* domains with real analytic boundaries such that the Bergman projection does not even map $C_0^\infty(\Omega)$ into $L^\infty(\Omega)$ [$C_0^\infty(\Omega)$ denotes, as usual, smooth functions with compact support in Ω]. However, when Ω is strictly pseudoconvex, Λ does preserve $C^\infty(\overline{\Omega})$ as was shown by Kohn [K 1, K 2]. His analogue of Spencer's formula for the Bergman projection is

$$P\varphi = \varphi - \bar{\partial}^* N\bar{\partial}\varphi$$

where N is the so called $\bar{\partial}$-Neumann operator. The best results to date on the regularity properties of the $\bar{\partial}$-Neumann operator have been obtained by Catlin [Ca 1,2,3] who uses the machinery of D'Angelo [D'A] to measure the degree of *weak* pseudoconvexity. Kohn's article [K 3] contains an excellent survey of this subject.

The ideas in the proof sketched earlier of Theorem 4.6 have other applications. We shall show how they can be used to prove the following fundamental theorem of Alexander [Al 1]; for a very elegant, direct proof, see the book of Rudin [Ru].

In the rest of this article, we denote by B^n the unit ball $\{z \in \mathbb{C}^n | \sum_{j=1}^n |z_j|^2 < 1\}$ in $\mathbb{C}^n$ and by Δ^n the unit polydisc $\{z \in \mathbb{C}^n | \, |z_j| < 1, j = 1, \ldots, n\}$.

Theorem 4.7. *A proper holomorphic map $f : B^n \to B^n$ is necessarily biholomorphic if $n > 1$.*

Sketch of Proof. Let $B = B^n$ and $\rho(z) = |z|^2 - 1$ the standard defining function of B. If $f : B \to B$ is proper, then, Theorem 4.6 and the Hopf lemma argument given in Theorem 4.5 show that $\rho \circ f$ is also a defining function for B. For any function φ, denote by H_φ the $(n+1) \times (n+1)$ augmented Hessian determinant

$$H_\varphi = \det \begin{bmatrix} 0 & \partial\varphi/\partial\bar{z}_j \\ \partial\varphi/\partial z_k & \partial^2\varphi/\partial z_k \partial\bar{z}_j \end{bmatrix} .$$

The strict convexity of B implies that H_φ has no zeros on ∂B for any defining function φ. Further, the chain rule gives

$$H_{\rho\circ f} = |u|^2 H_\rho \circ f$$

where $u = \det(f')$. Since $H_{\rho\circ f} \neq 0$ on ∂B, it follows that $u \neq 0$ on ∂B. But, if $n > 1$, Hartogs' theorem shows that $u \neq 0$ on B. Thus f is an unramified covering, and since B is simply connected, the theorem follows.

This argument can be generalized to proper self maps of strictly pseudoconvex (smooth) domains. However, a much better approach was found by Pinčuk [P 1], who used Alexander's theorem directly, together with certain non-homogeneous dilations to prove the following general result.

Theorem 4.8. *A proper holomorphic map of a bounded strictly pseudoconvex domain in $\mathbb{C}^n$ with C^2 boundary into itself is necessarily biholomorphic if $n > 1$.*

This has been generalized to domains on certain manifolds by Burns-Shnider [BuSh]. Also, the argument given above to prove Alexander's theorem was generalized by Bedford and Bell [BB] to prove the following theorem.

Theorem 4.9. *Let Ω be a bounded weakly pseudoconvex domain in $\mathbb{C}^n$ with real analytic boundary. If $n > 1$, then any proper holomorphic self map of Ω is biholomorphic.*

Theorem 4.1 implies, of course, that the unit ball and the unit polydisc are *not* biholomorphic if $n > 1$ (proved, at least for $n = 2$, by Poincaré). One sometimes adds the misleading comment that there is no Riemann mapping theorem in several variables. We next describe two positive mapping theorems. It is not hard to see that the only plane domains $\Omega \neq \mathbb{C}$ for which the automorphism group is transitive are those biholomorphic to the disc. Bun Wong [Wo] showed that the corresponding statement is true for strictly pseudoconvex domains in $\mathbb{C}^n$, $n > 1$. Later, Rosay [Ro] generalized it to the following statement.

Theorem 4.10. *A bounded domain in $\mathbb{C}^n$ with C^2 boundary whose automorphism group is transitive is biholomorphic to the unit ball in $\mathbb{C}^n$; i.e. a bounded homogeneous domain with C^2 boundary is biholomorphic to the ball.*

It should be remarked that bounded homogeneous domains are always (without any assumptions on the boundary) domains of holomorphy, i.e. pseudoconvex. There are, of course, many such domains without smooth boundary.

The next theorem, due to B. Fridman [Fri], may claim to be an approximate Riemann mapping theorem.

Theorem 4.11. *Let D be a bounded pseudoconvex domain in $\mathbb{C}^n$ diffeomorphic to the unit ball B^n. Given $\varepsilon > 0$, there exist domains $\Omega_1 \subset D, \Omega_2 \subset B^n$ whose boundaries are within ε of those of D and B^n respectively such that Ω_1 and Ω_2 are biholomorphically equivalent.*

Let us mention also a recent result of Bedford and Pinčuk [BP] which confirms, at least partially, a conjecture of Greene and Krantz [GK].

Theorem 4.12. *A (weakly) pseudoconvex domain in $\mathbb{C}^2$ with real analytic boundary whose automorphism group is non-compact is biholomorphic to a domain of the form $\{(z,w) \in \mathbb{C}^2 \mid |z|^2 + |w|^{2p} < 1\}, p > 0$ an integer.*

We have described above the approach to boundary regularity based on the Bergman projection and results which can be proved using related techniques. Different methods have been found. Pinčuk has used the method of non-homogeneous dilations which he introduced in connection with Theorem 4.8 to give a new and elementary proof of the Wong-Rosay theorem. Further, he and Hasanov [PH] have used it in conjunction with a C^∞ version of the "edge of the wedge" theorem to give a direct elementary proof of Fefferman's theorem which gives results about extensions in domains with C^k boundary ($k \leq \infty$). Lempert [Lem 2] obtained these results independently using his study of extremal discs for the Kobayashi metric. Other noteworthy elementary proofs of Fefferman's theorem were given by Nirenberg, Webster and Yang [NWY] and by Lempert [Lem 1]. The survey article of Forstnerič [FF 1] contains an excellent summary of these various methods.

If we consider the questions of extension of holomorphic maps *past* the boundary, the Schwarz reflection principle gives the definitive answer in one variable. In several variables, the simple geometric picture of reflection is lost; it has been restored only for strictly pseudoconvex domains. The first, and most basic, theorem was proved by Lewy [Le] and Pinčuk [P 2].

Theorem 4.13. *Let Ω_1, Ω_2 be bounded strictly pseudoconvex domains in $\mathbb{C}^n$ with real analytic boundaries and let $f : \Omega_1 \to \Omega_2$ be a proper holomorphic map. Then, there exist domains Ω_1', Ω_2' with $\overline{\Omega}_j \subset \Omega_j' (j = 1, 2)$ such that f extends to a proper holomorphic map of Ω_1' to Ω_2'.*

Webster [W 1] has given a very nice geometric interpretation of this result in terms of the geometry of strictly pseudoconvex domains and the edge of the wedge theorem.

Baouendi, Jacobowitz and Trèves [BJT] proved this theorem when Ω_1, Ω_2 are only weakly pseudoconvex but assuming that f is biholomorphic. Baouendi, Bell and Rothschild [BBR] extended this to proper mappings f when $n = 2$, and finally Baouendi and Rothschild [BR 1] and Diederich and Fornaess [DF 3] proved this for arbitrary n (when the Ω_j are weakly pseudoconvex and f is proper). Local versions of the extension theorem appear to be more difficult. Recently, Baouendi and Rothschild [BR 2] have obtained what appears to be the optimal theorem in $\mathbb{C}^2$.

We shall conclude this article by describing some of the main results concerning proper maps into the ball B^N or the polydisc Δ^N. The general area of proper maps of a bounded domain into one of larger dimension is currently very active, and we refer the reader to the survey of Forstnerič [FF 1] for a thorough treatment of the subject as it stands today. This area was initiated by the following theorem of Webster [W 3].

Theorem 4.14. *If $n \geq 3$, then every proper holomorphic map f from B^n to B^{n+1} which is at least C^3 up to the boundary is equivalent to the linear embedding: there exist automorphisms σ_1 of B^n and σ_2 of B^{n+1} such that $\sigma_2 \circ f \circ \sigma_1$ is the map $z \mapsto (z, 0)$.*

When $n = 2$, this result is no longer true as it stands. J. Faran [Fa] proved the following theorem.

Theorem 4.15. *Let f be a proper holomorphic map of B^2 to B^3 which is C^3 up to the boundary. Then, f is equivalent (in the same sense as in Theorem 4.14) to one of the following four maps:*

$$(z, w) \mapsto (z, w, 0)$$
$$(z, w) \mapsto (z^2, \sqrt{2}\, zw, w^2)$$
$$(z, w) \mapsto (z^3, \sqrt{3}\, zw, w^3)$$
$$(z, w) \mapsto (z^2, zw, w)$$

Cima and Suffridge [CS 1] reduced the smoothness requirement at the boundary from C^3 to C^2. Rather surprisingly, some smoothness at the boundary turns out to be necessary in these results; mere continuity is certainly not sufficient. As for maps into balls with larger codimension, Forstnerič [FF 2] has recently proved the following general theorem.

Theorem 4.16. *Let $N > n > 1$. Then, any proper holomorphic map from B^n to B^N which is C^{N-n+1} up to the boundary is a rational map.*

Moreover, Cima and Suffridge [CS 2] have shown that rational proper holomorphic maps cannot have singular points on the boundary, and so extend holomorphically past the boundary.

It is a natural question, and one of very great interest, to find analogues of the embedding theorems for Stein manifolds (Sect. 3) in the case of bounded domains. Of course, there is no single standard domain into which one might attempt to map them; the two obvious candidates are B^N and Δ^N (the ball and the polydisc). Here, the situation is more complicated than embedding in $\mathbb{C}^N$; one has the following result of N. Sibony [Si].

Theorem 4.17. *There exists a bounded pseudoconvex Reinhardt domain in $\mathbb{C}^2$ having C^∞ smooth boundary which does not admit a proper holomorphic map into any bounded (euclidean) convex open set in $\mathbb{C}^N$ for any N.*

It turns out, again rather surprisingly, that strictly pseudoconvex domains do admit proper maps into both B^N and Δ^N. These positive results really go back to the construction by Aleksandrov [Al 1] (in 1982) of inner functions on the ball B^n; these are bounded holomorphic functions whose radial limits have absolute value 1 almost everywhere on ∂B^n; it had long been believed that inner functions did not exist when $n > 1$. A slightly weaker statement had been obtained by Hakim and Sibony [HM]. Using the ideas of these constructions, Aleksandrov [A 2] (for the ball) and Løw [Lø1] proved the following theorem.

Theorem 4.18. *Let Ω be a bounded strictly pseudoconvex domain in $\mathbb{C}^n$ with smooth boundary. Then Ω can be properly embedded in Δ^N for some N.*

Løw [Lø 2] and Forstneric [FF 3] have also shown that strictly pseudoconvex domains Ω can be embedded into B^N for some N. Some of these embeddings do not even extend continuously to $\partial\Omega$, while some are continuous but not differentiable. There have been several refinements of these theorems.

It is of course natural to ask about the smallest codimension that can be allowed in these theorems. The analogue of Bishop's theorem (Theorem 3.2 part 2) would be that there are proper maps in codimension 1 also in this context. One of the present authors long believed that such maps did not exist. However, Berit Stensønes [St] proved the following theorem.

Theorem 4.19. *Any bounded strictly pseudoconvex domain in $\mathbb{C}^n$ with smooth boundary admits a proper holomorphic map into Δ^{n+1}.*

The proof of Stensønes involves a very ingenious adaptation of the constructions of Hakim-Sibony and of Løw referred to earlier. Moreover, Dor [D] and Hakim [H] have shown that *strictly pseudoconvex domains in $\mathbb{C}^n$ admit proper holomorphic maps into B^{n+1} which are continuous but not C^1 up to the boundary.* Thus, as mentioned earlier, some regularity assumption is necessary in Webster's theorem (Theorem 4.14). Just how much regularity would force the maps to extend across the boundary (if the boundary is, say real-analytic) remains open. Do there exist, for example, proper maps of B^n into B^N (N large) which are C^1 but do not extend holomorphically past the boundary? Forstnerič's theorem (4.16) asserts that they do not exist if the map is C^{N-n+1}.

The theorem of Stensønes implies that the analogue of the standard embedding theorem holds: *strictly pseudoconvex domains in $\mathbb{C}^n$ with smooth boundary can be properly embedded in Δ^{2n+1}.* Since the analogy with Stein manifolds seems rather strong, one would expect to be able to lower the codimension. Is it, for instance, possible to embed B^2 in Δ^3? (This seems unlikely.) Also, the theorem of Løw-Forstnerič (referred to after 4.18) furnishes embeddings of a strictly pseudoconvex domain Ω into B^N for an N depending on Ω. Can one choose N independent of Ω (depending only on the dimension of Ω)? This problem is probably more delicate than that of embeddings in Δ^N. This circle of ideas is certainly worthy of further study.

References

[AM] Abhyankar, S.S., Moh, T.T.: Embeddings of the line in the plane. J. reine u. angew. Math. **276** (1975) 148–166

[ABT] Acquistapace, F., Broglia, F., Tognoli, A.: A relative embedding theorem for Stein spaces. Ann. Scuol. Norm. Sup. Pisa **2** (1975) 507–522

[A 1] Aleksandrov, A.B.: The existence of inner functions in the ball. Mat. Dokl. Akad. Nauk SSSR **118** (1982) 147–163

[A 2] Aleksandrov, A.B.: Proper holomorphic mappings from the ball into a polydisc. Dokl. Akad. Nauk SSSR **286** (1986) 11–15 [Engl. transl. Sov. Math. Dokl. **33** (1986) 1–5]

[Al 1] Alexander, H.: Proper holomorphic mappings in $\mathbb{C}^n$. Indiana Math. J. **26** (1977) 137–146

[Al 2] Alexander, H.: Explicit imbedding of the (punctured) disc into $\mathbb{C}^2$. Comm. Math. Helv. **52** (1977) 539–544

[AF] Andreotti, A., Frankel, T.: The Lefschetz theorem on hyperplane sections. Ann. Math. **69** (1959) 713–717

[BS] Bănică, C., Stănăşilă, O.: Algebraic methods in the global theory of complex spaces. John Wiley, New York 1976

[BBR] Baouendi, M.S., Bell, S., Rothschild, L.P.: Mappings of three dimensional *CR* manifolds and their holomorphic extension. Duke Math. J. **56** (1988) 503–530

[BJT] Baouendi, M.S., Jacobowitz, H., Trèves, F.: On the analyticity of *CR* mappings. Ann. Math. **122** (1985) 365–400
[BR 1] Baouendi, M.S., Rothschild, L.P.: Germs of *CR* maps between real analytic hypersurfaces. Invent. math. (to appear)
[BR 2] Baouendi, M.S., Rothschild, L.P.: A general reflection principle in $\mathbb{C}^2$. Preprint
[Ba] Barrett, D.: Irregularity of the Bergman projection on a smooth bounded domain in $\mathbb{C}^n$. Ann. Math. **119** 431–436
[BCW] Bass, H., Connell, E.H., Wright, D.: The Jacobian conjecture. Bull. Am. Math. Soc. **7** (1982) 287–330
[Bed] Bedford, E.: Proper holomorphic mappings. Bull. Am. Math. Soc. **10** (1984) 157–175
[BB] Bedford, E., Bell, S.: Proper self maps of weakly pseudoconvex domains. Math. Ann. **261** (1982) 47–49
[BP] Bedford, E., Pinčuk, S.: Domains in $\mathbb{C}^2$ with noncompact holomorphic morphism groups. Math. Sb. **135** (1988) 147–157
[Be 1] Bell, S.: Non-vanishing of the Bergman kernel function at boundary points of certain domains in $\mathbb{C}^n$. Math. Ann. **244** (1979) 69–74
[Be 2] Bell, S.: Biholomorphic mappings and the $\bar{\partial}$-problem. Ann. Math. **114** (1981) 103–113
[BC] Bell, S., Catlin, D.: Boundary regularity of proper holomorphic mappings. Duke Math. J. **49** (1982) 385–396
[BK] Bell, S., Krantz, S.: Smoothness at the boundary of conformal maps. Rocky Mountain J. Math. **17** (1987) 23–40
[BL] Bell, S., Ligocka, E.: A simplification and extension of Fefferman's theorem on biholomorphic mappings. Invent. math. **57** (1980) 283–289
[Berg] Bergman, S.: The kernel function and conformal mapping. American Mathematical Society, Providence 1970
[Bi] Bishop, E.: Mappings of partially analytic spaces. Am. J. Math. **83** (1961) 209–242
[BM] Bochner, S., Martin, W.T.: Several complex variables. Princeton University Press, Princeton 1948
[Bo] Borel, A., et. al.: Algebraic $\mathscr{D}$-modules. Academic Press, New York 1987
[BuSh] Burns, D., Shnider, S.: Real hypersurfaces in complex manifolds. Proc. Symp. Pure Mathematics, vol. 30. American Mathematical Society 1977, pp. 141–168
[C 1] Cartan, H.: Sur les fonctions de plusieurs variables complexes. L'itération des transformations intérieures d'un domaine borné. Math. Zeit. **35** (1932) 760–773
[C 2] Cartan, H.: Quotients of complex analytic spaces. Int. Coll. on Function Theory, Tata Inst. Bombay, 1960, pp. 1–15
[C 3] Cartan, H.: Familles d'espaces complexes et fondements de la géométrie analytique. Sém. ENS, 1960–61
[Ca 1] Catlin, D.: Necessary conditions for subellipticity of the $\bar{\partial}$-Neumann problem. Ann. Math. **117** (1983) 147–171
[Ca 2] Catlin, D.: Boundary invariants of pseudoconvex domains. Ann. Math. **120** (1984) 529–586
[Ca 3] Catlin, D.: Subelliptic estimates for the $\bar{\partial}$-Neumann problem on pseudoconvex domains. Ann. Math. **126** (1987) 131–191
[CM] Chern, S., Moser, J.: Real analytic hypersurfaces in complex manifolds: Acta math. **133** (1974) 219–271
[CS 1] Cima, J., Suffridge, T.: A reflection principle with applications to proper holomorphic mappings. Math. Ann. **265** (1983) 489–500
[CS 2] Cima, J., Suffridge, T.: Boundary behavior of rational proper maps. Preprint

[D'A] D'Angelo, J.: Real hypersurfaces, orders of contact, and applications. Ann. Math. **115** (1982) 615–637

[DF 1] Diederich, K., Fornaess, J.E.: Pseudoconvex domains: Bounded strictly plurisubharmonic exhaustion functions. Invent. math. **39** (1977) 129–141

[DF 2] Diederich, K., Fornaess, J.E.: Boundary regularity of proper holomorphic mappings. Invent. math. **67** (1982) 363–384

[DF 3] Diederich, K., Fornaess, J.E.: Proper holomorphic mappings between real-analytic pseudoconvex domains in $\mathbb{C}^n$. Preprint

[D] Dor, A.: Proper holomorphic maps from strongly pseudonvex domains in $\mathbb{C}^2$ to the unit ball in $\mathbb{C}^3$... Ph.D. Thesis, Princeton 1987

[DV] Douady, A., Verdier, J.L.: Sém. de géométrie analytique de l'ENS. Astérisque **16** (1971–72)

[Fa] Faran, J.J.: Maps from the two-ball to the three-ball. Invent. math. **68** (1982) 441–475

[F] Fefferman, C.: The Bergman kernel and biholomorphic mappings of pseudoconvex domains. Invent. math. **26** (1974) 1–65

[FS] Fornaess, J.E., Stensønes, B.: Lectures on counterexamples in several complex variables. Princeton Math. Notes **8** (1987)

[Fo 1] Forster, O.: Plongements de variétés de Stein. Comm. Math. Helv. **45** (1970) 170–184

[Fo 2] Forster, O.: Power series methods in deformation theory. Proc. Symp. Pure Math., vol. 2. American Mathematical Society 1977, pp. 199–217

[FK] Forster, O., Knorr, K.: Ein Beweis des Grauertschen Bildgarbensatzes nach Ideen von B. Malgrange. Manuscripta math. **5** (1971) 19–44

[FR] Forster, O., Ramspott, K.J.: Homotopieklassen von Idealbasen in Steinschen Algebren. Invent. math. **5** (1968) 255–276

[FF 1] Forstnerič, F.: Proper holomorphic mappings: A survey. (Princeton Lecture Notes in Mathematics). Princeton University Press, Princeton 1990

[FF 2] Forstnerič, F.: Extending proper holomorphic mappings of positive codimension. Invent. math. **95** (1989) 31–62

[FF 3] Forstnerič, F.: Embedding strictly pseudoconvex domains into balls. Trans. Am. Math. Soc. **295** (1986)

[Fri] Fridman, B.: An approximate Riemann mapping theorem in $\mathbb{C}^n$. Math. Ann. **257** (1986) 49-55

[Fris] Frisch, J.: Points de platitude d'un morphisme d'espaces analytiques complexes. Invent. math. **4** (1967) 118–138

[Fu] Fulton, W.: Intersection theory. (Ergebnisse der Mathematik, vol. 2). Springer, Berlin Heidelberg 1984

[FuL] Fulton, W., Lang, S.: Riemann-Roch algebra. (Grundlehren der mathematischen Wissenschaften, vol. 277). Springer, Berlin Heidelberg 1985

[GM] Goresky, M., MacPherson, R.: Stratified morse theory. (Ergebnisse der Mathematik, vol. 14). Springer, Berlin Heidelberg 1988

[G 1] Grauert, H.: Charakterisierung der holomorph-vollständigen komplexen Räume. Math. Ann. **129** (1955) 233–259

[G 2] Grauert, H.: Analytische Faserungen über holomorph-vollständigen Räumen. Math. Ann. **135** (1958) 263–273. (See also the two earlier papers in Math. Ann. **133** (1957) 139–159, 450–472

[G 3] Grauert, H.: Ein Theorem der analytischen Garbentheorie und die Modulräume komplexer Strukturen. Publ. Math. IHES **5** (1960) 233–292

[G 4] Grauert, H.: Über Modifikationen und exzeptionelle analytische Mengen. Math. Ann. **146** (1962) 331–368

[G 5] Grauert, H.: Set theoretic complex equivalence relations. Math. Ann. **265** (1983) 137–148

[GR 1] Grauert, H., Remmert, R.: Theorie der Steinschen Räume. (Grundlehren der mathematischen Wissenschaften, vol. 227). Springer, Berlin Heidelberg 1977

[GR 2] Grauert, H., Remmert, R.: Coherent analytic sheaves. (Grundlehren der mathemtischen Wissenschaften, vol. 265). Springer, Berlin Heidelberg 1984

[GR 3] Grauert, H., Remmert, R.: Komplexe Räume. Math. Ann. **136** (1958) 245–318

[GK] Greene, R., Krantz, S.: Biholomorphic self maps of domains. (Lecture Notes in Mathematics, vol. 1276). Springer, Berlin Heidelberg 1987, pp. 136–207

[Gr] Gromov, M.: Partial differential relations. (Ergebnisse der Mathematik, 3. Folge, vol. 9). Springer, Berlin Heidelberg 1986

[GE] Gromov, M., Eliashberg, J.: Nonsingular maps of Stein manifolds. Funct. Anal. Appl. **5** (1971) 82–83

[GN] Gunning, R.C., Narasimhan, R.: Immersion of open Riemann surfaces. Math. Ann. **174** (1967) 103–108

[H] Hakim, M.: Applications holomorphes propres continues de domaines strictement pseudoconvexes de $\mathbb{C}^n$ dans la boule unité de $\mathbb{C}^{n+1}$. Duke Math. J. **60** (1990) 115–133

[HS] Hakim, M., Sibony, N.: Fonctions holomorphes bornées sur la boule unité de $\mathbb{C}^n$. Invent. math. **67** (1982) 213–222

[Ham] Hamm, H.: Zum Homotopietyp Steinscher Räume. J. reine angew. Math. **338** (1983) 121–135 [See also the correction in: Hamm: Zum Homotopietyp q-vollständiger Räume, *same journal* **364** (1986), 1–9].

[Har] Hartshorne, R.: Residues and duality. (Lecture Notes in Mathematics, vol. 20). Springer, Berlin Heidelberg 1966

[Hop] Hopf, E.: Elementare Bemerkungen über die Lösungen partieller Differentialgleichungen zweiter Ordnung vom elliptischen Typus. Sitz. Preuss. Akad. Wiss. Berlin, Math.-Phys. Kl. **19** (1927) 147–152

[Hö] Hörmander, L.: An introduction to complex analysis in several variables. North-Holland, 1973

[Ka] Karchyauskas, K.: Homotopy properties of complex algebraic sets (Studies in Topology). Steklov Inst., Leningrad 1979

[KV] Kiehl, R., Verdier, J.L.: Ein einfacher Beweis des Kohärenzsatzes von Grauert. Math. Ann. **195** (1971) 24–50

[Kn] Knorr, K.: Der Grauertsche Projektionssatz. Invent. math. **12** (1971) 118–172

[Kod] Kodaira, K.: Complex manifolds and deformation of complex structures. (Grundlehren der mathematischen Wissenschaften, vol. 283). Springer, Berlin Heidelberg 1986

[K 1] Kohn, J.J.: Harmonic integrals on strongly pseudoconvex manifolds, I. Ann. Math. **78** (1963) 112–148

[K 2] Kohn, J.J.: Harmonic integrals on strongly pseudoconvex manifolds, II. Ann. Math. **79** (1964) 450–472

[K 3] Kohn, J.J.: A survey of the $\bar{\partial}$-Neumann problem. Proc. Pure Math. **41** (1984) 137–145

[L] Laufer, H.B.: Imbedding annuli in $\mathbb{C}^2$. J. Analyse Math. **26** (1973) 187–215

[Lem 1] Lempert. L.: La métrique de Kobayashi et la représentation des domaines sur la boule. Bull. Soc. Math. France **109** (1981) 427–474

[Lem 2] Lempert. L.: A precise result on the boundary regularity of biholomorphic mappings. Math. Zeit. **193** (1986) 559–579

[Le] Lewy, H.: On the boundary behavior of holomorphic mappings. Acad. Naz. Lincei **35** (1977) 1–8

[Li] Ligocka, E.: How to prove Fefferman's theorem without use of differential geometry. Ann. of Polish Math. **39** (1981) 117–130

[Lø 1] Løw, E.: The ball in $\mathbb{C}^n$ is a closed complex submanifold of a polydisc. Invent. math. **83** (1986) 405–410

[Lø 2] Løw, E.: Embeddings and proper holomorphic maps of strictly pseudoconvex domains into polydiscs and balls. Math. Zeit. **190** (1985) 401–410

[Mi] Milnor, J.: Morse theory. Ann. Math. Studies, Princeton 1963

[M] Mumford, D.: Abelian varieties. Oxford University Press 1970

[N 1] Narasimhan, R.: Imbedding of holomorphically complete complex spaces. Am. J. Math. **82** (1960) 917–934

[N 2] Narasimhan, R.: On imbeddings and immersions of Stein manifolds. Ist. Naz. Mat. Rome, Symp. Math. (1968) 297–301

[NWY] Nirenberg, L., Webster, S., Yang, P.: Local boundary regularity of holomorphic mappings. Comm. Pure Appl. Math. **33** (1980) 306–338

[OTT] O'Brian, N.R., Toledo, D., Tong, Y.L.L.: Grothendieck-Riemann-Roch for complex manifolds. Bull. Am. Math. Soc. **5** (1981)

[Pa] Painlevé, S.: Sur les lignes singulières des functions analytiques. Thèse. Gauthier-Villars, Paris 1887

[P 1] Pinčuk, S.: On proper holomorphic mappings of strictly pseudoconvex domains. Siberian Math. J.**15** (1974) 909-917

[P 2] Pinčuk, S.: On the analytic continuation of holomorphic mappings. Math. USSR Sb. **27** (1975) 375–392

[PH] Pinčuk, S., Hasanov, S.: Asymptotically holomorphic functions. Mat. Sb. **134** (1987) 546–555

[RR] Ramis, J.P., Ruget, G.: Complexes dualisants et théorèmes de dualité en géométrie analytique complexe. Publ. Math. IHES **38** (1970) 77–91

[RRV] Ramis, J.P., Ruget, R., Verdier, J.L.: Dualité relative en géométrie analytique complexe. Invent. math. **13** (1971) 261–283

[Ra] Range, R.E.: The Carathéodory metric and holomorphic maps on a class of weakly pseudoconvex domains. Pacific J. Math. **78** (1978) 173–189

[R 1] Remmert, R.: Projektionen analytischer Mengen. Math. Ann. **130** (1956) 410–441

[R 2] Remmert, R.: Sur les espaces analytiques holomorphiquement séparables et holomorphiquement convexes. C. R. Acad. Sci., Paris **243** (1956) 118–121

[RS 1] Remmert, R., Stein, K.: Über die wesentlichen Singularitäten analytischer Mengen. Math. Ann. **126** (1953) 263–306

[RS 2] Remmert, R., Stein, K.: Eigentliche holomorphe Abbildungen. Math. Zeit. **73** (1960) 159–189

[Ri] Rischel, H.: Holomorphe Überlagerungskorrespondenzen. Math. Scand. **15** (1964) 49–63

[Ro] Rosay, J.-P.: Sur une characterization de la boule parmi les domains de $\mathbb{C}^n$ par son groupe d'automorphismes. Ann. Inst. Four. Grenoble **29** (1979) 91-97

[Ru] Rudin, W.: Function theory in the unit ball of $\mathbb{C}^n$. (Grundlehren der mathematischen Wissenschaften, vol. 241). Springer, Berlin Heidelberg 1980

[Sch] Schafft, U.: Einbettungen Steinscher Mannigfaltigkeiten. Manuscripta math. **47** (1984) 175–186

[Si] Sibony, N.: Sur le plongement des domaines faiblement pseudoconvexes dans des domaines convexes. Math. Ann. **273** (1986) 209–214

[Sk] Skoda, H.: Applications de techniques L^2 à la théorie des idéaux d'une algèbre de fonctions holomorphes avec poids. Ann. Sci. E.N.S., Paris **5** (1972) 545–575

[S] Stein, K.: Analytische Zerlegungen komplexer Räume. Math. Ann. **132** (1956) 63–93

[St] Stensønes, B.: Proper holomorphic mappings from strongly pseudoconvex domains in $\mathbb{C}^n$ to the unit polydisc in $\mathbb{C}^{n+1}$. Preprint

[SZ] Stout, E.L., Zame, W.R.: A Stein manifold topologically but not holomorphically equivalent to a domain in $\mathbb{C}^n$. Adv. Math. **60** (1986) 154–160

[W 1] Webster, S.M.: On the reflection principle in several complex variables. Proc. of the Am. Math. Soc. **71** (1978) 26–28

[W 2] Webster, S.M.: Biholomorphic mappings and the Bergman kernel off the diagonal. Invent. math. **51** (1979) 155–169

[W 3] Webster, S.M.: On mapping an n-ball into an $(n+1)$-ball in complex space. Pacific J. Math. **81** (1979) 267–272

[Wh] Whitney, H.: Complex analytic varieties. Addison-Wesley, 1972

[Wo] Wong, B.: Characterization of the ball in $\mathbb{C}^n$ by its automorphism group. Invent. math. **41** (1977) 253–257

An Overview of Recent Advances in Hodge Theory

Jean-Luc Brylinski and Steven Zucker

Contents

Introduction

Our aim, in writing this article, is to give a survey of the progress in Hodge theory over the past 15 years. Of course, it was first necessary to decide what we mean by "Hodge theory"! This is by no means an easy question, and our answer to it shaped the contents of the article. Indeed, the nature of the subject has evolved over the years.

The classical antecedents of Hodge theory lie in the discussion of the periods of the integrals of the first kind on algebraic plane curves over $\mathbb{C}$. In more modern terms, one is describing where the holomorphic 1-forms sit in its cohomology with $\mathbb{C}$-coefficients, and how this changes with parameters – in other words, a variation of Hodge structure of weight one. The crowning achievement of this era, due to Poincaré and Lefschetz (see [Lef]), was the inversion of so-called normal functions (Abel's theorem with parameter and degeneration) to characterize the cohomology classes on a non-singular algebraic surface that are represented by algebraic 1-cycles. That was the state of the subject, say, 60 years ago.

Hodge's own work on the theory of harmonic integrals [Ho] resulted in the well-known *Hodge Decomposition Theorem* (cf. our Proposition 3.7): for a compact Kähler manifold X, its cohomology groups decompose:

$$H^m(X, \mathbb{C}) = \oplus_{p+q=m} H^{p,q}(X) \tag{0-1}$$

in a natural way, with $\overline{H^{p,q}(X)} = H^{q,p}(X)$. This work led to the famous Hodge Conjecture, for X a projective algebraic manifold, which now asserts that every rational class in $H^{p,p}(X)$ is represented by a codimension p algebraic cycle with $\mathbb{Q}$-coefficients. This is a beguilingly simple statement, about which distressingly little is understood beyond the case $p = 1$. If, in addition, we replace Hodge's use of integral equations by L_2 methods (see [Gaf 2]), we get the state of Hodge theory 35 years ago.

It is fair to say that the modern era of Hodge theory began in the late 1960's with the work of Griffiths [Gr 1–Gr 4]. This was a neoclassical development, looking back at the work of Poincaré and Lefschetz from the point of view of a (then) modern complex geometer, and generalizing it to higher dimensional varieties and higher codimension cycles. This produced the notions of variation of Hodge structure and classifying spaces for Hodge structures; the intermediate Jacobians (a modification of Weil's) and Abel-Jacobi mappings (see also [Li]). Although the algebro-geometric aspects of the program never achieved full success, it has had a lasting influence on modern Hodge theory and algebraic geometry.

Shortly after Griffiths' work appeared, Deligne's work on mixed Hodge theory [D 4] introduced a set of new ideas and techniques into the subject. He introduced the notion of a mixed Hodge structure, which is, roughly speaking, an iterated extension of Hodge structures of increasing weight. The cohomology groups of *any* complex algebraic variety (not necessarily compact, not necessarily smooth) carry mixed Hodge structures. These are functorial for morphisms of varieties; and in the compact smooth case, one recovers the usual Hodge structure 0-1. This work was heavily influenced by Grothendieck's philosophy of motives, in particular by his idea that there should exist "mixed motives", associated to the cohomology groups of general algebraic varieties. Such mixed motives should be successive extensions of ordinary (or *pure*) motives. A key insight of Deligne was to conjecture the existence of a natural increasing *weight filtration* $W_\bullet$ on a mixed motive, hence on any cohomology group of an algebraic variety.

What led Deligne to the idea of an *increasing* filtration is the fact that there are non trivial extensions of abelian varieties by tori (in the category of algebraic groups) – to wit, Rosenlicht's generalized Jacobians –, but not the other way around. In characteristic $p > 0$, this weight filtration may be read from the eigenvalues of the Frobenius automorphism on l-adic cohomology (using the Weil conjecture, which was later proved [D 5]). In characteristic 0, the weight filtration W on any $H^m(X, \mathbb{Q})$ was constructed in [D 4]; Deligne showed that any graded piece $\mathrm{Gr}_p^W H^m(X, \mathbb{Q})$ is pure of weight p. The weight filtrations in characteristic zero and in positive characteristic have been shown to coincide under comparison between cohomology theories (see [D 4] and [B-B-D: § 6]).

The impact of Deligne's work on Griffiths' program was already clear in [Gr 4]. The first big payoff came soon, in the form of Schmid's results on the degeneration of Hodge structures. This is reported in the survey article [Gr-S], which also reported the consequences of the new de Rham theory for degenerating geometric Hodge structures, that eventually appeared in [Cl 2] and [St 1]. If only for the reason that such an article appeared at that time, we will push our stopwatch and declare that to represent the state of the subject 15 years ago, from the point of view of Griffiths. We feel, by the way, that [Gr-S] helped to pave the way for the broad-scale acceptance of Grothendieck-style algebraic geometry among the coming generation of transcendental algebraic geometers.

This article can be regarded as a report on the progress in Hodge theory since 1975. We consider Hodge theory to be an independent and mature branch of modern mathematics, which has many relations with many other parts of mathematics, from which it draws techniques and inspiration, and to which it contributes in return. Hodge theory can no longer be regarded as a subset of algebraic geometry or of complex analysis. It has become a vast and intricate field, and in many ways a rather difficult one. The progress made in the last 15 years is certainly impressive, and this blossoming should continue in the years to come.

Before we describe the contents of this report, we should mention that because of limitations of time, space and energy, we were forced to make some choices. We have given preference to those active topics that are internal to Hodge theory, and which are heavily based on complex analysis (as is fitting for a volume in honor of Remmert). We regret that we have had to omit, or give inadequate mention to, the following topics: geometric applications of Hodge theory (the Hodge conjecture, Torelli problems, Hodge structures and moduli, see [TTAG: Ch. VIII]; infinitesimal variation of Hodge structure, see [TTAG: Ch. III]; vanishing theorems); Deligne's theory of absolute Hodge cycles [D-M-O-S]; number-theoretic aspects (special values of L-functions [D 9], [Be 1], the Taniyama group and its extension [An], p-adic Hodge theory [F 2]); automorphic forms and representations, Shimura varieties (see the review article [Mi]).

In the first section, we review Deligne's construction of mixed Hodge structures on the cohomology of all algebraic varieties, and simplicial varieties, over $\mathbb{C}$, for it sets the tone of much of the work that followed. A simplified account of this already appeared in [Gr-S], at least for smooth varieties and for varieties with normal crossings, but we emphasize the use of the notions of mixed Hodge complexes and cohomological mixed Hodge complexes, which were introduced in [D 4], and have proven important in many subsequent developments.

Much of Sect. 2 also contains review material, namely the theorems of Schmid [Sc] on degeneration of Hodge structures *in the abstract*, i.e., in the absence of any hypothesis that the variation of Hodge structure arises from a family of smooth projective varieties. These are the nilpotent orbit theorem and the SL(2)-orbit theorem; at that time, the latter was proved only with one parameter.

One of the recent advances is its extension to several variables [C-K-S 1] (see our Theorem 2.16), which is based on work of Cattani and Kaplan [C-K 2] on the special properties of the commuting nilpotent monodromy logarithms. One consequence of the (one-variable) SL(2)-orbit theorem is the existence of "limit mixed Hodge structures". In Sect. 2, we also run through the construction of the latter in the geometric case by de Rham theoretic methods by Steenbrink [St 1] and Clemens [Cl 2] (this is also discussed in [TTAG: Ch. VII]). This affords us a natural opportunity to bring in the construction of the sheaves of "vanishing cycles" and "nearby cycles" (2.21), which are fundamental in the work of M. Saito (discussed in Sect. 4). We then present the generalization by Navarro Aznar (Theorem 2.25). We conclude the section with the relation with the asymptotics of integrals at an isolated singularity, work of Varchenko and others – which, incidentally, is very close to the spirit of the classical era, and to Griffiths' original point of view as well.

Sect. 3 is about L_2-cohomology. Lying at the heart of the proof of the Hodge Theorem (by Hilbert space methods), it is an integral part of the subject. The importance of this topic has emerged during the past 15 years, after the work of Cheeger [Ch 1] and Zucker [Z 1], [Z 3]. There is a general relation between L_2-cohomology and harmonic forms (see formula 3-5, and Proposition 3.5), which gives a key motivation for its introduction: it was the most familiar way to establish the existence of useful Hodge decompositions. The following two general questions have proven to be very interesting:

(i) Given a Riemannian manifold M, is there a compactification $\overline{M}$ for which the L_2-cohomology becomes some known topological invariant?

(ii) Given a projective variety X, is there a Kähler metric on its regular locus X^0 for which the L_2-cohomology becomes the middle-perversity intersection cohomology of [G-M]?

By a conjecture of Cheeger, Goresky and MacPherson (see our 3.20), the restriction of the Fubini-Study metric from projective space would work in (ii), but verifying this has proven very difficult for several reasons. In the other direction, the Zucker Conjecture, settled affirmatively three years ago ([Lo], [Sap-S]), is an illustration of (i).

It is easy to extend the definition of L_2-cohomology to allow coefficients in any metrized local system $\mathbb{V}$ on the manifold M. When the local system underlies a variation of Hodge structure over a smooth variety X, metrized by its Hodge metric, there is on the one hand a natural bigradation for a Hodge decomposition (see our 3-7) given by Deligne; on the other hand, the asymptotics of the Hodge metric are known, as a consequence of the SL(2)-orbit theorem (see Theorem 2.18). If, now, X is given a Poincaré metric with respect to a good compactification $\overline{X}$, the L_2-cohomology of X with coefficients in $\mathbb{V}$ satisfies the requisite local conditions for giving the intersection cohomology of $\overline{X}$ with the same coefficients (Theorems 3.12 and 3.15); this is a result of class (i) above.

Ironically, it was not by L_2-cohomology that Hodge structures for the intersection homology groups of singular projective varieties were finally produced. Instead, it came out of work in algebraic analysis (i.e., **D**-modules), a subject that also has blossomed during the past 15 years, as we report in Sect. 4. The starting point is the so-called *Riemann-Hilbert correspondence* of [K-K 1] and [Me 1], which asserts that taking the de Rham complex sets up an equivalence of categories between holonomic **D**-modules with regular singularities and perverse sheaves (and likewise for their derived categories). The idea is to equip such **D**-modules with Hodge (and weight) filtrations, so that they induce (mixed) Hodge structures on hypercohomology.

In his theory of polarizable Hodge modules [Sa 3] and mixed Hodge modules [Sa 11], Morihiko Saito inductively defines categories of filtered **D**-modules that have this property, using conditions on the vanishing and nearby cycle sheaves associated to local holomorphic functions, and compatibility with the so-called V-filtration of Kashiwara-Malgrange. For this reason, it is often difficult to understand the objects explicitly. In fact, a main part of the theory is to verify that there are non-trivial objects in these categories! Fortunately, Saito was able to show that the filtered **D**-modules associated to variations of polarized Hodge structure, and more generally admissible variations of mixed Hodge structure (see our 7.2), do belong to them, so one gets Hodge theory for intersection cohomology, and mixed Hodge structures on cohomology with coefficients in an admissible variation of mixed Hodge structure, as expected. Mixed Hodge modules attain, though perhaps in an unexpected way, Grothendieck's hope for a theory of "Hodge sheaves".

Section 5 is devoted to Deligne cohomology and to its generalization, by Beilinson, to noncompact varieties. This theory, which involves a mixture of integral cohomology and truncated de Rham cohomology, was invented by Deligne around 1970; it unified the Hodge cohomology groups with Griffiths' intermediate Jacobians (see [Bl 1] or [E-Z]). Around 1976-77, Bloch found remarkable applications of Deligne cohomology to Chern classes for unitary local systems on algebraic varieties. In particular, he found a regulator map from $K_2(X)$ to $H^1(X,\mathbb{C}^*)$, for X a curve. This regulator map has a beautiful interpretation in terms of products in Deligne cohomology (see Theorem 5.9), which was found by Bloch, Beilinson and Deligne.

Beilinson [Be 1, Be 2] generalized Deligne cohomology to noncompact varieties and constructed general regulator maps from $K_m(X)$ to Deligne cohomology (see 5.19). He formulated several conjectures, some of a geometric nature, which we discuss in 5.21 and 5.22, others of a number-theoretic nature (special values of L-functions), which we do not discuss here. Beilinson [Be 2] gives a remarkable interpretation of Deligne cohomology in terms of extensions in the derived category of mixed Hodge structures. This is related to work of Carlson and Hain [C-H]. We discuss this in 5.22–5.25. These ideas are connected to recent progress in the theory of Grothendieck motives [B-M-S], [H-M]; we comment briefly on these at the end of Sect. 5.

Several approaches have been used to put mixed Hodge structures on homotopy groups of algebraic varieties, discussed in Sect. 6. The method of Morgan [Mo] is based on Sullivan's theory of minimal models for differential graded algebras [Su]. It consists in putting filtrations on the minimal model. In the simply-connected case, Morgan obtains mixed Hodge structures on the homotopy groups (Theorem 6.13). In the projective case, Deligne, Griffiths, Morgan and Sullivan showed that the minimal model is "formal", which means that the rational homotopy groups are a formal consequence of the cohomology ring [D-G-M-S] (Theorem 6.14). In the non-simply-connected case, Morgan also obtains interesting results, but his Hodge filtration is only well-defined up to inner conjugation.

The method of Hain [H 1] is based on Chen's method of iterated integrals, which gives a map of complexes from a bar complex to the algebra of differential forms on the free loop space of a manifold, or on the path space. It is a very useful topological tool in the study of the path fibration. Hain shows that the bar construction preserves mixed Hodge complexes. Since on the pointed loop space, the rational homotopy is the primitive part of the rational homology, Hain obtains mixed Hodge structures on the completion of the group algebra of the fundamental group with respect to its augmentation ideal (Theorem 6.23). Geometric applications are discussed in 6.25. Alternative approaches to mixed Hodge theory on homotopy groups, due Deligne and to Navarro Aznar [NA 1], are only briefly mentioned.

The notion of a variation of mixed Hodge structure is a very natural generalization of that of a variation of Hodge structure. The idea, naturally, is that a variation of mixed Hodge structure on X must at least yield a filtered local system $(\mathbb{V}, W)$ on X, and that the graded local systems $\mathrm{Gr}_k^W \mathbb{V}$ should be honest polarized variations of Hodge structure. The delicate problem is to find a good notion which gives sufficient control on the extensions, when X is non-compact. In Sect. 7, we present such a notion, that of *admissible* variations of mixed Hodge structure. The definition is due to Steenbrink and Zucker [St-Z] in the case of curves, and to Kashiwara [K 6] in the general case (by a curve test); it is given in 7.2. In the case where the base manifold is a disc, one imposes two constraints: first, the Hodge filtration should admit a well-behaved limit; second, the relative monodromy weight filtration of Deligne should exist.

There are two justifications for this notion. First, any geometric variation is admissible (Theorem 7.3). The proof of this statement is a natural extension of the techniques of Steenbrink [St 1], which we discussed in Sect. 2. Second, if the base manifold is quasi-projective, the cohomology groups with coefficients in a variation of mixed Hodge structure acquire a functorial mixed Hodge structure (Theorem 7.4). We also discuss a general rigidity theorem for variation of mixed Hodge structure (Theorem 7.12). At the end of Sect. 7, we present the results of Hain and Zucker [H-Z] on unipotent variations of mixed Hodge structure (see Definition 7.16), which are a very natural application of Hain's homotopy mixed Hodge theory.

Section 8 is devoted to the question of determining which local systems on a given quasi-projective manifold underlie a variation of Hodge structure. We primarily discuss very recent work of C. Simpson, which, in our opinion, promises to have profound repercussions in Hodge theory. He uses the non-linear P.D.E. methods of differential geometry to obtain a correspondence between irreducible vector bundles on a compact Kähler manifold and stable Higgs bundles with vanishing total Chern class (see Theorem 8.9). A Higgs bundle is a vector bundle $\mathscr{E}$, together with an operator-valued one-form θ on $\mathscr{E}$ of square 0. $\mathbb{C}^*$ acts on Higgs bundles in the obvious way, by dilating θ.

The remarkable fact is that the fixed points of this $\mathbb{C}^*$-action correspond exactly to complex variations of Hodge structure (Corollary 8.10), and real variations correspond to fixed points which give self-dual Higgs bundles (see Proposition 8.12). Some striking applications are presented in 8.15 and 8.16. Simpson is presently working on the non-compact case; his most recent results (new at the time of this writing) are briefly discussed at the end of Sect. 8.

It is probably already clear that the eight sections are not all independent from one another. It is the case that whenever $p < q$, Sect. p does not use results from Sect. q, although occasionally we may allude to material to be covered later. In most sections, for reason of space, we give precious few proofs or even comments on the proofs. Some sections are inherently more technical than others, but we hope that in each case it is possible for the reader to get a good idea of the main ideas without undue suffering.

It is a pleasure to thank our many colleagues who helped us throughout the preparation of this report. We are grateful to Pierre Deligne for many valuable suggestions. We thank Pierre Deligne, Phillip Griffiths, Nicolas Katz for discussions about the history of the subject. We want to express our gratitude to Carlos Simpson for patiently explaining his work to us. We thank the many people who sent us their articles and preprints. We are very grateful to Pierre Deligne, Fouad El Zein, Richard Hain, Morihiko Saito, Wilfried Schmid and Joseph Steenbrink for reading large portions of the first version of this report, and for pointing out mistakes and suggesting improvements in the presentation. We would also like to express appreciation to Raghavan Narasimhan for the kind encouragement he has given us.

1. Review of Hodge-Deligne Theory

We will review the main results and constructions of Deligne's Hodge theory, since not only the theorems obtained by Deligne (for instance, the existence of a functorial mixed Hodge structure on the cohomology of an arbitrary complex algebraic variety) but also the concepts and methods introduced by him have been essential to most of the recent developments of Hodge theory.

To start with, we review the notions of Hodge structures and mixed Hodge structures. Let V be a finite-dimensional vector space over $\mathbb{R}$. A *Hodge structure of weight m* on V is a decreasing filtration $\ldots \subset F^p(V_{\mathbb{C}}) \subset F^{p-1}(V_{\mathbb{C}}) \subset \ldots$ of the complexified vector space $V_{\mathbb{C}} = V \otimes_{\mathbb{R}} \mathbb{C}$, such that for $\overline{F}^{\bullet}$ the complex-conjugate filtration and $V^{p,q} := F^p \cap \overline{F}^q$, one has the Hodge decomposition:

$$V_{\mathbb{C}} = \oplus_{p+q=m} V^{p,q}.$$

In the terminology of [D4: §2], the filtrations $F^{\bullet}$ and $\overline{F}^{\bullet}$ are said to be *m-opposed.*

A famous theorem of Hodge [Ho] asserts that for X a compact Kähler manifold, the group $H^m(X, \mathbb{R})$ carries a Hodge structure of weight m. The complex vector space $V^{p,q}$ may be identified with the space of harmonic differential forms of type (p, q). The Hodge filtration has the following complex-analytic description. The complex cohomology of X is equal to the hypercohomology of the holomorphic de Rham complex $\Omega_X^{\bullet}$. Let $F^p\Omega_X^{\bullet}$ be the "tronqué bête" (in Deligne's terminology).

$$F^p\Omega_X^{\bullet} : 0 \to \ldots 0 \to \Omega_X^p \xrightarrow{d} \Omega_X^{p+1} \to \ldots \Omega_X^n. \tag{0-1}$$

This gives a decreasing filtration of $\Omega_X^{\bullet}$. The corresponding spectral sequence for hypercohomology degenerates at E_1 and the induced filtration on $H^m(X, \mathbb{C})$ is the Hodge filtration.

To allow for direct sums of Hodge structures of different weights, one introduces the algebraic group S over $\mathbb{R}$, defined as $S = R_{\mathbb{C}/\mathbb{R}}\mathbb{G}_{\mathrm{m}}$, the restriction of scalars, à la Weil, from $\mathbb{C}$ to $\mathbb{R}$, of the multiplicative group $\mathbb{G}_{\mathrm{m}}$, or in other words, $G = \mathbb{C}^*$ viewed as an algebraic group over $\mathbb{R}$. Deligne defines a Hodge structure as a real representation $\rho : S \to \mathrm{GL}(V)$. Since $S(\mathbb{C}) = \mathbb{C}^* \times \mathbb{C}^*$, a complex representation of S on a complex vector space $V_{\mathbb{C}}$ amounts to a bigrading $V_{\mathbb{C}} = \oplus_{p,q} V^{p,q}$, where $V^{p,q} := \{v \in V_{\mathbb{C}} : (z_1, z_2)v = z_1^p z_2^q v\}$. Since complex conjugation acts on $\mathbb{C}^* \times \mathbb{C}^*$ as $(z_1, z_2) \mapsto (\overline{z}_2, \overline{z}_1)$, when $V_{\mathbb{C}}$ has a real structure, a representation of S as above is defined over $\mathbb{R}$ if and only if the condition $\overline{V^{p,q}} = V^{q,p}$ is verified.

The algebraic torus $\mathbb{G}_m$ is a subgroup of S. A real representation V of S decomposes as $V = \sum_m V_m$, where V_m is the subspace of V on which $a \in \mathbb{G}_m$ acts as a^m. The bigrading on $(V_m)_{\mathbb{C}}$ determines on it a pure Hodge structure of weight m. Hence Hodge structures of weight m correspond to representations of S on which $a \in \mathbb{G}_m$ acts as a^m. The language of S-representations is useful in many contexts, in particular as regards the Mumford-Tate group [D-M-O-S: I.3]. The automorphism of $V_{\mathbb{C}}$ induced by $i \in S(\mathbb{R}) = \mathbb{C}^*$ is denoted by C.

For A a subring of $\mathbb{R}$, an *A-Hodge structure* of weight m is a A-module V_A of finite type, together with a Hodge structure of weight m on $V_{\mathbb{R}} := V_A \otimes_A \mathbb{R}$. The *Hodge structure of Tate* $\mathbb{Z}(1)$ is the unique $\mathbb{Z}$-Hodge structure of weight -2, with $\mathbb{Z}(1) := 2\pi i\mathbb{Z} \subset \mathbb{C}$; the action of S is as follows: (z_1, z_2) acts on $\mathbb{C}(1) = \mathbb{C}$ by multiplication by $z_1^{-1}z_2^{-1}$. For $n \in \mathbb{Z}$, the Hodge structure $\mathbb{Z}(n)$ is defined as the tensor power $\mathbb{Z}(n) = \mathbb{Z}(1)^{\otimes n}$.

A *polarization* of a Hodge structure V of weight m is a homomorphism $(\ ,\ ) : V \otimes V \to \mathbb{Z}(-m)$ of Hodge structures, which is $(-1)^m$-symmetric and such that

$$\langle x, y\rangle = (2\pi i)^m (x, Cy)$$

is a positive-definite symmetric bilinear form on V.

For X a smooth projective algebraic variety over $\mathbb{C}$, of dimension n, let $L \in H^2(X, \mathbb{Z}(1))$ be the class of a hyperplane section. For $0 \le j \le n$, the *primitive part* $P^j(X)$ of $H^j(X)$ is the kernel of $L^{n-j+1} : H^j(X,\mathbb{Z}) \to H^{2n+2-j}(X, \mathbb{Z}(n-j+1))$. The morphism of Hodge structures

$$\begin{aligned} H^j(X,\mathbb{Z}) \otimes H^j(X,\mathbb{Z}) &\to \mathbb{Z}(-j) \\ (\alpha, \beta) &\mapsto \langle \alpha \cup \beta \cup L^{n-j}, [X]\rangle \end{aligned}$$

for $[X] \in H^{2n}(X, \mathbb{Z}(n))$ the orientation class, is a polarization of the Hodge structure $P^j(X)$ of weight j.

Definition 1.1. *Let A be a subring of $\mathbb{R}$. An A-*mixed Hodge structure *(often abbreviated M.H.S.) consists of the following data:*

(1) An A-module V_A of finite type;

(2) An increasing filtration $\ldots \subset W_n \subset W_{n+1} \subset \ldots$ of the $A \otimes \mathbb{Q}$-module $V_{A\otimes\mathbb{Q}} := V_A \otimes \mathbb{Q}$;

(3) A decreasing filtration F^p of $V_{\mathbb{C}} := V_A \otimes_A \mathbb{C}$.

These data should satisfy the following requirement: for each $j \ge 0$, the filtrations induced by F and $\overline{F}$ on $\mathrm{Gr}_j^W(V_{\mathbb{C}})$ are j-opposed (hence determine an A-Hodge structure of weight j on $\mathrm{Gr}_j^W(V_A)$).

A morphism $f : V_A \to V'_A$ of M.H.S. is an A-linear map $f : V_A \to V'_A$ which is compatible with the filtrations W and F.

Theorem 1.2 (Deligne, [D 4: II, Théorème 2.3.5]). *(1) The category of A-mixed Hodge structures is abelian; the kernels and cokernels of morphisms of M.H.S. are endowed with the induced filtrations.*

(2) Assume $A \otimes \mathbb{Q}$ is a field. Then every morphism of A-M.H.S. is strictly compatible with the filtrations $W_\bullet$ and $F^\bullet$.

(3) The functor $V \mapsto \mathrm{Gr}_j^W(V)$ is an exact functor from the category of A-M.H.S. to the category of A-Hodge structures of weight j.

(4) The functor $V \mapsto \mathrm{Gr}_F^p(V)$ is exact.

For V a M.H.S., one sets $V^{p,q} = (\mathrm{Gr}_{p+q}^W(V))^{p,q}$. The *Hodge numbers* $h^{p,q}$ of V are the integers $h^{p,q} = \dim_{\mathbb{C}}(V^{p,q})$.

We recall next some constructions essential to Deligne's mixed Hodge structures on the cohomology groups of complex algebraic varieties. One first treats the case of a smooth variety X. From Hironaka's theorem on resolution of singularities [Hi], one may find a compactification $X \overset{j}{\hookrightarrow} \overline{X}$ such that the complement $D = \overline{X} - X$ is a divisor with normal crossings. Such a compactification will be called *good*. One may choose a good compactitication in such a way that $D = \cup_i D_i$ with D_i smooth; we then fix an order on the set of components of D.

Deligne introduces in [D 2] and [D 4] the *logarithmic de Rham complex* $\Omega^\bullet_{\overline{X}}(\log D)$ (this is nowadays the standard notation; the same complex was denoted by $\Omega^\bullet_{\overline{X}}\langle D\rangle$ in [D 2]). $\Omega^\bullet_{\overline{X}}(\log D)$ is a complex of sheaves on $\overline{X}$. First, $\Omega^1_{\overline{X}}(\log D)$ is the $\mathcal{O}_{\overline{X}}$-submodule of $j_*(\Omega^1_X)$ generated by the $\frac{dz_i}{z_i}$, for z_i a local equation of a component D_i of D. Then let $\Omega^p_{\overline{X}}(\log D) = \wedge^p_{\mathcal{O}_{\overline{X}}} \Omega^1_{\overline{X}}(\log D)$. The sheaves $\Omega^p_{\overline{X}}(\log D)$ are locally free over $\mathcal{O}_{\overline{X}}$.

Let $W_n(\Omega^p_{\overline{X}}(\log D))$ be the submodule spanned by products $\alpha \wedge \frac{dz_{i_1}}{z_{i_1}} \wedge \ldots \wedge \frac{dz_{i_m}}{z_{i_m}}$, for $m \leq n$ and α holomorphic. This gives a subcomplex of sheaves, and one has

$$W_n(\Omega^p_{\overline{X}}(\log D)) \wedge W_m(\Omega^q_{\overline{X}}(\log D)) \subset W_{n+m}(\Omega^{p+q}_{\overline{X}}(\log D)).$$

Let D^n (resp. $\tilde{D}^n$) denote the union inside $\overline{X}$ (resp. the disjoint union) of the intersections $D_{i_1} \cap \ldots \cap D_{i_n}$ of n components of D. Let $i_n : \tilde{D}^n \to \overline{X}$ be the canonical map. Recall that for $K^\bullet$ a complex and for $n \in \mathbb{Z}$, the complex $K^\bullet[n]$ is defined by $K[n]^p = K^{n+p}$ (translation of $K^\bullet$ to the left by n).

Proposition 1.3. *The map* $\alpha \wedge \frac{dz_{i_1}}{z_{i_1}} \wedge \ldots \wedge \frac{dz_{i_n}}{z_{i_n}} \mapsto \alpha|_{D_{i_1} \cap \ldots \cap D_{i_n}}$ *gives an isomorphism of complexes of sheaves on* $\overline{X}$

$$\mathrm{Res} : \mathrm{Gr}_n^W(\Omega^\bullet_{\overline{X}}(\log D)) \cong i_{n*}\Omega^\bullet_{\tilde{D}^n}[-n].$$

Now let $\tau_{\leq n}(K^\bullet)$, for any complex $K^\bullet$, denote the canonical filtration of Deligne

$$\tau_{\leq n}(K^\bullet) : \quad \ldots \to K^{n-2} \to K^{n-1} \overset{d_{n-1}}{\to} \ker d_n \to 0 \to 0 \ldots$$

such that the induced map on cohomology $H^p\tau_{\leq n}(K^\bullet) \to H^p(K^\bullet)$ is an isomorphism for $p \leq n$ and is 0 for $p > n$. It has the property that $\mathrm{Gr}_n^\tau(K^\bullet)$ is quasi-isomorphic to $H^n(K^\bullet)[-n]$.

Recall that a filtered morphism $\phi : (A^\bullet, W_\bullet) \to (B^\bullet, W_\bullet)$ between filtered complexes is a *filtered quasi-isomorphism* if for every j, the induced morphism of complexes $\mathrm{Gr}_j^W : \mathrm{Gr}_j^W(A) \to \mathrm{Gr}_j^W(B)$ is a quasi-isomorphism. Equivalently, this means that ϕ induces an isomorphism on the E_1-terms of the spectral sequences for these filtered complexes; hence, if the filtrations are exhaustive and separated, ϕ itself is a quasi-isomorphism. Then we have:

Theorem 1.4 (Deligne, [D 4: II, Proposition 3.1.8]). *The morphisms of filtered complexes*

$$(\Omega^\bullet_{\overline{X}}(\log D), W) \leftarrow (\Omega^\bullet_{\overline{X}}(\log D), \tau) \to (j_*\Omega^\bullet_X, \tau) \cong (Rj_*\mathbb{C}, \tau)$$

are filtered quasi-isomorphisms. Hence the Leray spectral cohomology for j in complex cohomology is identified with the hypercohomology spectral sequence for the filtered complex of sheaves $(\Omega^\bullet_{\overline{X}}(\log D), W)$ on $\overline{X}$.

The Hodge filtration $F^\bullet$ of the log complex is just the filtration bête:

$$\text{(1-2)} \qquad F^p\Omega^\bullet_{\overline{X}}(\log D) : \quad 0 \to \ldots \to \Omega^p_{\overline{X}}(\log D) \to \Omega^{p+1}_{\overline{X}}(\log D) \to \ldots$$

The filtrations $W_\bullet$ and $F^\bullet$ on the complex of sheaves $\Omega^\bullet_{\overline{X}}(\log D)$ on $\overline{X}$ induce filtrations $W_\bullet$ and $F^\bullet$ on the hypercohomology $H^p(\overline{X}, \Omega^\bullet_{\overline{X}}(\log D)) = H^p(X, \mathbb{C})$.

In the following theorem, the conventions for the spectral sequence of a filtered complex are the ones introduced by Deligne [D 4: II, § 3.2], which differ from the classical ones in [Go]. In the following, it will be more convenient to deal with a decreasing filtration $W^\bullet$ rather than with an increasing filtration $W_\bullet$. One goes from one to the other simply by the rule $W_n = W^{-n}$.

To describe Deligne's conventions, assuming $W^\bullet$ is an increasing filtration on a complex $K^\bullet$ in any abelian category, introduce

$$Z_r^{p,q} = \mathrm{Ker}(d : W^p(K^{p+q}) \to K^{p+q+1}/W^{p+r}(K^{p+q+1}))$$

and

$$B_r^{p,q} = W^{p+1}(K^{p+q}) + d(W^{p-r+1}(K^{p+q-1})) \subset K^{p+q} .$$

Then the E_r-term of the spectral sequence for $(K^\bullet, W)$ is $E_r^{p,q} := Z_r^{p,q}/(B_r^{p,q} \cap Z_r^{p,q})$. The differential d_r, which maps $E_r^{p,q}$ to $E_r^{p+r,q-r-1}$, is induced by the differential of K, as usual, and the cohomology of the complex $(E_r^{\bullet,\bullet}, d_r)$ is identified with $E_{r+1}^{\bullet,\bullet}$.

It is worthwhile to point out an elegant reformulation of these definitions, which is due to El Zein [E 7: § 2.8]. He shows that $E_r^{p,q}$ is equal to $E_\infty^{p,q}(W^{p-r+1}K/W^{p+r}K) = \mathrm{Gr}_W^p H^{p+q}(W^{p-r+1}K/W^{p+r}K)$. This makes the self-dual character of Deligne's conventions immediately visible.

We also recall Deligne's construction of the *shifted filtration* Dec(W) (in French, "filtration décalée"). Put Dec$(W)^p K^n := Z_1^{p+n,-p}$. Then, by construction, the complex $(E_0(\mathrm{Dec}(W)), d_0)$ is quasi-isomorphic to the complex $(E_1(W), d_1)$. Hence $E_r(\mathrm{Dec}(W)) \cong E_{r+1}(W)$ for $r \geq 1$ [D 4: II, Proposition 1.3.4].

Theorem 1.5 (Deligne, [D 4: II, Théorème 3.2.5]). *(1) The filtration W on $H^p(X, \mathbb{C})$ comes from a filtration W of $H^p(X, \mathbb{Q})$. This filtration, as well as the filtration F on $H^p(X, \mathbb{C})$, is independent on the choice of the compactification $\overline{X}$ of X.*

(2) The filtrations $W[p]$ and F define on $H^p(X, \mathbb{Z})$ a mixed Hodge structure; this mixed Hodge structure is functorial with respect to algebraic maps.

(3) The spectral sequence for the filtration W has E_1-term ${}_WE_1^{p,q} = H^{2p+q}(\tilde{D}^p, \mathbb{Q})$. This spectral sequence degenerates at E_2. The Leray spectral sequence for $j : X \hookrightarrow \overline{X}$ degenerates at E_3.

(4) The spectral sequence for the filtration F

$$ {}_FE_1^{p,q} = H^q(\overline{X}, \Omega^p_{\overline{X}}(\log D)) \Rightarrow H^{p+q}(X, \mathbb{C}) $$

degenerates at E_1.

(5) The spectral sequence for the sheaf $\Omega^p_{\overline{X}}(\log D)$, endowed with the filtration W, degenerates at E_2. Its E_1-term is $E_1^{-n,k+n} = H^k(\tilde{D}^n, \Omega^{p-n}_{\tilde{D}^n})$.

(6) The Hodge numbers $h^{p,q}$ of the M.H.S. $H^n(X, \mathbb{Z})$ can only be non-zero when $p \leq n$, $q \leq n$, and $p + q \geq n$.

Note for statement (3) that the spectral sequence for the filtered complex $(Rj_*\mathbb{C}, \tau)$ is obtained by "décalage" from the Leray spectral sequence.

The proof of this theorem uses several ingredients, one of which is classical Hodge theory for the smooth projective varieties $\tilde{D}^n$; another one is a theorem on complexes endowed with two filtrations. This theorem is rather technical but important. If $K^\bullet$ is a complex in some abelian category, endowed with two filtrations W and F, the terms $E_r(K, W)$ of the spectral sequence relative to the W-filtration are equipped with three filtrations. To describe them, assuming $W^\bullet$ is an increasing filtration, recall $Z_r^{p,q} = \mathrm{Ker}(d : W^p(K^{p+q})) \to K^{p+q+1}/W^{p+r}(K^{p+q+1})$ and $B_r^{p,q} = W^{p+1}(K^{p+q}) + d(W^{p-r+1}(K^{p+q-1})) \subset K^{p+q}$.

(1) Since $E_r^{p,q} = Z_r^{p,q}/(B_r^{p,q} \cap Z_r^{p,q})$, there is a natural filtration on $E_r^{p,q}$ as a quotient of the subobject $Z_r^{p,q}$ of K^{p+q}, filtered by F. This is the *first direct filtration*, denoted F_d.

(2) Since, dually, $E_r^{p,q} = \mathrm{Ker}(K^{p+q}/B_r^{p,q} \to K^{p+q}/(Z_r^{p,q} + B_r^{p,q}))$, $E_r^{p,q}$ receives a filtration as a subobject of the quotient $K^{p+q}/B^{p,q}$ of K^{p+q}, filtered by F. This is the *second direct filtration*, denoted F_{d^*}.

(3) The *recursive filtration* (in French, *filtration récurrente*) F_r is defined on $E_r^{p,q}$ by induction on $r \geq 0$. For $r = 0$, $F_r = F_d = F_{d^*}$ on $E_0^{p,q} = H^{p+q}(\mathrm{Gr}^p_W(K))$. On $E_{r+1}^{p,q}$, the recursive filtration is the one induced on $E_{r+1}^{p,q}$ as a subquotient of $E_r^{p,q}$.

One always has $F_d \subset F_r \subset F_{d^*}$ on $E_r^{p,q}$. The following result gives a criterion for equality of these filtrations.

Theorem 1.6 (Deligne, [D 4: II, Théorème 1.3.16] and [D 4: III, Proposition 7.2.8]). *(1) Let the complex K be equipped with two filtrations W and F, with F exhaustive and separating. Let r_o be an integer such that for $0 \leq r \leq r_o$, the differentials of the complex $E_r(K, W)$ are strictly compatible with the recursive filtration F_r. Then for $r \leq r_o + 1$, the three filtrations F_d, F_{d^*} and F_r coincide on $E_r^{p,q}$.*

(2) If for every $r \geq 0$ the differentials d_r are strictly compatible with the recursive filtrations of $E_r^{p,q}$, then the filtrations F_d, F_{d^*} and F_r coincide on E_∞, and coincide with the filtration induced by F on the subquotients $\mathrm{Gr}_j^W H^\bullet(K^\bullet)$ of $H^\bullet(K^\bullet)$.

(3) Under the assumption of (2), the spectral sequence $E(K,F)$ degenerates at E_1, and one has an isomorphism of spectral sequences:

$$\mathrm{Gr}_F^p(E_r(K,W)) \cong E_r(\mathrm{Gr}_F^p K, W).$$

This theorem applies to the bi-filtered complex of global sections of a bifiltered acyclic complex of sheaves on $\overline{X}$ resolving the complex $\Omega^\bullet_{\overline{X}}(\log D)$ of Theorem 1.5. On the terms ${}_W E_r^{p,q}$ of the spectral sequence for W in hypercohomology, it is not hard to see that the three filtrations coincide ([D 4: II, Théorème 3.2.5]).

Theorem 1.5 was axiomatized by Deligne in the notion of *cohomological mixed Hodge complexes,* which we will presently define. The definition involves the concept of derived categories, due to Verdier [Ve 2, G-M, Bo 2]. A reader who is not familiar with this notion may think of an object of the derived category of complexes as a complex defined up to quasi-isomorphism. For example, for $\underline{K}^\bullet$ a complex of sheaves on a space X, the complex $R\Gamma(X, \underline{K}^\bullet)$ is defined as $\Gamma(X, I^\bullet)$, for $I^\bullet$ an acyclic resolution of $\underline{K}^\bullet$.

Definition 1.7. *(1) Let A be a subring of $\mathbb{R}$, such that $A \otimes \mathbb{Q}$ is a field. An* A-Hodge complex of weight *m consists of the following:*

a) a complex K_A in the derived category of complexes of A-modules that are bounded below, such that each $H^p(K_A)$ is a A-module of finite type;

b) a bounded below filtered complex $(K_{\mathbb{C}}, F)$ of $\mathbb{C}$-vector spaces, and an isomorphism $\alpha : K_{\mathbb{C}} \xrightarrow{\sim} K_A \otimes \mathbb{C}$.

The following conditions must be fulfilled.

(CH 1) The differential d of $K_{\mathbb{C}}$ is strictly compatible with F; equivalently, the spectral sequence for the filtered complex $(K_{\mathbb{C}}, F)$ degenerates at E_1.

(CH 2) For each $p \geq 0$, the filtration F on $H^p(K_{\mathbb{C}}) = H^p(K_A) \otimes \mathbb{C}$ determines an A-Hodge structure of weight $m + p$.

(2) Let X be a topological space. An A-cohomological Hodge complex of weight *m on X consists of the following*

a) A complex K_A in the derived category of complexes of sheaves of A-modules on X that are bounded below.

b) An object $(K_{\mathbb{C}}, W)$ in the derived category of filtered complexes of sheaves of $\mathbb{C}$- vector spaces over X that are bounded below.

c) An isomorphism $\alpha : K_{\mathbb{C}} \xrightarrow{\sim} K_A \otimes \mathbb{C}$ in the derived category of complexes of sheaves on X that are bounded below.
The following axiom must be verified.

(CHC) The triple $(R\Gamma(X, K_A), (R\Gamma(X, K_{\mathbb{C}}), F), R\Gamma(X, \alpha))$ is a Hodge complex of weight m.

For example, a cohomological Hodge complex is associated to any compact Kähler manifold X. Namely, let $A = \mathbb{Z}$, $K_{\mathbb{Z}} = \mathbb{Z}$ (constant sheaf), $K_{\mathbb{C}}$ the holomorphic de Rham complex $\Omega_X^\bullet$ with the "filtration bête" F. With $\alpha : \mathbb{C} \to K_{\mathbb{C}} = \Omega_X^\bullet$ the inclusion, the triple $(K_{\mathbb{Z}}, (K_{\mathbb{C}}, F), \alpha)$ is a cohomological Hodge complex of weight 0.

Definition 1.8. *(1) An* A-mixed Hodge complex *consists of the following.*

a) A complex K_A of A-modules (in the derived category), that is bounded below, such that $H^p(K_A)$ is an A-module of finite type for all p.

b) A filtered complex $(K_{A\otimes\mathbb{Q}}, W)$ of $A \otimes \mathbb{Q}$-vector spaces that is bounded below, and an isomorphism $K_{A\otimes\mathbb{Q}} \xrightarrow{\sim} K_A \otimes \mathbb{Q}$ in the derived category.

c) A bifiltered complex $(K_{\mathbb{C}}, W, F)$ of $\mathbb{C}$-vector spaces, and a filtered isomorphism $\alpha : (K_{\mathbb{C}}, W) \xrightarrow{\sim} (K_{A\otimes\mathbb{Q}}, W) \otimes \mathbb{C}$ in the filtered derived category.
The following axiom must be verified.

(CHM) For every n, the triple $(\mathrm{Gr}_n^W(K_{A\otimes\mathbb{Q}}), \mathrm{Gr}_n^W(K_{\mathbb{C}}), \mathrm{Gr}_n^W(\alpha))$ is an $A\otimes\mathbb{Q}$-Hodge complex of weight n.

(2) An A-cohomological mixed Hodge complex *on a space X consists in the following.*

a) A complex K_A of sheaves of A-modules over X, that is bounded below, and such that the cohomology groups $H^p(X, K_A)$ are A-modules of finite type, and that the map $H^*(X, K_A) \otimes \mathbb{Q} \to H^*(X, K_A \otimes \mathbb{Q})$ is an isomorphism.

b) A filtered complex $(K_{A\otimes\mathbb{Q}}, W)$ of sheaves of $A\otimes\mathbb{Q}$-modules that is bounded below, (where W is an increasing filtration), and an isomorphism $K_{A\otimes\mathbb{Q}} \cong K_A \otimes \mathbb{Q}$ in the derived category.

c) A bifiltered complex $(K_{\mathbb{C}}, W, F)$ of sheaves of $\mathbb{C}$-vector spaces on X (with W increasing and F decreasing) that is bounded below, and an isomorphism $\alpha : (K_{\mathbb{C}}, W) \xrightarrow{\sim} (K_{A\otimes\mathbb{Q}}, W) \otimes \mathbb{C}$ in the filtered derived category.
The following axiom must be verified.

(CHMC) For all n, the triple $(\mathrm{Gr}_n^W(K_{A\otimes\mathbb{Q}}), \mathrm{Gr}_n^W(K_{\mathbb{C}}), \mathrm{Gr}_n^W(\alpha))$ is an $A\otimes\mathbb{Q}$-cohomological Hodge complex of weight n.

It is easy to see that for $(K_A, (K_{A\otimes\mathbb{Q}}, W), (K_{\mathbb{C}}, W, F))$ a cohomological mixed Hodge complex on X, the triple

$$(R\Gamma(X, K_A), R\Gamma(X, (K_{A\otimes\mathbb{Q}}, W)), R\Gamma(X, (K_{\mathbb{C}}, W, F))$$

is an A-mixed Hodge complex.

The main theorem of [D 4: II] (Theorem 1.5) may be restated as follows.

Theorem 1.9 [D 4: Scholie 8.1.8]. *Let $\overline{X}$ be a smooth projective variety over $\mathbb{C}$ and $D \subset \overline{X}$ a divisor with normal crossings. Set $X = \overline{X} - D$. Denote by j the inclusion $X \hookrightarrow \overline{X}$. Let W be the canonical filtration $W_n Rj_*\mathbb{Q} = \tau_{\leq n} Rj_*\mathbb{Q}$ on $Rj_*\mathbb{Q}$. Let the log complex $\Omega^\bullet_{\overline{X}}(\log D)$ be endowed with the filtrations W and F. The isomorphism α in the derived category $\Omega^\bullet_{\overline{X}}(\log D) \xrightarrow{\sim} Rj_*\Omega^\bullet_X \cong Rj_*\mathbb{C} = (Rj_*\mathbb{Q}) \otimes \mathbb{C}$ is a filtered isomorphism in the derived category.*

Then the triple $(Rj_*\mathbb{Z}, (Rj_*\mathbb{Q}, W), (\Omega^\bullet_{\overline{X}}(\log D), W, F))$ is a $\mathbb{Z}$-cohomological Hodge complex on $\overline{X}$.

The following theorem shows that the conditions that define a mixed Hodge complex have strong consequences.

Theorem 1.10 [D 4: III, Scholie 8.1.9]. *Let K be a mixed Hodge complex.*

(1) On the terms ${}_WE_r^{p,q}$ of the spectral sequence for $(K_\mathbb{C}, W)$, the two direct filtrations and the recursive filtration defined by F coincide.

(2) The filtration $W[n]$ of $H^n(K_{A\otimes\mathbb{Q}}) = H^n(K_A) \otimes \mathbb{Q}$ and the filtration F of $H^n(K_\mathbb{C}) = H^n(K_A) \otimes \mathbb{C}$ determine an A-M.H.S. on $H^n(K_A)$.

(3) The differentials ${}_Wd_1$ for ${}_WE_1^{p,q}$ preserve the Hodge bigrading.

(4) The spectral sequence of $(K_{A\otimes\mathbb{Q}}, W)$ degenerates at E_2.

(5) The spectral sequence for $(K_\mathbb{C}, F)$ degenerates at E_1.

(6) The spectral sequence for $(\mathrm{Gr}_F^p(K_\mathbb{C}), W)$ degenerates at E_2.

Deligne's general construction of a mixed Hodge structure on $H^p(X, \mathbb{Z})$, for X a separated scheme over $\mathbb{C}$, rests on the existence of a smooth simplicial scheme $X_\bullet = (X_n)_n$ over $\mathbb{C}$ together with an augmentation $a : X_\bullet \to X$ and a good compactification $X_\bullet \hookrightarrow \overline{X}_\bullet$. The simplicial space $X_\bullet$ should be a "proper hypercovering" of X, i.e. it should satisfy the following property. For each $n \geq -1$, the natural map $X_{n+1} \to \cos q_n^X(sq_n\ X_\bullet)$ is proper and surjective, where $sq_n\ X_\bullet$ denotes the n-skeleton of $X_\bullet$ – a n-truncated augmented simplicial space – and $cosq_n^X$ is the "coskeleton functor" from n-truncated augmented simplicial spaces to augmented simplicial spaces (this functor is right adjoint to the skeleton functor $sq_n = sq_n^X$ for augmented simplicial spaces, cf. [SGA 4]). The significance of this condition is that it ensures that a is "universally of cohomological descent", in particular $H^*(X, \mathbb{Z}) \cong H^*(X_\bullet, \mathbb{Z})$. One needs therefore an adequate generalization of A-mixed Hodge complexes to cosimplicial complexes or to DG (differential graded) complexes. A DG complex is a complex of complexes, or equivalently a double complex, where the first degree is that of the complex, the second degree that of each DG object.

Let us give an example of such a simplicial scheme X_n. For X an algebraic curve over $\mathbb{C}$, let $\pi : X_0 \to X$ be its normalization, let $g : X' \to X$ be the curve, homeomorphic to X, obtained by contracting each finite subset $\pi^{-1}(s)$ of X_0 to a point (the regular functions on X' are those regular functions on X which take the same value on every point of a fiber $\pi^{-1}(x)$). Let $X_1 = X_0 \times_{X'} X_0$, $X_2 = X_0 \times_{X'} X_0 \times_{X'} X_0$, etc.

The face maps are the natural projections, and the degeneracy maps are deduced from the diagonal inclusion $X_0 \hookrightarrow X_1 = X_0 \times_X X_0$. Then each X_n is smooth, as it is the disjoint union of X_0 and of a finite set. $X_\bullet$ thus defined is a proper hypercovering of X. Let $X_n \hookrightarrow \overline{X}_n$ be the natural smooth compactification (note each component of $\overline{X}_n$ has dimension 0 or 1). Then we have a smooth proper simplicial scheme $\overline{X}_\bullet$.

Definition 1.11. *An* A-DG mixed Hodge complex *consists of the following.*

(1) A complex K_A in the derived category of complexes bounded from below of bounded from below DG A-modules (i.e. K_A is a double complex which is bounded below both horizontally and vertically).

(2) A filtered complex $(K_{A\otimes\mathbb{Q}}, W)$ of DG $(A \otimes \mathbb{Q})$-vector spaces, and an isomorphism $K_A \otimes \mathbb{Q} \xrightarrow{\sim} K_{A\otimes\mathbb{Q}}$.

(3) A bifiltered complex $(K_\mathbb{C}, W, F)$ of DG $\mathbb{C}$-vector spaces, with a filtered quasi-isomorphism $(K_{A\otimes\mathbb{Q}}, W) \otimes \mathbb{C} \xrightarrow{\sim} (K_\mathbb{C}, W)$.

The following condition must be satisfied. For each n, the triple $(K_A^n, (K_{A\otimes\mathbb{Q}}^n, W), (K_\mathbb{C}^n, W, F))$ is an A-mixed Hodge complex.

Proposition 1.12 [D4: III, 8.1.19]. *Let $X_\bullet \hookrightarrow \overline{X}_\bullet$ be a good compactification of a smooth simplicial variety $X_\bullet$, with $D_n = \overline{X}_n - X_n$ a divisor with normal crossings. Put $K_\mathbb{Z}^n := R\Gamma(\overline{X}_n, Rj_*\mathbb{Z})$, with the usual differential induced by the simplicial structure. Let $K_\mathbb{Q}^n := R\Gamma(\overline{X}_n, Rj_*\mathbb{Q})$ with the filtration $W_k = \tau_{\leq k}$. Let $K_\mathbb{C}^n := R\Gamma(\overline{X}_n, \Omega^\bullet_{\overline{X}_n}(\log D_n))$, equipped with the usual filtrations W and F. Then $(K_\mathbb{Z}^\bullet, (K_\mathbb{Q}^\bullet, W), (K_\mathbb{C}^\bullet, W, F))$ is a $\mathbb{Z}$-DG mixed Hodge complex.*

For $K^\bullet$ an A-DG mixed Hodge complex, let $sK^\bullet$ be the associated simple complex. The filtration L is the filtration by the DG degree. The diagonal filtration $\delta(W, L)$ on $sK^\bullet$ is defined by $\delta(W,L)_n(sK^\bullet) = \oplus_{p,q} W_{n+p}(K^{p,q})$. Let $\mathrm{Dec}_1 W$ be the filtration induced by "décalage" of the filtration W on $K^\bullet_{A\otimes\mathbb{Q}}$. The following theorem shows that the associated simple complex of a DG mixed Hodge complex is an A-mixed Hodge complex.

Proposition 1.13 [D4: III, Théorème 8.1.15]. *Let $K^\bullet$ be an A-DG mixed Hodge complex.*

(1) The associated simple complex, with the filtrations $\delta(W, L)$ and F, is an A-mixed Hodge complex.

(2) The E_1-term $E_1^{a,b}$ of the spectral sequence for $(sK \otimes \mathbb{Q}, \delta(W, L))$ is isomorphic to $\sum_{a=\gamma-\beta, b=\alpha+\beta} H^\alpha(\mathrm{Gr}_\beta^W K^{\bullet,\gamma})$. The d_1-differential is the sum of the connecting homomorphism $H^\alpha(\mathrm{Gr}_\beta^W K^{\bullet,\gamma}) \to H^{\alpha+1}(\mathrm{Gr}_{\beta-1}^W K^{\bullet,\gamma})$ and of the simplicial differential $H^\alpha(\mathrm{Gr}_\beta^W K^{\bullet,\gamma}) \to H^\alpha(\mathrm{Gr}_\beta^W K^{\bullet,\gamma+1})$.

(3) On the ${}_LE_r$-terms of the spectral sequence defined by $(sK^\bullet, L)$, the two direct filtrations and the recursive filtration defined by $\mathrm{Dec}_1(W)$ *coincide; the same holds for the three filtrations defined by F. These filtrations will be denoted by W and F.*

(4) For each $r \geq 1$, $({}_LE_r, W, F)$ is an A-mixed Hodge structure. The differentials d_r are morphisms of M.H.S.

Applying this theorem to the geometric situation in Proposition 1.12, Deligne obtains:

Theorem 1.14. *Under the assumptions of Proposition 1.12, one has:*

(1) There is a natural M.H.S. on $H^\bullet(X_\bullet, \mathbb{Z})$, which is functorial in the pair $(X_\bullet, \overline{X}_\bullet)$.

(2) In rational cohomology, the spectral sequence for $(sK^\bullet \otimes \mathbb{Q}, \delta(W, L))$ has E_1-term $E_1^{-a,b} = \oplus_{p+2r=b, q-r=-a} H^p(\tilde{D}_q^r, \mathbb{Z})$. This spectral sequence degenerates at E_2.

(3) The Hodge numbers $h^{p,q}$ verify:

$$h^{p,q} \neq 0 \Rightarrow 0 \leq p, q \leq n.$$

(4) If $D_\bullet = \emptyset$, then

$$h^{p,q} \neq 0 \Rightarrow 0 \leq p, q \leq n \quad \text{and} \quad p + q \leq n.$$

Since, for all cohomological purposes, every separated scheme over $\mathbb{C}$ may be replaced by a smooth simplicial scheme, Deligne deduces:

Theorem 1.15 [D 4: III, § 8.2] *(1) There is a functorial M.H.S. on $H^n(X, \mathbb{Z})$, for X a separated scheme over $\mathbb{C}$. The Künneth isomorphism and the cup-product are morphisms of M.H.S.*

(2) If the Hodge number $h^{p,q}$ of $H^n(X, \mathbb{Z})$ is non-zero, then $0 \leq p, q \leq n$. If $N = \dim(X)$ and $n \geq N$, then we have the more precise inequalities $n - N \leq p, q \leq N$. If X is proper, then $p + q \leq n$. If on the other hand X is smooth, then $p + q \geq n$.

El Zein [E 8] has found another method to construct the Deligne mixed Hodge structure on $H^n(Y, \mathbb{Z})$, which does not use simplicial methods, but is based instead on his technique of “mixed cone”. Say Y is a closed subvariety of the smooth algebraic manifold X. Let $p : X' \to X$, with X' smooth, be a proper birational morphism such that p is an isomorphism outside of Y and that $Y' = p^{-1}(Y)$ is a divisor with normal crossings. Then one has an exact sequence in the category of M.H.S. [E 8: I]:

$$0 \to H^n(X', \mathbb{Q}) \to H^n(Y', \mathbb{Q}) \oplus H^n(X, \mathbb{Q}) \to H^n(Y, \mathbb{Q}) \to 0\,.$$

This means that $H^n(Y,\mathbb{Q})$, with its M.H.S., is expressible in terms of the cohomology of varieties with at worst normal crossing singularities. El Zein constructs a cohomological Hodge complex based on this geometric situation.

Deligne [D 4] proves the following result by inspection of weights.

Theorem 1.16 [D 4: III, § 8.2]. *(1) Let $f : Y \to X$ be a proper surjective algebraic morphism, with X proper and Y smooth. Then the top weight quotient $\mathrm{Gr}_n^W H^n(X,\mathbb{Q})$ of $H^n(X,\mathbb{Q})$ identifies with the image of $H^n(X,\mathbb{Q})$ in $H^n(Y,\mathbb{Q})$. Equivalently, the kernel of $f^*: H^n(X,\mathbb{Q}) \to H^n(Y,\mathbb{Q})$ is equal to $W_{n-1}H^n(X,\mathbb{Q})$. The sequence*

$$H^n(X,\mathbb{Q}) \to H^n(Y,\mathbb{Q}) \overset{p_1^\bullet - p_2^\bullet}{\to} H^n(Y \times_X Y,\mathbb{Q})$$

is exact.

(2) In the diagram of schemes

$$\tilde{X} \overset{\pi}{\to} X \overset{f}{\to} Y$$

assume• Y is smooth, X is proper, $\tilde{X}$ is proper and smooth and π is surjective. Then the kernels of f^* and of $(f\pi)^*$ inside $H^*(Y,\mathbb{Q})$ coincide.

Deligne [D 4: III, § 10] pursued a detailed study of the H^1 of an algebraic curve. He describes it in terms of a "1-motive"; this concept has proven very important in compactifying moduli spaces of curves, abelian varieties and $K3$-surfaces, and in number theory [Tat].

More generally, Deligne put a mixed Hodge structure on the cohomology of any separated simplicial scheme.

Theorem 1.17 [D 4: III, § 8.3]. *The cohomology of a separated simplicial scheme $X_\bullet$ carries a natural mixed Hodge structure. The spectral sequence $E_1^{p,q} = H^q(X_p,\mathbb{Z}) \Rightarrow H^{p+q}(X_\bullet,\mathbb{Z})$ is a spectral sequence in the category of mixed Hodge structures.*

In particular, for the classifying space B_G of a linear algebraic group G (a smooth simplicial scheme), the odd-dimensional rational cohomology is 0 and $H^{2n}(B_G,\mathbb{Q})$ is purely of type (n,n) ([D 4: III, Théorème 9.1.1]).

Deligne also established [D 4: III, Proposition 9.3.1] that for X a quasi-projective variety over $\mathbb{C}$, there exists a "complex of differential operators of order 1" $K^\bullet$ on X (i.e. each K^n is a coherent $\mathcal{O}_X$-module, and $d : K^n \to K^{n+1}$ is a differential operator of order 1), which is a resolution of the constant sheaf $\mathbb{C}_X$. Deligne deduces from the existence of such a complex that the natural map $H^\bullet(X,\mathbb{C}) \to H^\bullet(X,\Omega_X^\bullet)$ is an injection with a natural splitting. This statement was first proven by Bloom and Herrera [B-H].

The complex $K^\bullet$ is defined as $sR\varepsilon_*\Omega^\bullet_{X_\bullet}$, for $\varepsilon : X_\bullet \to X$ a proper hypercovering. There is a natural map $\Omega_X^\bullet \to K^\bullet$. The complex $K^\bullet$ also has the property that for X proper, the spectral sequence for the filtration F degenerates at E_1.

Du Bois [Du 1] extended the construction of $K^\bullet$ to separated schemes X over $\mathbb{C}$. He introduced the derived category $D_{\text{diff}}(X)$ of complexes of differential operators on X and showed that in $D_{\text{diff}}(X)$, the complex $K^\bullet$ is independent of the proper hypercovering chosen. He denoted this complex by $\underline{\Omega}^\bullet_X$. For $f : X \to S$ a morphism of separated schemes, there is a natural map $\underline{\Omega}^\bullet_S \to Rf_*\underline{\Omega}^\bullet_X$. There is also a version $\underline{\Omega}^\bullet_{\overline{S}}(\log \overline{S} - S)$of $\underline{\Omega}^\bullet_S$ with logarithmic singularities, which is a complex of sheaves on a compactification $\overline{S}$ of S.

Du Bois computed the complex $\underline{\Omega}^\bullet_X$ for a number of interesting examples. For instance, if X has normal crossings singularities, let $X_0 \xrightarrow{\varepsilon_0} X$ be the normalization of X and $X_1 = X_0 \times_X X_0 \xrightarrow{\varepsilon_1} X$, then $\underline{\Omega}^p_X = \ker(\varepsilon_{0*}\Omega^p_{X_0} \to \varepsilon_{1*}\Omega^p_{X_1})$.

For X a curve, let $\pi : X_0 \to X$ be its normalization, and let $g : X' \to X$ be the curve, homeomorphic to X, obtained by contracting each finite subset $\pi^{-1}(s)$ of X_0 to a point. Then $\underline{\Omega}^\bullet_X$ is isomorphic to the complex $0 \to g_*\mathcal{O}_{X'} \to \pi_*\Omega^1_{X_0} \to 0$.

It should be emphasized that complexes like $\Omega^\bullet_{\overline{X}}(\log D)$ and the Deligne-Du Bois complexes (with logarithmic singularities) can be defined in the Zariski topology, and that according to the GAGA theorem of Serre [Se], the hypercohomology of the Zariski complex is identical to the hypercohomology of the complex-analytic one. On the other hand, the constructions of Deligne work equally well for Kähler manifolds, if one considers only Kähler manifolds which admit a Kähler compactification; so for instance one obtains M.H.S. on the cohomology of smooth simplicial Kähler manifolds.

Deligne's theory [D 4], as well as El Zein's treatment [E 8], also give M.H.S. on compactly-supported cohomology and on Borel-Moore homology. Quite recently, Deligne's Hodge theory has been reformulated and extended to the homotopy groups of algebraic varieties by Navarro Aznar [NA 1]. Deligne also obtained similar results. One of the main tools in [NA 1] is the construction of a direct image functor for derived categories of sheaves of DG-$\mathbb{Q}$-algebras. See Theorem 2.25 for the results of Navarro Aznar on M.H.S. for vanishing cycles.

2. Degeneration of Hodge Structures

We start with the notion of *polarized variation of A-Hodge structure*, for A a subring of $\mathbb{R}$. This is due to Griffiths [Gr 1] and Deligne [D 3]. Recall that for $\mathbb{V}_{\mathbb{C}}$ a local system of complex vector spaces over a complex manifold X, the corresponding holomorphic vector bundle $\mathcal{V} := \mathcal{O}_X \otimes \mathbb{V}_{\mathbb{C}}$ is endowed with an integrable connection ∇, with space of horizontal sections $\mathbb{V}_{\mathbb{C}}$.

Definition 2.1 [Gr 1, D 3]. *Let A be a subring of $\mathbb{R}$, and X a complex manifold. A* polarized variation of *A*-Hodge structure of weight *m over X consists of a local system $\mathbb{V}_A$ over X of A-modules of finite rank, together with the following data:*

(1) a decreasing filtration $\ldots \mathscr{F}^{p+1} \subset \mathscr{F}^p \subset$ *of the holomorphic vector bundle* $\mathscr{V} = \mathscr{O}_X \otimes \mathbb{V}_A$ *by holomorphic sub-bundles;*

(2) a flat bilinear form $S : \mathbb{V}_A \times \mathbb{V}_A \to A(-m)$*;*

which satisfy:

(a) the Griffiths transversality condition $\nabla(\mathscr{F}^p) \subset \Omega^1_X \otimes_{\mathscr{O}_X} \mathscr{F}^{p-1}$*;*

(b) for every x in X, the A-module $\mathbb{V}_{A,x}$*, with the filtration* F^p_x *induced by* $\mathscr{F}^p$ *and the bilinear form* S_x*, is a polarized Hodge structure of weight m.*

The typical geometric situation where such objects appear arises from a smooth projective morphism $f : Y \to X$ between quasi-projective algebraic varieties over $\mathbb{C}$. Denote the fiber $f^{-1}(x)$ by Y_x. For each $m \geq 0$, let $\mathscr{P}^m_{\mathbb{Z}} \subset R^m f_* \mathbb{Z}$ consist of the primitive cohomology classes in the fibers of f. Let $\mathscr{F}^p_x = \oplus_{j \geq p} P^{j,m-j}(Y_x)$ define the sub-bundle $\mathscr{F}^p$; it is holomorphic since it may be described in terms of $R^m f_* F^p \Omega^\bullet_{Y/X}$, where the subcomplex $F^p\Omega^\bullet_{Y/X}$ of $\Omega^\bullet_{Y/X}$ is defined in 0-1. There is natural bilinear form S, described in Sect. 1, and one obtains a $\mathbb{Z}$-polarized variation of Hodge structure.

If we choose a base point x of X, the variation of Hodge structure $\mathbb{V}$ gives rise to a holomorphic mapping $\Phi : X \to M/\Gamma$, where M is the appropriate classifying space of polarized Hodge structures on the fiber $\mathbb{V}_x$, and Γ is the monodromy group, which acts on M as a subgroup of the Lie group G of automorphisms of $(\mathbb{V}_{\mathbb{R}}, S)$. Φ is the *period mapping* of Griffiths. More precisely, let $\rho : \pi_1(X, x) \to \mathrm{GL}(\mathbb{V}_{\mathbb{R},x})$ denote the monodromy homomorphism. Put $\Gamma = \mathrm{im}(\rho)$. Then, for $\tilde{X}$ a universal covering space of X, Φ admits the lifting $\tilde{\Phi} : \tilde{X} \to M$ such that $\tilde{\Phi}(\gamma z) = \rho(\gamma)\tilde{\Phi}(z)$ [Gr 1: III].

We need to say more about the classifying space M for polarized real Hodge structures of weight m with Hodge numbers $h^{p,q}$. Fix a real vector space V and a $(-1)^m$-symmetric bilinear form S on V. M is an open subset of the flag manifold $\widehat{M}$ (for the orthogonal group of S), consisting of the decreasing filtrations $\ldots \subset F^{p+1} \subset F^p \subset \ldots$ of $V_{\mathbb{C}}$ such that

(1) $\dim \mathrm{Gr}^p_F V_{\mathbb{C}} = h^{p,m-p}$;

(2) the space orthogonal to F^p (with respect to S) is F^{m+1-p}.

The complex semi-simple Lie group $G_{\mathbb{C}}$ of automorphisms of $(V_{\mathbb{C}}, S)$ acts transitively on $\widehat{M}$, with isotropy group a parabolic subgroup B. M itself is an orbit of the real group $G = G_{\mathbb{R}}$, with compact isotropy group $K = G \cap B$. Let $\mathfrak{g}$, $\mathfrak{g}_{\mathbb{R}}$, $\mathfrak{b}$ be the Lie algebras of $G_{\mathbb{C}}$, $G_{\mathbb{R}}$ and B, respectively. Any F in $\widehat{M}$ defines a filtration, also denoted by F, of $\mathrm{End}(V) = V^* \otimes V$ and of $\mathfrak{g}$, with $F^0(\mathfrak{g}) = \mathfrak{b}$. The *horizontal subbundle* $T_h\widehat{M}$ of the tangent bundle $T\widehat{M}$ has fiber at F equal to $F^{-1}(\mathfrak{g})/\mathfrak{b} \subset \mathfrak{g}/\mathfrak{b} = T_F(\widehat{M})$. This subbundle is clearly $G_{\mathbb{C}}$-equivariant. Griffiths transversality means that $\tilde{\Phi} : \tilde{X} \to M$ is horizontal, i.e. $d\tilde{\Phi}$ takes values in $T_h\widehat{M}$.

When $F \in M$, the above filtration on $\mathfrak{g}$, together with the Killing form on $\mathfrak{g}_{\mathbb{R}}$, define a polarized Hodge structure. The hermitian form induced on $\mathfrak{b}^{\perp} \cong T_F M$ by the polarization form gives a hermitian metric on M, which is actually Kähler.

The following important result was first proven by Landman [La] and

Clemens [Cl 1] for geometric variations of Hodge structure, then by Borel in the general case.

Proposition 2.2 (Borel, [D 3, Sc]). *Let $\mathbb{V}_{\mathbb{Z}}$ be a $\mathbb{Z}$-polarized variation of Hodge structure over the complex manifold X. Assume that there is an open embedding $X \hookrightarrow \overline{X}$, where $D := \overline{X} - X$ is a divisor with normal crossings. Then the monodromy of $\mathbb{V}_{\mathbb{Z}}$ around each local component of D is quasi-unipotent (i.e. all its eigenvalues are roots of unity).*

Most of this chapter will be devoted to the theory of degenerations of Hodge structures, mostly in a local context. In the geometric situation, a smooth projective morphism $f : Y \to X$ degenerates over the boundary D of some (partial) compactification $X \hookrightarrow \overline{X}$. This means that one cannot extend f to a smooth family over $\overline{X}$. To handle this situation, one needs to extend f in some way to a projective morphism $f : \overline{Y} \to \overline{X}$, and to use resolution of singularities to control the singularities of the fibers over D. When X has dimension 1, such methods are very successful. In case X has dimension bigger than 1, controlling the singularities becomes much more problematic.

In any case, it is of great interest to study the degeneration of the family in some way that does not depend on the choice of a particular extension of f over $\overline{X}$. The theory of degenerations of variations of Hodge structure works independently of any geometric assumptions and gives information on the asymptotic behavior of the Hodge structures on the cohomology of the fibers, hence also concrete information about the behavior of the period mapping.

The *nilpotent orbit theorem* of W. Schmid works in the generality of a real polarized variation of Hodge structure with quasi-unipotent monodromy (according to Borel's theorem, this covers the case of $\mathbb{Z}$-variations). This theorem is local near $D = \overline{X} - X$, so one may as well assume that X is a product of discs Δ and punctured discs Δ^*, say $X = (\Delta^*)^n \times \Delta^k \subset \overline{X} = \Delta^{n+k}$. Let $\gamma_1, \ldots, \gamma_n$ denote the monodromy transformations, acting on $\mathbb{V}$. Since each γ_j is quasi-unipotent, there exist nilpotent transformations N_j – the *monodromy logarithms* – and positive integers $l_1, \ldots, l_n$ such that

$$\gamma_j^{l_j} = \exp(N_j).$$

The universal covering space of $(\Delta^*)^n \times \Delta^k$ is $H^n \times \Delta^k$, where $H = \{z \in \mathbb{C} : \operatorname{Im}(z) > 0\}$ is the upper half-plane. Let $(\mathbf{z}, \mathbf{w})$ denote coordinates on the covering space, with $\mathbf{z} = (z_1, \ldots, z_n)$ and $\mathbf{w} = (w_1, \ldots, w_k)$. Let $s_i = e^{z_i}$, $\mathbf{s} = (s_1, \ldots, s_n)$. Then $(\mathbf{s}, \mathbf{w})$ are coordinates on $(\Delta^*)^n \times \Delta^k$. In this context, the mapping $\tilde{\Phi}$ satisfies $\tilde{\Phi}(\mathbf{z} + l_j \varepsilon_j, \mathbf{w}) = \gamma_j^{l_j} \cdot \tilde{\Phi}(\mathbf{z}, \mathbf{w})$, for ε_j the j-th standard basis vector of $\mathbb{C}^n$.

First one defines a nilpotent orbit.

Definition 2.3. *A* nilpotent orbit *is a holomorphic mapping $\theta : \mathbb{C}^n \to \widehat{M}$ of the form*

$$\theta(\mathbf{z}) = \exp\left(\sum z_j N_j\right) \cdot F,$$

where

(1) $F \in \widehat{M}$;

(2) $\{N_j\}_{j=1}^n$ *is a commuting set of nilpotent elements of* $\mathfrak{g}_{\mathbb{R}} \cap F^{-1}\mathfrak{g}$ *(hence* θ *is a horizontal mapping);*

(3) There exists $\alpha \in \mathbb{R}$ *such that* $\theta(\mathbf{z}) \in M$ *for* $\mathrm{Im}(z_j) > \alpha$.

In our situation, from the lifting $\tilde{\Phi} : H^n \times \Delta^k \to M$ of the period mapping, one can construct another mapping

$$\tilde{\psi}(\mathbf{z}, \mathbf{w}) := \exp(-\sum z_j N_j) \cdot \tilde{\Phi}(\mathbf{l} \cdot \mathbf{z}, \mathbf{w}),$$

for $\mathbf{l} = (l_1, \ldots, l_n)$.

This mapping is single-valued, i.e. factors through a holomorphic map $\psi : (\Delta^*)^n \times \Delta^k \to \widehat{M}$.

Theorem 2.4 (Schmid's Nilpotent Orbit Theorem) [Sc: Theorem 4.12]. *With* $\psi : (\Delta^*)^n \times \Delta^k \to \widehat{M}$ *defined above, we have:*

(1) ψ *extends holomorphically to* Δ^{n+k};

(2) Let $F(\mathbf{w}) = \psi(0, \mathbf{w})$ *and* $\theta_{\mathbf{w}}(\mathbf{z}) = \exp(\sum z_j N_j) \cdot F(\mathbf{w})$. *Then, for each* $\mathbf{w} \in \Delta^k$, $\mathbf{z} \to \theta_{\mathbf{w}}(\mathbf{z})$ *is a nilpotent orbit. Furthermore, for any G-invariant distance* d *on* M, *there are non-negative constants* α, β *and* K *such that for* $\mathrm{Im}(z_j) > \alpha$, $\theta_{\mathbf{w}}(\mathbf{z}) \in M$ *and*

$$d(\tilde{\Phi}(\mathbf{z}, \mathbf{w}), \theta_{\mathbf{w}}(\mathbf{z})) \leq K \sum_{j=1}^{n} (\mathrm{Im}(z_j))^{\beta} e^{-2\pi l_j^{-1} \mathrm{Im}(z_j)}.$$

The constants α, β, K *depend only on the moduli space* M, *through the corresponding Hodge numbers and weight, and on the unipotency indices* $l_1, \ldots, l_n$.

Remark 2.5. The distance estimate given here, due to Deligne, is sharper than the one given in [Sc], and is proven in [C-K-S 1: § 1].

In dimension one, the nilpotent orbit theorem is equivalent, for period mappings of geometric origin, to the regularity of the Gauss-Manin connection [Gr-Sc]. In arbitrary dimensions, the nilpotent orbit theorem implies the regularity of that connection.

Since the nilpotent orbit theorem states that near a point of D, a polarized variation of Hodge structure may be approximated quite closely by a nilpotent orbit, the study of nilpotent orbits is of paramount importance for the general theory of degenerating M.H.S. This is achieved by the SL(2)-*orbit theorem*, due to Schmid [Sc] in the one-variable case and to Cattani-Kaplan-Schmid [C-K-S 2] in the several variable case.

Before discussing the SL(2)-orbit theorem, we discuss splittings of a mixed Hodge structure. Let (V, W, F) be a M.H.S. over $\mathbb{R}$ (so V is a real vector space). A *splitting* of the M.H.S. (V, W, F) is a bigrading $V_{\mathbb{C}} = \sum V^{p,q}$ such that $W_l = \sum_{p+q \leq l} V^{p,q}$ and $F^p = \sum_{r \geq p} V^{r,s}$. Deligne constructed a canonical splitting of any M.H.S.

Definition and Proposition 2.6 (Deligne, see [Mo] and [C-K-S 1]). *For (V, W, F) a M.H.S. over $\mathbb{R}$, let*

$$I^{p,q} = F^p \cap W_{p+q} \cap ((\overline{F}^q \cap W_{p+q}) + \overline{U}^{q-1}_{p+q-2}),$$

where $U^a_b = \sum_{j \geq 0} F^{a-j} \cap W_{b-j}$.

(i) $(I^{p,q})_{p,q}$ is a functorial splitting of (V, W, F), which satisfies

(2-1) $$I^{p,q} \equiv \overline{I}^{q,p} \pmod{\oplus_{r<p,s<q} I^{r,s}}.$$

Furthermore this is the only splitting verifying (2-1).

(ii) This natural splitting establishes an equivalence of categories between the category of M.H.S. over $\mathbb{R}$ and the category of $\mathbb{R}$-vector spaces V equipped with a bigrading $V_{\mathbb{C}} = \sum_{p,q} I^{p,q}$ satisfying (2-1).

A mixed Hodge structure (V, W, F) is said *to be split over* $\mathbb{R}$ if it admits a splitting $(J^{p,q})_{p,q}$ such that $J^{p,q} = \overline{J^{q,p}}$. From Proposition 2.6, such a splitting, if it exists, is unique, and is given by $J^{p,q} = F^p \cap \overline{F}^q \cap W_{p+q}$. There is a canonical method to deform any real M.H.S. into one split over $\mathbb{R}$. To the M.H.S. (V, W, F) one associates the nilpotent Lie algebra

$$\mathfrak{l}^{-1,-1} := \{X \in \mathrm{End}(V_{\mathbb{C}}) : X(I^{p,q}) \subset \oplus_{r<p,s<q} I^{r,s}\}.$$

It is clear that $\overline{\mathfrak{l}}^{-1,-1} = \mathfrak{l}^{-1,-1}$, hence $\mathfrak{l}^{-1,-1}$ admits the real form $\mathfrak{l}^{-1,-1}_{\mathbb{R}} = \mathrm{End}(V_{\mathbb{R}}) \cap \mathfrak{l}^{-1,-1}$.

Proposition 2.7. *Given a M.H.S. (V, W, F) over $\mathbb{R}$, there exists a unique $\delta \in \mathfrak{l}^{-1,-1}_{\mathbb{R}}$ such that $(V, W, e^{-i\delta} \cdot F)$ is a M.H.S. split over $\mathbb{R}$. The functor $(V, W, F) \to ((V, W, e^{-i\delta} \cdot F), \delta)$ establishes an equivalence of categories between the category of M.H.S. over $\mathbb{R}$ and the category whose objects are pairs consisting of a M.H.S. (V, W, F) split over $\mathbb{R}$ and an element $\delta \in \mathfrak{l}^{-1,-1}_{\mathbb{R}}(W, F)$, and whose morphisms are morphisms of M.H.S. intertwining the endomorphisms δ.*

The geometry of the situation may be summarized as follows. For $(V_{\mathbb{R}}, W_\bullet)$ a fixed real vector space with a given filtration, let F^W be the variety of filtrations F of $V_{\mathbb{C}}$ such that (V, F, W) is a M.H.S. Let $F^W_{\mathbb{R}}$ be the subvariety of F such that the M.H.S. (V, F, W) is split over $\mathbb{R}$. Then the varieties F^W and $F^W_{\mathbb{R}}$ are smooth. Let U be the complex nilpotent Lie group with Lie algebra $\mathfrak{u} = \{T \in W_{-1}\mathrm{End}(V_{\mathbb{C}}) : T \equiv \overline{T} \bmod \mathfrak{l}^{-1,-1}\}$. Then U operates on F^W, and its operation leaves fixed the Hodge structures on $\mathrm{Gr}^W_j(V)$. We have $F^W = U \cdot F^W_{\mathbb{R}}$, and for $F \in F^W_{\mathbb{R}}$ with stabilizer subgroup $F^0(U)$, the tangent space to $F^W_{\mathbb{R}}$ at F is $W_{-1}\mathrm{End}(V_{\mathbb{R}}) + F^0(U)/F^0(U)$.

Now $U/F^0(U)$ is isomorphic to $\exp(\mathfrak{l}_{\mathbb{C}}^{-1,-1})$, as $\mathfrak{l}_{\mathbb{C}}^{-1,-1}$ is a complementary subspace for $F^0(\mathfrak{u})$ inside $\mathfrak{u}$. If we now divide $U/F^0(U)$ by $\exp(W_{-1}\mathrm{End}(V_{\mathbb{R}}))$, the corresponding double coset space is isomorphic to $\exp(\mathfrak{l}_{\mathbb{R}}^{-1,-1})\backslash\exp(\mathfrak{l}_{\mathbb{C}}^{-1,-1})$, hence to $\exp(i \cdot \mathfrak{l}_{\mathbb{R}}^{-1,-1})$. This is meant to explain the role of $\exp(i \cdot \mathfrak{l}_{\mathbb{R}}^{-1,-1})$ in Proposition 2.7.

We next discuss polarized M.H.S. For k an integer, a *k-bilinear form* on a real M.H.S. $(V_{\mathbb{R}}, W, F)$ is a non-degenerate bilinear form S on $V_{\mathbb{R}}$ with values in $\mathbb{R}(-k)$, which is $(-1)^k$-symmetric and satisfies $S(F^p, F^{k-p+1}) = 0$.

Definition 2.8 (Polarized Mixed Hodge Structure). *Let $(V_{\mathbb{R}}, W, F)$ be a M.H.S., and let S be a k-bilinear form on $(V_{\mathbb{R}}, W, F)$. We say that $(V_{\mathbb{R}}, W, F, k, S)$ is* polarized *by a nilpotent real endomorphism N of V if N satisfies*

(1) N preserves S;

(2) $W = W(N)[k]$, where the weight filtration *$W(N)$ of [SGA 7] is the unique filtration of V such that $N \cdot W_l(N) \subset W_{l-2}(N)$ and for every $j \geq 0$, N^j induces an isomorphism $N^j : \mathrm{Gr}^W_j(N) \xrightarrow{\sim} \mathrm{Gr}^W_{-j}(N)$;*

(3) $NF^p \subset F^{p-1}$;

(4) the Hodge structure of weight $k + l$ on the primitive part $P_{k+l} := \ker(N^{l+1} : \mathrm{Gr}^W_{k+l} \to \mathrm{Gr}^W_{k-l-2})$ is polarized by the bilinear form $(2\pi i)^{-l} S(\cdot, N^l \cdot)$.

A splitting of a polarized M.H.S. $(V, W, F, S; N)$ is a splitting of the M.H.S.

$$W_l = \sum_{p+q \leq l} J^{p,q} \ , \quad \text{and} \quad F^p = \sum_{r \geq p} J^{r,s}$$

which is compatible with N (i.e. $NJ^{p,q} \subset \oplus_{r \leq p-1, s \leq q-1} J^{r,s}$), and such that the splitting $H_l = \sum_{p+q=k+l} J^{p,q}$ is self-dual with respect to S, i.e. $S(H_l, H_m) = 0$ unless $l + m = -1$. The semi-simple endomorphism Y of V defined by $Y(u) = (p + q - k)u$ for $u \in J^{p,q}$ is compatible with S, W and F. In particular, Deligne's splitting $I^{p,q}$ is compatible with any polarization of a M.H.S.

The relation with nilpotent orbits is the following. Let $z \mapsto \exp(zN)F$ be a one-parameter nilpotent orbit (cf. Definition 2.3). Introduce the real filtration W as a suitable shift of the monodromy weight filtration $W(N)$; precisely $W = W(N)[-m]$, where m is the weight of the variation of Hodge structure.

Theorem 2.9 (Schmid, [Sc]). *For a one-parameter nilpotent orbit, $(V, W, F, S; N)$ is a polarized mixed Hodge structure.*

This theorem was obtained by Schmid as a corollary of his SL(2)-orbit theorem, to which we turn next. For variations of Hodge structure of geometric origin, it admits a geometric proof [St 1], which we discuss at the end of this section. A converse to the theorem is given in [C-K-S 1: Corollary 3.13]. This theorem may be rephrased in terms of the canonical extension $\overline{\mathcal{V}}$ of $\mathcal{V}$ to a holomorphic vector bundle over Δ^n, which is spanned by sections of the form

$$\tilde{v}(\mathbf{s}) = \exp\Big(\frac{-1}{2\pi i}\sum_{j=1}^{n}(\log s_j)N_j\Big)v, \tag{2-2}$$

for v a multivalued section of $\mathbb{V}$. Note that the connection ∇ has logarithmic poles with respect to $\overline{\mathcal{V}}$, i.e. it extends to give a $\mathcal{O}_{\Delta^n}$-linear map

$$\overline{\nabla} : \overline{\mathcal{V}} \to \Omega^1_{\Delta^n}(\log D) \otimes_{\mathcal{O}_{\Delta^n}} \overline{\mathcal{V}}.$$

Corollary 2.10. *(1) The filtration $\mathcal{F}^\bullet$ of $\mathcal{V}$ extends to a filtration of $\overline{\mathcal{V}}$ by subbundles.*

(2) Let $n = 1$. The vector space $V_0 := \overline{\mathcal{V}}_0$, endowed with the real structure induced by the isomorphism $v \mapsto \tilde{v}$ of (2-1) and the pair of filtrations (W, F), where W is the monodromy weight filtration centered at m (cf. Definition 2.8) and F is induced from $\mathcal{F}$, is a mixed Hodge structure, polarized by N.

Starting from a polarized M.H.S. (V, W, F, S, N), let $(V, W, \tilde{F})$ be the associated $\mathbb{R}$-split M.H.S. Write $F = e^{i\delta}\tilde{F}$, with $\delta \in \mathfrak{l}_{\mathbb{R}}^{-1,-1}$. Since N commutes with δ, it is also an endomorphism of the M.H.S. $(V, W, \tilde{F})$ of type $(-1,-1)$. Let $(\tilde{I}^{r,s})_{r,s}$ be the $\mathbb{R}$-splitting of $(V, W, \tilde{F})$, with $\tilde{I}^{p,q}$ complex-conjugate to $\tilde{I}^{q,p}$. Let $\tilde{Y}$ be the corresponding semisimple transformation:

$$\tilde{Y}(v) = (p+q+k)v \quad \text{for } v \in \tilde{I}^{p,q}.$$

The pair $(\tilde{Y}, N)$ can be completed, in a unique way, to an $\mathfrak{sl}(2)$-triple $(\tilde{N}^+, \tilde{Y}, N)$. We have the commutation relations:

$$[\tilde{N}^+, N] = \tilde{Y}, \quad [\tilde{Y}, \tilde{N}^+] = 2\tilde{N}^+, \quad [\tilde{Y}, N] = -2N.$$

It is clear that $\tilde{N}^+$ is a nilpotent endomorphism of $(V, W, \tilde{F})$ of type $(1, 1)$. We have a Lie algebra homomorphism $\tilde{\rho} : \mathfrak{sl}(2, \mathbb{R}) \to \mathfrak{g}$, which maps the standard generators of $\mathfrak{sl}(2, \mathbb{R})$ to $(\tilde{N}^+, \tilde{Y}, N)$. It lifts to a Lie group homomorphism $\tilde{\rho} : \mathrm{SL}(2, \mathbb{C}) \to G_{\mathbb{C}}$; $\tilde{\rho}$ is defined over $\mathbb{R}$. Since $\tilde{Y}$ and $\tilde{N}^+$ fix the filtration $\tilde{F}$, the map $g \mapsto \tilde{\rho}(g) \cdot \tilde{F}$ induces an equivariant embedding

$$\mathbb{CP}^1 \cong \mathrm{SL}(2, \mathbb{C})/B^- \hookrightarrow \widehat{M},$$

where $B^- = \left\{\begin{pmatrix} a & 0 \\ b & a^{-1}\end{pmatrix}\right\}$ is a Borel subgroup. Hence the nilpotent orbit $z \mapsto \exp(zN)\tilde{F}$ is SL(2)-equivariant. The SL(2)-orbit theorem gives a precise comparison between this orbit and the original nilpotent orbit (with F instead of $\tilde{F}$). To state it, it is convenient to put a connection on the $\tilde{B}$-principal bundle $G_{\mathbb{R}} \to G_{\mathbb{R}}/\tilde{B} \cong M$, where $\tilde{B}$ is the stabilizer of $e^{iN}\tilde{F}$ in $G_{\mathbb{R}}$. As $\tilde{B}$ preserves positive-definite hermitian forms on the Hodge spaces of $\mathrm{Gr}^W_\bullet(V, e^{iN}\tilde{F})$, it is compact. So we have the orthogonal decomposition $\mathfrak{g}_{\mathbb{R}} = \tilde{\mathfrak{b}} \oplus \tilde{\mathfrak{b}}^{\perp}$ with respect to the Killing form. This $\tilde{B}$-invariant decomposition defines a $G_{\mathbb{R}}$-invariant connection on the principal bundle.

Theorem 2.11 (Schmid's SL(2)-orbit Theorem [Sc]). *Let $(V, W, F, S; N)$ be as in Theorem 2.9. Let α be such that $\exp(zN)F$ and $\exp(zN)\tilde{F}$ are both in M for $\mathrm{Im}(z) > \alpha$. Then there exists a real-analytic $G_{\mathbb{R}}$-valued function $\tilde{g}(y)$, defined for $y > \alpha$, such that*

(1)

$$\exp(iyN)F = \tilde{g}(y)\exp(iyN)\tilde{F},$$

for $y > \max(0, \alpha)$ (i.e. $\tilde{g}(y)$ transforms one nilpotent orbit into the other);

(2) Putting $\tilde{h}(y) = \tilde{g}(y)\exp(-\frac{1}{2}\log y\tilde{Y})$, the $G_{\mathbb{R}}$-valued curve $y \mapsto \tilde{h}(y)$ is horizontal with respect to the connection on the principal bundle $G_{\mathbb{R}} \to G_{\mathbb{R}}/\tilde{B} \cong M$; equivalently, $\tilde{h}(y)$ satisfies the differential equation $\tilde{h}(y)^{-1}\tilde{h}'(y) \in \tilde{\mathfrak{b}}^{\perp}$;

(3) $\tilde{g}(y)$ has a convergent Taylor expansion around ∞:

$$\tilde{g}(y) = \tilde{g}(\infty)(1 + \tilde{g}_1 y^{-1} + \tilde{g}_2 y^{-2} + \ldots);$$

(4) $\tilde{g}(\infty) \in \exp(\mathfrak{l}_{\mathbb{R}}^{-1,-1} \cap \ker(\mathrm{ad}N))$.

These properties determine $\tilde{g}(y)$ uniquely. Let

$$\tilde{g}(y)^{-1} = (1 + \tilde{f}_1 y^{-1} + \ldots)\tilde{g}(\infty)^{-1}$$

denote the Taylor series for $\tilde{g}(y)^{-1}$. Then we have in addition:

(5) $(\mathrm{ad}N)^{k+1}\tilde{g}_k = 0$ and $(\mathrm{ad}N)^{k+1}\tilde{f}_k = 0$;

(6) $\tilde{g}_k$ and $\tilde{f}_k$ belong to the linear span of the $\mathrm{End}(V)^{p,q}$ for $p + q \leq k - 1$.

We note that the significance of $\tilde{h}(y)$ is that it conjugates $e^{iN}\tilde{F}$ into $\exp(iYN)F$: we have $\exp(iyN)F = \tilde{h}(y)e^{iN}\tilde{F}$.

An important corollary of the SL(2)-orbit theorem is the Hodge norm estimate:

Corollary 2.12 (Schmid, [Sc]). *Let $(\mathbb{V}, \mathcal{F}, S)$ be a polarized variation of $\mathbb{R}$-Hodge structure of weight m over the punctured disc Δ^*, with coordinate s. Assume $\mathbb{V}$ has unipotent monodromy. Let $W = W(N)[m]$ be the shifted monodromy weight filtration. A flat section v of $\mathbb{V}$ lies in $W_l - W_{l-1}$ if and only if, on any small sector around 0, one has*

$$\langle v, v \rangle \sim (\log|s|)^l.$$

The Hodge filtration $\tilde{F}_0 := \tilde{g}(\infty)\tilde{F}$ is of particular importance. Like $\tilde{F}$, it is related to F by a (non-real) element of the nilpotent group $\exp(\mathfrak{l}^{-1,-1} \cap \ker \mathrm{ad}N)$. Since $\tilde{g}(\infty) \in G_{\mathbb{R}}$, $(V, W, \tilde{F}_0)$ is an $\mathbb{R}$-split M.H.S., polarized by N. A remarkable fact is that $\tilde{F}_0$ depends only on (V, W, F), not on N. One can restate the SL(2)-orbit theorem in terms of $\tilde{F}_0$. Introducing $N^+ := Ad(\tilde{g}(\infty))\tilde{N}^+$,

$Y := Ad(\tilde{g}(\infty))\tilde{Y}$, one obtains an SL(2)-triple (N^+, Y, N). With the new $G_{\mathbb{R}}$-valued curves $g(y) = \tilde{g}(y)\tilde{g}(\infty)^{-1}$, $h(y) = \tilde{h}(y)\tilde{g}(\infty)^{-1} = g(y)\exp(-\frac{1}{2}(\log y)Y)$, one has:

$$\exp(iyN)F = g(y)\exp(iyN)\tilde{F}_0 = h(y)e^{iN}\tilde{F}_0.$$

The coefficients g_j (resp. f_j) of the Taylor expansions of $g(y)$ (resp. $g(y)^{-1}$) satisfy conditions (5) and (6) of Theorem 2.10.

Proposition 2.13 [C-K-S 1: Proposition 3.28]. *Let $\delta = \sum_{p,q\geq 1}\delta_{-p,-q}$. The linear transformations g_k and f_k are expressed as universal non-commutative polynomials in the $\delta_{-p,-q}$ and in $ad\, N^+$. They satisfy bounds*

$$||g_k||, ||f_k|| \leq C(1 + ||N^+|| + ||Y|| + ||N||)^n (\max_{p,q} ||\delta_{-p,-q}||^m)^k.$$

The generalization of the SL(2)-theorem to nilpotent orbits in several variables proved extremely difficult. Even what its statement should be remained an enigma for several years. The first step was accomplished by Cattani and Kaplan, who, inspired by the analogous result of Deligne in [D 5: II], proved that the monodromy weight filtration is constant on an open cone. To state fully their result, we need to introduce another concept from [D 5].

Proposition and Definition 2.14 (Deligne, [D 5: II], see also [St-Z]). *Let $(V, W_\bullet)$ be a filtered vector space, and let N be a nilpotent endomorphism of V, preserving W. A weight filtration of N relative to W is a filtration M of V such that*

(1) $NM_j \subset M_{j-2}$;

(2) The filtration induced by M on $\mathrm{Gr}_k^W(V)$ is the weight filtration for $(\mathrm{Gr}_k^W(V), W)$, centered at k.

If M exists, it is unique.

We refer to [St-Z: § 2] for a detailed discussion of the relative weight filtration M. Note for instance [St-Z: Proposition 2.6] that if W has length 2, then M exists if and only if N^l is a strict endomorphism of (V, W) for each $l > 0$.

Proposition 2.15 [C-K 2: § 3]. *(1) Let $\theta(\mathbf{z}) = \exp\left(\sum_{j=1}^n z_j N_j\right) \cdot F$, $\mathbf{z} \in \mathbb{C}^n$ be a nilpotent orbit (cf. Definition 2.3). Let $C := \{\sum_{j=1}^n \lambda_j N_j : \lambda_j \in \mathbb{R}, \lambda_j > 0\}$. Then the weight filtration $W(N)$ associated to $N \in C$ is independent of N. Call it $W(C)$.*

For J a subset of $\{1, \ldots, n\}$, let $C_J = \{\sum_{j\in J}\lambda_j N_j : \lambda_j \in \mathbb{R}, \lambda_j > 0\}$.

(2) Let $N \in C_J$ and $N' \in C_{J'}$. For $N'' \in C_{J\cup J'}$, $W(N'')$ is the weight filtration of N relative to $W(N')$.

We point out that assertion (2) is slightly misstated in [C-K 2] (see [St-Z: Theorem 3.12]).

From Theorem 2.9, for every $N \in C$, $(V, W(C)[-m], F, N)$ is a polarized mixed Hodge structure. The following constructions depend on a fixed ordering of the variables in $\mathbb{C}^n$. For $1 \le r \le n$, let $C_r := \{\sum_{j=1}^r \lambda_j N_j : \lambda_j \in \mathbb{R}, \lambda_j > 0\}$. Every element of C_r defines the same weight filtration, say $W(C_r)$; set $W^r = W(C_r)[-m]$. Define Hodge filtrations $\tilde{F}_r \in \widehat{M}$, for $1 \le r \le n$ by descending induction on r as follows.

Let $\tilde{F}_n$ be the Hodge filtration of the $\mathbb{R}$-split M.H.S. canonically associated to (V, W^n, F) by the SL(2)-orbit theorem (this filtration was denoted by $\tilde{F}_0$ in the discussion following Theorem 2.11). Now $(V, W^n, \tilde{F}_n)$ is polarized by every $N \in C = C_n$, hence the mapping $(z_1, \dots, z_{n-1}) \mapsto \exp(\sum_{j=1}^{n-1} z_j N_j) \cdot (e^{iN_n} \tilde{F}_n)$ is a nilpotent orbit in $(n-1)$ variables, which takes values in M for $\mathrm{Im}(z_j) > 0$. In particular, $(V, W^{n-1}, e^{iN_n} \tilde{F}_n)$ is a M.H.S., polarized by every $N \in C_{n-1}$. Let $\tilde{F}^{n-1}$ be the Hodge filtration of the $\mathbb{R}$-split M.H.S. associated to it by the SL(2)-orbit theorem. Continuing, one obtains $\tilde{F}_r \in \widehat{M}$, such that for each r, $(V, W^r, \tilde{F}^r)$ is the $\mathbb{R}$-split M.H.S. canonically associated to $(V, W^r, e^{iN_{r+1}} \tilde{F}_{r+1})$. $(V, W^r, \tilde{F}^r)$ is polarized by every $N \in C_r$.

For $t_1, \dots t_{r-1}$ positive real numbers, the mapping $z \mapsto \big(\exp z N_r + \sum_{j=1}^{r-1} t_j N_j)\big) \cdot (e^{iN_{r+1}} \tilde{F}_{r+1})$ is a one-variable nilpotent orbit. Let $h_r(t_1, \dots t_{r-1}; y)$ be the $G_{\mathbb{R}}$-valued function associated to it (with $(t_1, \dots, t_{r-1})$ as parameters).

Theorem 2.16 (SL(2)-orbit Theorem in n Variables, Cattani-Kaplan-Schmid [C-K-S 1, C-K 3]). *Let $\theta : \mathbb{C}^n \to \widehat{M}$ be a nilpotent orbit $\theta(\mathbf{z}) = \exp\big(\sum_{j=1}^n z_j N_j\big) \cdot F$. Then there exists a unique Lie group homomorphism $\rho : \mathrm{SL}(2, \mathbb{C})^n \to G_{\mathbb{C}}$ with the following properties*

(1) $\rho_* : \mathfrak{sl}(2, \mathbb{C})^n \to \mathfrak{g}$ is a morphism of M.H.S. for the M.H.S. on $\mathfrak{g}$ induced by $e^{iN_1} \cdot \tilde{F}_1$;

Denote by $(\widehat{N}_j^+, \widehat{Y}_j, \widehat{N}_j^-)$ the images under ρ of the standard generators of the j-th copy of $\mathfrak{sl}(2, \mathbb{R})$. Let $\widehat{N}_{\mathbf{r}}^{\pm} = \sum_{j=1}^r \widehat{N}_j^{\pm}$, $\widehat{Y}_{\mathbf{r}} = \sum_{j=1}^r \widehat{Y}_j$. Then

(2) $\tilde{F}_r = e^{-i\widehat{N}_{\mathbf{r}}^-} \big(e^{iN_1} \tilde{F}_1\big)$;

(3) $\widehat{Y}_{\mathbf{r}}$ is the semi-simple endomorphism associated to the split M.H.S. $(W^r, \tilde{F}_r)$ and, consequently $W(\widehat{N}_{\mathbf{r}}^-) = W(C_r)$;

(4) $\widehat{N}_r^-$ is the component of N_r in the subspace $\cap_{j=1}^{r-1} \ker(\mathrm{ad}\ \widehat{Y}_{\mathbf{j}})$ relative to the joint spectral decomposition of the commuting semi-simple endomorphisms $ad(\widehat{Y}_{\mathbf{j}})_{j=1}^{r-1}$;

(5) $\sum_{s=1}^r y_s y_r^{-1} N_s = \mathrm{Ad}\big(\prod_{j=r-1}^1 h_j(y_1 y_j^{-1}, \dots, y_{j-1} y_j^{-1}; y_j y_{j+1}^{-1})\big) \widehat{N}_{\mathbf{r}}^-$.

Moreover, there exist $G_{\mathbb{R}}$-valued functions $g_r(y_1, \dots, y_r)$, defined for $y_j > 0$ if $1 \le r \le n-1$ and for $y_j > \alpha$ for some α if $r = n$, such that

(6) For $j < r \le n$, $g_j(y_1, \dots, y_j)$ commutes with $\widehat{Y}_r$ and fixes $\tilde{F}_r \in M$. Hence it induces a $(0,0)$-endomorphism of the M.H.S. $(W^r, \tilde{F}_r)$.

(7) $\sum_{s=1}^r y_s N_s = \mathrm{Ad}\big(\prod_{j=r-1}^1 g_j(y_1 y_{j+1}^{-1}, \dots, y_j y_{j+1}^{-1})\big) \sum_{s=1}^r y_s \widehat{N}_s^-$;

(8) $\exp(i \sum_{j=1}^n y_j N_j) F = \big(\prod_{r=n}^1 (g_r(y_1 y_{r+1}^{-1}, \dots, y_r y_{r+1}^{-1}))\big) \exp(i \sum_{j=1}^n y_j \widehat{N}_j^-) \cdot \tilde{F}_n$;

(9) The functions $g_r(y_1, \ldots, y_r)$ and their inverses admit power series expansions in non-negative powers of $y_1 y_2^{-1}, y_2 y_3^{-1}, \ldots, y_r$ with constant term 1, which converge in a region of the form $y_1 y_2^{-1} > \beta, \ldots, y_{r-1} y_r^{-1} > \beta, y_r > \beta$.

The proof is a delicate induction on the number of variables. The commuting endomorphisms $\widehat{Y}_j$ of $\mathbb{V}$ yield a flat multigrading $\mathbb{V} = \oplus \mathbb{V}_{l_1, \ldots l_n}$. For each j, the subbundles $\mathbb{V}_l^j := \oplus_{l_j = l} \mathbb{V}_{l_1, \ldots, l_n}$ grade the weight filtration W^j. One has $\mathbb{V}_{l_1, \ldots, l_n} \cong \mathrm{Gr}_{l_n}^{W^n}(\ldots(\mathrm{Gr}_{l_1}^{W^1}(\mathbb{V}))$. This multigrading of $\mathbb{V}$ is very useful to obtain growth estimates for sections in sectors of $(\Delta^*)^n$ of the type $\frac{\log|s_j|}{\log|s_{j+1}|} > \varepsilon$. To treat other sectors, one performs a permutation of the indices. Let $e : (\Delta^*)^n \to G_{\mathbb{R}}$ be characterized by $e(\mathbf{s})v = \prod_j \left(\log|s_j| \cdot \log|s_{j+1}|^{-1}\right)^{\frac{l_j}{2}} v$ for $v \in V_{l_1, \ldots l_n}$. Then we have the following norm estimate.

Proposition 2.17 [C-K-S 1]: Proposition 5.24, [K 5]. *Assume the monodromy is unipotent and let h denote the Hodge metric induced by the polarization. Let v_1, v_2 be multivalued flat sections of $\mathbb{V}$ and $\tilde{v}_i(\mathbf{s}) = \exp(-\sum \frac{\log s_j}{2\pi} N_j) \cdot v_i$ the corresponding holomorphic sections of $\mathcal{V}$.*

Set

$$h(v_1, v_2) =: \langle e \cdot v_1, e \cdot v_2 \rangle$$
$$\tilde{h}(v_1, v_2) =: \langle e \cdot \tilde{v}_1, e \cdot \tilde{v}_2 \rangle$$

Then in any locally flat frame of V, the matrices h and $\tilde{h}$, as well as their inverses, are bounded in any sector of the form $\frac{\log|s_j}{\log|s_{j+1}|} > \varepsilon, \ldots, -\log|s_n| > \varepsilon$, $1 \leq j \leq n$.

This can be rephrased in terms of norm estimates for flat sections of $\mathbb{V}$, which generalize Corollary 2.12.

Proposition 2.18 [C-K-S 1: Theorem 5.21]. *A flat section v lies in $\cap_j W_{l_j}^j$ but not $\sum_j (W_{l_1}^1 \cap \ldots \cap W_{l_j - 1}^j \cap \ldots \cap W_{l_n}^n)$ if and only if on any sector of the form $\frac{\log|s_1|}{\log|s_2|} > \varepsilon, \ldots, -\log|s_n| > \varepsilon$, one has*

$$\langle v, v \rangle \sim \prod_j \left(\frac{\log|s_j|}{\log|s_{j+1}|} \right)^{l_j}.$$

These estimates are crucial in the comparison theorem between L_2-cohomology with coefficients in $\mathbb{V}$ and intersection cohomology, which we will discuss in Sect. 2.

To return to one-variable variations of Hodge structure, a way they arise geometrically is from proper holomorphic mapping $f : X \to \Delta$, with f smooth over the punctured disc Δ^*. One fixes the coordinate function s on Δ. The constructions are coordinate dependent. One may as well assume, using resolution of singularities, that $D := f^{-1}(0)$ is a divisor with normal crossings. We will follow Steenbrink's treatment [St 1]. Very similar results were obtained by Clemens [Cl 2]. Steenbrink makes use of the relative version $\Omega^\bullet_{X/\Delta}(\log D)$ of Deligne's log complex. One has $\Omega^1_{X/\Delta}(\log D) = \Omega^1_X(\log D)/f^*\Omega^1_\Delta(\log 0)$ and $\Omega^p_{X/S}(\log D) = \wedge^p\Omega^1_{X/\Delta}(\log D)$.

Proposition 2.19 [St 1: Theorem 2.18]. *For each $m \geq 0$, the higher direct image sheaf $\mathscr{V}^m := R^m f_*(\Omega^\bullet_{X/\Delta}(\log D))$ is locally free on Δ and for all s in Δ, the canonical map $R^m f_*(\Omega^\bullet_{X/\Delta}(\log D)) \otimes_{\mathcal{O}_\Delta} \kappa(s) \to H^m(X_s, \Omega^\bullet_{X/\Delta}(\log D) \otimes_{\mathcal{O}_X} \mathcal{O}_{X_s})$ is an isomorphism.*

For all $s \in \Delta^*$, the vector space $H^m(X_s, \Omega^\bullet_{X/\Delta}(\log D) \otimes_{\mathcal{O}_X} \mathcal{O}_{X_s})$ is isomorphic to $H^m(X_s, \mathbb{C})$. Hence the fiber $\mathscr{V}^m_0$ is a good candidate to describe the "limit as $s \to 0$ of the cohomology groups $H^m(X_s, \mathbb{C})$". Indeed, the complex $\Omega^\bullet_{X/\Delta}(\log D) \otimes_{\mathcal{O}_X} \mathcal{O}_D$ is quasi-isomorphic to the complex of nearby cycles $\psi_f(\mathbb{C}_X)$ described in 2.21 below (see [St 1: § 2]); if D is reduced, it admits an intrinsic description on D [Fr: (3.2)].

Proposition 2.20 [St 1: Proposition 2.20]. *The vector bundle with meromorphic connection $\mathscr{V}^m$ is a canonical extension in the sense of Deligne* [D 2], *i.e. it satisfies:*

(1) the connection ∇ on $\mathscr{V}^m$ has a pole of order ≤ 1 at 0;

(2) every eigenvalue α of the residue $\mathrm{Res}_0(\nabla) \in \mathrm{End}(\mathscr{V}^m_0)$ satisfies: $\alpha \in \mathbb{Q}$ and $0 \leq \alpha < 1$.

We now need to recall the definition of the spaces and sheaves of nearby cycles and of vanishing cycles, due to Grothendieck and Deligne [SGA 7]. At this point we only need to consider the case of the constant sheaf on X, but for future use (in Sect. 4), we consider an arbitrary bounded complex of sheaves $K^\bullet$ on X. We have the holomorphic mapping $f : X \to \Delta$. Let $\tilde{\Delta}^*$ be a universal covering space of Δ^* and let $\pi : \tilde{\Delta}^* \to \Delta$ be the natural map. Consider the cartesian diagram

$$\begin{array}{ccccc} \tilde{X}^* & \xrightarrow{k} & X & \xleftarrow[i]{} & D \\ \downarrow & & f\downarrow & & \downarrow \\ \tilde{\Delta}^* & \xrightarrow{\pi} & \Delta & \longleftarrow & \{0\} \end{array}$$

Definition 2.21 [SGA 7]. *For $K^\bullet$ a bounded complex of sheaves on X^*, the complex $\psi_f(K^\bullet)$ of nearby cycles for $K^\bullet$ is the object $\psi_f(K^\bullet) = i^*Rk_*k^*(K^\bullet)$ of the derived category of complexes of sheaves on D. For $K^\bullet$ a complex of sheaves on X, the complex of vanishing cycles $\phi_f(K^\bullet)$ is defined as the cone of the morphism of complexes $i^*K^\bullet \to \psi_f(K^\bullet)$. Denote the canonical morphism $\psi_f \to \phi_f$ by* can. *Let T be the monodromy operation, an automorphism of $\psi_f(K^\bullet)$ and of $\phi_f(K^\bullet)$. There exist a canonical morphism* Var : *$\phi_f(K^\bullet) \to \psi_f(K^\bullet)$ such that $T - 1 = \mathrm{can} \circ \mathrm{Var}$ on ϕ_f and $T - 1 = \mathrm{Var} \circ \mathrm{can}$ on ψ_f.*

Let us illustrate these definitions in the case of $K^\bullet = \mathbb{C}_X$. The complexes of sheaves $\psi_f(\mathbb{C}_X)$ and of $\phi_f(\mathbb{C}_X)$ are constructible. The cohomology of their stalks at $y \in D$ is as follows: $\mathscr{H}^j\psi_f(\mathbb{C}_X)_y = H^j(X(s,y),\mathbb{C})$, where $X(s,y)$ is the "Milnor fiber", and $\mathscr{H}^j\phi_f(\mathbb{C}_X)_y = \tilde{H}^j(X(s,y),\mathbb{C})$ (reduced cohomology of the Milnor fiber). The hypercohomology group $H^j(D,\psi_f(\mathbb{C}_X))$ is equal to $H^j(\tilde{X}^*,\mathbb{C})$, hence to the cohomology $H^j(X_s,\mathbb{C})$ of a general fiber X_s.

We return to our situation $f : X \to \Delta$, with f proper and smooth over $X^* = X - D$, D a divisor with normal crossings. We assume that the monodromy automorphism T on $H^p(X_s,\mathbb{Q}) = H^p(\tilde{X}^*,\mathbb{Q})$ is *unipotent* and that D is an algebraic variety. Steenbrink constructs a cohomological mixed Hodge complex on D as follows. Let $A^\bullet_{\mathbb{Z}} = \psi_{f,1}(\mathbb{Z}_X)$, where the subscript 1 denotes the maximal subcomplex of $\psi_f(\mathbb{Z})$ on which T is unipotent. Then $H^p(Y,A^\bullet_{\mathbb{Z}})) \cong H^p(\tilde{X}^*,\mathbb{Z}) = H^p(X_s,\mathbb{Z})$.

The construction of $(A^\bullet_{\mathbb{Q}}, W_\bullet)$ is more complicated. Let $j : X^* \hookrightarrow X$ be the inclusion; for $k \geq 0$, let

$$H^k_{\mathbb{Q}} := i^*Rj_*\mathbb{Q}_{X^*}(k+1)[k+1]/(\tau_{\leq k}i^*Rj_*\mathbb{Q}_{X^*}(k+1)[k+1]\ .$$

There is a natural map $\theta : H^k \to H^{k+1}$, the cup-product with the generator θ of $H^1(\Delta^*,\mathbb{Q}(1))$. Using a sheafification of Sullivan's d.g.a.'s of polynomial differential forms on a triangulation, one may realize the H^k by actual complexes of sheaves, and arrange that $\theta \circ \theta = 0$. Steenbrink then proves that the sequence

$$0 \to \psi_{f,1}(\mathbb{Q}_X) \xrightarrow{\theta} H^0_{\mathbb{Q}} \xrightarrow{\theta} H^1_{\mathbb{Q}} \to \dots$$

is exact. Hence, defining $A^\bullet_{\mathbb{Q}}$ to be the associated simple complex to the double complex $H^0_{\mathbb{Q}} \xrightarrow{\theta} H^1_{\mathbb{Q}} \to \dots$, there is a quasi-isomorphism $\theta : A^\bullet_{\mathbb{Z}} \otimes_{\mathbb{Z}} \mathbb{Q} \tilde{\to} A^\bullet_{\mathbb{Q}}$. The filtration $W_\bullet$ of $A^\bullet_{\mathbb{Q}}$ is defined as follows. First let $W_rH^k_{\mathbb{Q}}$ be the image of $\tau_{\leq r+2k+1}(i^*Rj_*\mathbb{Q}_{X^*}(k+1))[k+1]$ in $H^k_{\mathbb{Q}}$ (we refer to Sect. 1 for the relation between the weight filtration and the canonical filtration). Then let $W_rA^\bullet_{\mathbb{Q}}$ be the simple complex associated to the double complex $W_rH^0_{\mathbb{Q}} \to W_rH^1_{\mathbb{Q}} \to \dots$

To construct $A^\bullet_{\mathbb{C}}$, one starts with the double complex

$$A^{p,q} := \Omega^{p+q+1}_X(\log D)/W_q\Omega^{p+q+1}_X(\log D)\ ,$$

with horizontal differential induced by exterior derivation of differential forms, and vertical differential induced by $\theta = \frac{ds}{s}$. Let $A^\bullet_{\mathbb{C}}$ be the associated simple complex. The map $\theta : \Omega^\bullet_{X/\Delta} \otimes_{\mathcal{O}_X} \mathcal{O}_{D^{red}} \to A^\bullet_{\mathbb{C}}$ is a quasi-isomorphism with respect to the filtrations F. The filtration W of $A^\bullet_{\mathbb{C}}$ is induced by $W_rA^{p,q} = W_{2q+r+1}\Omega^{p+q+1}_X(\log D)/W_q\Omega^{p+q+1}_X(\log D)$. The filtration F is given by $F^pA^\bullet_{\mathbb{C}} = \oplus_{r\geq p}A^{r,s}$.

Theorem 2.22 (Steenbrink, [St 1: Theorem 4.19]). *The above complexes, filtrations and quasi-isomorphisms determine a cohomological mixed Hodge complex on D. In particular, the cohomology group $H^p(\tilde{X}^*,\mathbb{Q})$ carries a M.H.S., T is an automorphism of this M.H.S. and N is an endomorphism of type* $(-1,-1)$.

Steenbrink establishes the filtered quasi-isomorphisms (with the notations of Sect. 1)

$$\mathrm{Gr}_r^W(A^\bullet_{\mathbb{Q}}) \cong \oplus_{k\geq 0,-r}\,(a_{r+2k+1})_*\mathbb{Q}_{\tilde{D}^{r+2k+1}}(-r-k)[-r-2k] \tag{2-3}$$

and

$$\mathrm{Gr}_r^W(A^\bullet_{\mathbb{C}}) \cong \oplus_{k\geq 0,-r}(a_{r+2k+1})_*\Omega^\bullet_{\tilde{D}^{r+2k+1}}(-r-k)[-r-2k]\ . \tag{2-4}$$

The fact that one has a cohomological Hodge complex is then deduced from Deligne-Hodge theory (see Sect. 1).

Some of the functorial properties of the M.H.S. on $H^*(\tilde{X}^*, \mathbb{Q})$ are summarized in the following

Theorem 2.23 (Steenbrink, [St 1: §4 and §5]; see also [E 7: I.3.11 and II.3.18] and [Sa-Z: 2.3]). *Endow $H^p(X)$ with the M.H.S. induced by the isomorphism $H^p(X) \cong H^p(D)$.*

(1) $H^\bullet(X^*)$ and $H^\bullet_D(X)$ carry natural M.H.S., and the exact sequence of cohomology with supports

$$\ldots \to H^p(D) \xrightarrow{\alpha} H^p(X^*) \xrightarrow{\beta} H^{p+1}(X) \to H^{p+1}(D) \to \ldots$$

is an exact sequence in the category of M.H.S.

(2) The Wang sequence

$$\ldots \to H^p(X^*) \to H^p(\tilde{X}^*) \xrightarrow{N} H^p(\tilde{X}^*) \to H^{p+1}(X^*) \to \ldots$$

is an exact sequence in the category of M.H.S.

(3) If f is a projective morphism of relative dimension n, the hyperplane class L induces, for any $r \geq 0$, an isomorphism of M.H.S. $L^r : H^{n-r}(\tilde{X}^*) \cong H^{n+r}(\tilde{X}^*)(r)$.

(4) If f is projective, the filtration W of $H^p(\tilde{X}^*)$ coincides with the monodromy weight filtration. The endomorphism N is induced by an endomorphism v of $A^\bullet$; it induces a polarization of the M.H.S. $H^p(\tilde{X}^*)$, and the sequence

$$H^p(D, \mathbb{Q}) \to H^p(\tilde{X}^*, \mathbb{Q}) \xrightarrow{N} H^p(\tilde{X}^*, \mathbb{Q})(-1)$$

is exact (i.e. the local invariant cycle theorem holds).

Remark 2.24. The local invariant cycle theorem was proven by transcendental methods by Griffiths and Schmid [Gr-S] and by Clemens and Schmid [Cl 2], and by l-adic methods by Katz [SGA 7]. A generalization is given in [Z 2].

We point out that the same methods also work in the case of degeneration of Kähler manifolds, cf. [Gr-S] and [Cl 2].

In [St 2], Steenbrink generalizes these results by getting rid of the assumption that the monodromy is unipotent. Since the monodromy is quasi-unipotent anyway, one may perform the base change

$$\begin{array}{ccc} \tilde{Z} & \xrightarrow{\tilde{f}} & \Delta \\ {\scriptstyle\pi}\downarrow & & {\scriptstyle\sigma}\downarrow \\ X & \xrightarrow{f} & \Delta \end{array}$$

with $\sigma(s) = s^e$, for suitable e. Now $\tilde{f}$ has unipotent monodromy, and $\tilde{Z}$ has at worst quotient singularities, hence by a generalization of his constructions to this case, $H^{\bullet}(\tilde{Z}^{\bullet})$ has a M.H.S. But $H^{\bullet}(\tilde{X}^{\bullet})$ is isomorphic to $H^{\bullet}(\tilde{Z}^{\bullet})$. We refer to [St 2] for a detailed treatment and for many interesting examples. We refer to [R-Z] for a "formal" version of Steenbrink's double complex $\ldots \to H^p_{\mathbb{Q}} \to H^{p+1}_{\mathbb{Q}} \to \ldots$.

In [St-vD], van Doorn and Steenbrink show that in the case where $f : B \to \Delta$ (B a small ball in $\mathbb{C}^n$) has an isolated singularity at 0, if T acting on $H^{n-1}(X_s, \mathbb{C})$ has a Jordan block of size n (necessarily for an eigenvalue $\neq 1$), then it also has a Jordan block of size $n - 1$ for the eigenvalue 1. This is a complement to the monodromy theorem of Landman [La], Brieskorn [Br], Lê [Le]. For $n = 2$, one recovers a result of Lê (the monodromy for an irreducible plane curve singularity is of finite order).

In [St 3], the case of vanishing cycles for arbitrary isolated singularities is treated. Very general results have recently been obtained by Navarro-Aznar [NA 1] by a combination of simplicial scheme techniques, of Steenbrink's techniques and of "Thom-Whitney" complexes.

Theorem 2.25 [NA 1]. *Let X be a complex-analytic space, $f : X \to \Delta$ a non-constant analytic mapping of relative dimension n, V a Zariski open subset of a subspace $\overline{V}$ of $D := f^{-1}(0)$ which is both compact and algebraic. Then the cohomology groups $H^p(V, \psi_f(\mathbb{Z}))$ and $H^p(V, \phi_f(\mathbb{Z}))$ carry M.H.S. such that*

(1) The exact sequence of vanishing cycles is an exact sequence of M.H.S.

(2) The cup-products on $H^{\bullet}(V, \psi_f(\mathbb{Z}))$ and on $H^{\bullet}(V, \phi_f(\mathbb{Z}))[-1]$ are morphisms of M.H.S.

(3) The semi-simple part T_s of the monodromy operator T is an automorphism of M.H.S. Hence the unipotent subspaces $H^p(V, \psi_f(\mathbb{Z})_1)$ and $H^p(V, \phi_f(\mathbb{Z})_1)$ are also M.H.S.

(4) The logarithm N of the unipotent part of T is a morphism of M.H.S. of type $(-1, -1)$, and the Wang sequence (cf. Theorem 1.22) is an exact sequence of M.H.S.

(5) If $X - D$ is smooth and $V = \{x\}$, where x is an isolated rational singularity, the Hodge filtration satisfies $\mathrm{Gr}^0_F(H^p(\psi_f(\mathbb{Q}))_x)_1 = 0$ for all $p > 0$.

A very interesting line of investigation, concerning the geometric or analytic significance of the Hodge filtration of Steenbrink's mixed Hodge structure, was started by Varchenko, and continued by Scherk, Steenbrink, Pham and M. Saito, and also Barlet. Varchenko [Va 1, Va 2] introduced the *asymptotic Hodge filtration* of $\mathcal{V} := \mathcal{V}^{n-1}$ in the case $f : B \to \Delta$ has relative dimension $n - 1$, B is a small ball in $\mathbb{C}^n$ around the isolated singular point 0 of D. For $\omega \in \Omega^n(X)$ a holomorphic n-form, let $\sigma[\omega]$ denote the section $\sigma[\omega] := \frac{\omega}{df}$ of

$\mathscr{V}$ over Δ^*. Let Λ be the set of eigenvalues of the monodromy T acting on $H^{n-1}(X_s, \mathbb{C})$. Then $\sigma[\omega]$ admits an asymptotic expansion of the type

$$\sigma[\omega] = \sum_{\lambda \in \Lambda} \sum_{\alpha \in \Lambda(\lambda)} \sum_{k=0}^{n-1} \frac{1}{k!} A_{k,\alpha} s^{\alpha} (\log s)^k,$$

where $\Lambda(\lambda) = \{\alpha \in \mathbb{Q} : \alpha > -1 \text{ and } \exp(-2\pi i \alpha) = \lambda\}$ and each $A_{k,\alpha}$ is a flat multivalued section. For $|s|$ sufficiently small, the series converges in any angular sector around 0.

For given $\omega \in \Omega^n(X)$, let $\alpha(\omega)$ be the minimum of the numbers $\alpha \in \cup\Lambda(\lambda)$ such that some $A_{k,\alpha}$ is non-zero. The principal part $\sigma_{\max}[\omega]$ is defined as

$$\sigma_{\max}[\omega] = \sum_{k=0}^{n-1} \frac{1}{k!} A_{k,\alpha(\omega)} s^{\alpha(\omega)} (\log s)^k .$$

Definition 2.26 (Varchenko). *The asymptotic Hodge filtration of* $\mathscr{V}$ *is defined by* $\mathscr{F}^p = \operatorname{span}(\{\sigma_{\max}[\omega] : \alpha(\omega) \leq n-1-p\})$. *It is a decreasing filtration by holomorphic subbundles, with* $\mathscr{F}^n = 0$ *and* $\mathscr{F}^0 = \mathscr{V}$.

Proposition 2.27 [Va 2]. *(1) For each* $s \in \Delta^*$, *the filtrations* $W_\bullet$ *and* $F^\bullet$ *on* $\mathscr{V}_s$ *define a M.H.S.*

(2) The asymptotic Hodge filtration and the Hodge filtration of Steenbrink induce the same filtration on $\mathrm{Gr}_\bullet^W$.

We note that the asymptotic Hodge filtration is in general different from the Hodge filtration of Steenbrink [Sch]. The asymptotic Hodge filtration has been extended to non-isolated singularities by Barlet [Ba 1]. Varchenko and Barlet obtained some remarkable relations between the zeroes of the Bernstein-Sato polynomial for f and the M.H.S. on vanishing cohomology.

Scherk and Steenbrink [Sch-St] found a very interesting description of the Hodge filtration on $H^{n-1}(X_s)$, which is closely related to the work of Varchenko. Their work was made more precise by Pham [P] and by M. Saito [Sa 1, Sa 2]. Let S be the extension of $\mathscr{V}$ to a holomorphic vector bundle over Δ, such that ∇ has pole of order at most one with respect to S and the residue of ∇ has all its eigenvalues in $(-1, 0]$.

Let H'' be the *Brieskorn extension* [Br] of $\mathscr{V}$, which is the image of $\Omega^n(B)$ under the map $\omega \mapsto \frac{\omega}{df}$. Let $F^p(S) = S \cap (\nabla_{\frac{d}{dz}})^{n-p} \cdot H''$, and denote by $F^p(S_0)$ the induced filtration on the fiber S_0. Then Scherk and Steenbrink, Pham and Saito prove [Sc-St, P, Sa 1, Sa 2] that this coincides with Steenbrink's Hodge filtration on vanishing cohomology. Their proof consists in compactifying the situation, and using Griffiths' description [Gr 1] of the Hodge filtration for a smooth projective hypersurface in terms of residues of meromorphic differential forms.

M. Saito has developed a theory of Hodge modules and mixed Hodge modules, based on filtered **D**-modules, which we will discuss in Sect. 4. In particular, he has a general method to study vanishing cycles, based on the **D**-module construction of vanishing cycles. We just mention here two results on vanishing cycles which are obtained by these methods.

The first one, due to M. Saito is the proof of a conjecture of Steenbrink about the "spectrum" of a hypersurface singularity. For T an automorphism of finite order of a M.H.S. V, let the *spectrum* $\mathrm{Sp}\,(H,T)$ be the Laurent polynomial $\sum_{\alpha\in\mathbb{Q}} n_\alpha t^\alpha$, with $n_\alpha = \dim \mathrm{Gr}^p_F V_{\mathbb{C},\lambda}$, for $\lambda = e^{-2\pi i\alpha}$, $-p < \alpha < p+1$. For $X \hookrightarrow \mathbb{C}^n$ an analytic space and $f : X \cap B \to \mathbb{C}$, Steenbrink defines the spectrum $\mathrm{Sp}\,(f,0)$ of the germ of f at 0 as $\mathrm{Sp}\,(f,0) = \sum_j (-1)^{n-j}\mathrm{Sp}\,(H^j\psi_f(\mathbb{R}_B)_0)(n), T_s)$. Steenbrink's conjecture, proven by Saito [Sa 8: Theorem 3.2.1] describes the spectrum of $f + l^m$, for f a linear form and $m >> 0$ in terms of $\mathrm{Sp}\,(f)$ and of the geometry of the singular set of f.

The second result, due to M. Saito and Zucker [Sa-Z], gives the degeneration at E_3 of the spectral sequence with E_2-term $H^p(\Delta, \mathcal{H}^q\psi_f(\mathbb{Q}_X))$, for $f : X \to \Delta$ a proper morphism, and X a Kähler manifold. This proves a conjecture of Illusie, and improves upon the methods of [Z 2].

3. L_2-Cohomology

Let M be a Riemannian manifold, and $\mathbb{E}$ a metrized local system on M (i.e. $\mathbb{E}$ is the sheaf of horizontal sections of a flat complex vector bundle, which is equipped with an Hermitian inner product, not necessarily flat). If φ is an $\mathbb{E}$-valued differential form on M, then its pointwise length $|\varphi|$ is a function on M; by integrating this against the Riemannian volume density, one obtains the induced L_2 semi-norm:

$$\|\varphi\|^2 = \int_M |\varphi|^2 dV_M \, .$$

We introduce the following notation:

(3-1) $A^\bullet(M,\mathbb{E})$ = the C^∞ de Rham complex of M with coefficients in $\mathbb{E}$,
$L^\bullet(M,\mathbb{E})$ = the domain of d (defined weakly) in the context of locally L_2 forms with measurable coefficients,
$A^\bullet_{(2)}(M,\mathbb{E})$ = the domain of d for L_2C^∞ forms,
$L^\bullet_{(2)}(M,\mathbb{E})$ = the domain of d for L_2 forms measurable coefficients.

For those seeking clarification of the definitions in 3-1, we point out that the notions "measurable", "set of measure zero" and "locally integrable" are intrinsically defined on any C^∞ manifold, independent of Riemannian structure. A locally integrable i-form φ defines a *current* (distribution defined on smooth forms of compact support) T_φ of codimension i by the formula

$$T_\varphi(\psi) = \int_M \varphi \wedge \psi \, ;$$

conversely, this determines φ almost everywhere. In case φ is smooth, one has by Stokes' theorem

$$T_\varphi(\psi) = \int_M d\varphi \wedge \psi = (-1)^{i+1} \int_M \varphi \wedge d\psi ,$$

so one defines, for any current T of codimension i, the codimension $i+1$ current dT by:

$$(dT)(\psi) = (-1)^{i+1} T(d\psi) .$$

This is the (weakly-defined) exterior derivative of a current. By definition, a locally L_2 i-form φ belongs to $L^i(M, \mathbb{E})$ if and only if the current dT_φ is given by a locally L_2 $(i+1)$-form; etc.

These complexes fit into a diagram:

$$\text{(3-2)} \qquad \begin{array}{ccc} A^\bullet(M,\mathbb{E}) & \xrightarrow{\alpha} & L^\bullet(M,\mathbb{E}) \\ \uparrow{\scriptstyle \iota} & & \uparrow{\scriptstyle \kappa} \\ A^\bullet_{(2)}(M,\mathbb{E}) & \xrightarrow{\beta} & L^\bullet_{(2)}(M,\mathbb{E}) \end{array}$$

Then α and β are quasi-isomorphisms (see [Ch 1: § 8]); when M is compact, ι and κ are isomorphisms. We remark that if M the interior of a manifold-with-corners $\overline{M}$, then $A^\bullet(M, \mathbb{E})$ is also quasi-isomorphic to its subcomplex $A^\bullet(\overline{M}, \mathbb{E})$, consisting of forms that are smooth up to the boundary, and likewise for $A^\bullet_{(2)}$.

Definition 3.1. *The* L_2-cohomology *of M with coefficients in $\mathbb{E}$, $H^\bullet_{(2)}(M, \mathbb{E})$, is the cohomology of the complex $L^\bullet_{(2)}(M, \mathbb{E})$.*

Remark 3.2. From the diagram above, there is a canonical mapping

$$\text{(3-3)} \qquad H^\bullet_{(2)}(M, \mathbb{E}) \to H^\bullet(M, \mathbb{E}) ,$$

that is an isomorphism when M (or $\overline{M}$ as above) is compact. It is an important feature that we are distinguishing L_2-cohomology from its image under this mapping.

It is clear that all of the constructions in (3-1) define presheaves on M, whose associated sheaves we denote with the corresponding script letter: $\mathscr{A}^\bullet(M, \mathbb{E})$, etc. Note, however, that $\mathscr{A}^\bullet_{(2)}(M, \mathbb{E}) = \mathscr{A}^\bullet(M, \mathbb{E})$, and likewise for $\mathscr{L}^\bullet$. The sheaves $\mathscr{A}^\bullet(M, \mathbb{E})$ and $\mathscr{L}^\bullet(M, \mathbb{E})$ are fine sheaves, for they are $\mathscr{A}^0(M, \mathbb{C})$-modules. Their hypercohomology is just $H^\bullet(M, \mathbb{E})$.

To represent L_2-cohomology as the hypercohomology of a complex of sheaves, it is necessary to utilize some suitable compactification $j : M \hookrightarrow M^*$ to store the global L_2 condition. Given M^*, one can define a subsheaf $\mathscr{L}^\bullet_{(2)}(M^*, \mathbb{E})$ of $j_*\mathscr{L}^\bullet(M, \mathbb{E})$, determined by the presheaf

$$\text{(3-4)} \qquad U \text{ open in } M^* \longmapsto L^\bullet_{(2)}(U \cap M, \mathbb{E}) .$$

Whether this gives a complex of fine sheaves depends on the relation between the metric on M and the boundary of M in M^*: there must exist cut-off functions with bounded differential (see [Z 3: p. 175]).

Proposition 3.3. *When the latter condition is satisfied,*

$$H^\bullet_{(2)}(M,\mathbb{E}) \simeq H^\bullet\left(M^*, \mathscr{L}^\bullet_{(2)}(M^*,\mathbb{E})\right) .$$

For a metrized complex vector bundle on an Hermitian complex manifold, one can define the L_2 Dolbeault cohomology in a parallel fashion, with the $\overline{\partial}$-operator replacing d in the above discussion. However, we will stress only L_2 d-cohomology here.

The main point of L_2-cohomology is the Hodge theorem, which can be stated in the following very general formulation. Let

$$Z^i_{(2)} = Z^i_{(2)}(M,\mathbb{E})$$

denote the (Hilbert) space of closed L_2 i-forms, and put

$$B^i_{(2)} = dL^{i-1}_{(2)}(M,\mathbb{E}) .$$

By definition, we have

$$H^i_{(2)} = H^i_{(2)}(M,\mathbb{E}) = Z^i_{(2)}/B^i_{(2)} .$$

Let $h^i_{(2)} = h^i_{(2)}(M,\mathbb{E})$ denote the orthogonal complement of $B^i_{(2)}$ in $Z^i_{(2)}$. We see at once

(3-5) $$H^i_{(2)} \simeq h^i_{(2)} \oplus \left(\overline{B^i_{(2)}}/B^i_{(2)}\right) ,$$

where the "bar" here indicates the closure in $Z^i_{(2)}$.

In order to give a more useful description of $h^i_{(2)}$, we recall that the densely-defined *adjoint* d^* of d is the operator determined by the following rule. For an L_2 i-form φ, and an L_2 $(i-1)$-form η, $d^*\varphi = \eta$ means that the operator

$$T(\psi) = \langle\varphi, d\psi\rangle \qquad \psi \in L^{i-1}_{(2)}(M,\mathbb{E})$$

is bounded in L_2-norm, so is represented by an L_2 $(i-1)$-form, namely η. It is well-known that d^* is the closure of its restriction δ to smooth forms of compact support, whose formula can be deduced from Stokes' theorem:

$$\delta\varphi = (-1)^i *^{-1} d* ,$$

where $*$ is the star-operator determined by the metrics of M and $\mathbb{E}$:

$$\langle\varphi,\psi\rangle = \int_M \varphi \wedge (*\overline{\psi}) .$$

Then

(3-6) $$h^i_{(2)} = \left\{\varphi \in L^i_{(2)}(M,\mathbb{E}) : d\varphi = 0 \,,\; d^*\varphi = 0\right\} .$$

For very general reasons, we have

Proposition 3.4. *i)* $h^i_{(2)} \subset \{L_2$ *solutions of the Laplace equation for* $\mathbb{E}$*-valued i-forms*$\}$.

ii) $\overline{B^i_{(2)}}/B^i_{(2)}$ *is an infinite dimensional vector space whenever it is non-zero.*

From this, we can assert:

Proposition 3.5 (Hodge Theorem). *If* $B^i_{(2)}$ *is closed in* $Z^i_{(2)}$, *then* $H^i_{(2)} \simeq h^i_{(2)}$. *This holds, in particular, if* $H^i_{(2)}$ *is finite dimensional.*

If M is complete, then it is not hard to show that *(i)* in Proposition 3.4 is an equality ([Gaf 1, A-V]). This gives:

Corollary 3.6. *If* M *is complete, and* $B^i_{(2)}$ *is closed, then* $H^i_{(2)}$ *is isomorphic to the space of* L^2 *harmonic* $\mathbb{E}$*-valued i-forms.*

Finally, we consider the case where M is a complex Kähler manifold, and $\mathbb{E} = \mathbb{C}$. Since the Laplacian then takes (p,q)-forms to (p,q)-forms, one obtains:

Proposition 3.7 (Hodge Decomposition Theorem). *Let* M *be a complete Kähler manifold with finite-dimensional* L_2*-cohomology. Then the decomposition of harmonic forms into components of pure bidegree induces a Hodge structure*

$$H^i_{(2)}(M,\mathbb{C}) \simeq \bigoplus_{p+q=i} H^{p,q}_{(2)}(M) .$$

In view of Remark 3.2, the above contains classical Hodge theory, i.e. the case where M is compact (recall Sect. 1).

Deligne extended Proposition 3.7 to variations of Hodge structure (see [Z 1: § 1–2]), as we now describe. Let X be an Hermitian complex manifold, and $\mathbb{V}$ the metrized local system underlying a polarized variation of Hodge structure of weight m over X. We let $A^{p,q;r,s}_{(2)}(X,\mathbb{V})$ be the space of L_2 (p,q)-forms with values in the (r,s) Hodge bundle ov $\mathbb{V}$. Then

$$A^i_{(2)}(X,\mathbb{V}) = \bigoplus_{\substack{p+q=i\\ r+s=m}} A^{p,q;r,s}_{(2)}(X,\mathbb{V})$$

(orthogonal direct sum). Thinking in terms of total holomorphic degree, we put

$$A^{(P,Q)}_{(2)}(X,\mathbb{V}) = \bigoplus_{\substack{p+r=P\\ q+s=Q}} A^{p,q;r,s}_{(2)}(X,\mathbb{V}) .$$

If the metric on X is Kähler, the Laplacian for $\mathbb{V}$-valued forms respects the bidegree (P,Q). This yields, under the assumptions of Proposition 3.7, a Hodge structure of weight $i+m$:

$$H^i_{(2)}(X,\mathbb{V}) \simeq \bigoplus_{P+Q=i+m} H^{(P,Q)}_{(2)}(X,\mathbb{V}) \,. \tag{3-7}$$

When X is compact, this admits a rather nice description (compare (1-1)). The holomorphic de Rham complex of $\mathbb{V}$, $\Omega^\bullet_X \otimes_{\mathbb{C}} \mathbb{V}$, resolves $\mathbb{V}$, and admits the decreasing filtration F, in which F^p is the subcomplex

$$\mathscr{F}^p \to \Omega^1_X \otimes \mathscr{F}^{p-1} \to \Omega^2_X \otimes \mathscr{F}^{p-2} \to \dots \,, \tag{3-8}$$

so that Gr^p_F is the $\mathscr{O}_X$-linear complex

$$\mathscr{F}^p/\mathscr{F}^{p+1} \to \Omega^1_X \otimes \left(\mathscr{F}^{p-1}/\mathscr{F}^p\right) \to \Omega^2_X \otimes \left(\mathscr{F}^{p-2}/\mathscr{F}^{p-1}\right) \to \dots \tag{3-9}$$

Proposition 3.8. *Let X be a compact Kähler manifold.*

i) The pair $K_A = \mathbb{V}_A$, $K_C = \left(\Omega^\bullet_X \otimes_A \mathbb{V}_A, F\right)$ defines a cohomological Hodge complex of weight m (1.7).

ii) The filtration induced by F on $H^i(X,\mathbb{V})$ is the Hodge filtration associated to (3-7).

In connection with Proposition 3.8, it is natural to ask for more in the geometric case. Let $f : Z \to X$ be a smooth projective morphism. One has the associated Leray spectral sequence

$$E_2^{p,q} = H^p(X, R^q f_* \mathbb{Q}) \Longrightarrow H^{p+q}(Z,\mathbb{Q}) \,. \tag{3-10}$$

Up to a change of indices, this is the spectral sequence of the canonical filtration τ on $\mathbf{R}f_*\mathbb{Q}_Z$. A nice consequence of classical Hodge theory is

Theorem 3.9 [D 1] (see also [Gr 1: III, § 3]). *The spectral sequence* (3-10) *degenerates at E_2.*

This gives

$$\mathrm{Gr}^\tau_m H^k(Z,\mathbb{Q}) \simeq H^{k-m}(X, R^m f_* \mathbb{Q}) \,. \tag{3-11}$$

It is reasonable to insist upon

Proposition 3.10 (Deligne, see [Z 1: (2.16)]). *Assume that X is compact. Then:*

i) The filtration τ of $H^k(Z,\mathbb{Q})$ is a filtration by Hodge substructures.

ii) When the right-hand side is given the Hodge structure of Proposition 3.8, *the isomorphism* (3-11) *is one of Hodge structures.*

Remark 3.11. A filtration of a Hodge structure by Hodge substructures is split by a polarization. Thus the $\mathbb{Q}$-Hodge structure $H^{k-m}(X, R^m f_* \mathbb{Q})$ is a direct factor of $H^k(Z,\mathbb{Q})$.

In the more usual situation where X only admits a good Kähler compactification, it was expected (see end of Introduction of [D 4: III]) that there would be a natural mixed Hodge structure on $H^\bullet(X,\mathbb{V})$, with an analogue of the previous proposition holding. The case where X is a curve already presented considerable difficulties, and the final result appears in [Z 1]. There, the mixed Hodge structure is induced from filtrations F and W on the logarithmic de Rham complex $\Omega^\bullet_{\overline{X}}(\log D)\otimes\overline{\mathscr{V}}$, a subcomplex of $j_*\left(\Omega^\bullet_X\otimes_{\mathbb{C}}\mathbb{V}\right)$ that is quasi-isomorphic to $\mathbf{R}j_*\mathbb{V}$. The filtration F is easy to describe; it is the restriction of the one induced from (3-8) by direct image, viz. F^p is

$$\overline{\mathscr{F}}^p \to \Omega^1_{\overline{X}}(\log D)\otimes\overline{\mathscr{F}}^{p-1}\ .$$

The discussion of the weight filtration W leads one to considerations of L_2-cohomology (in the actual development, this came before the mixed Hodge theory). For this, one makes use of a Poincaré metric on X, i.e., any metric that in a disc Δ about a point of D has the asymptotics of the Poincaré metric of Δ^*:

$$r^{-2}\log^{-2}r\left[dr^2+(rd\theta)^2\right]\quad\text{in polar coordinates } (r,\theta)\ .$$

Theorem 3.12. *For X a curve with Poincaré metric and $\mathbb{V}$ metrized by the Hodge metric:*

i) $\mathscr{L}^\bullet_{(2)}(\overline{X},\mathbb{V})$ is a resolution of $j_*\mathbb{V}$. Thus

$$H^i(\overline{X},j_*\mathbb{V})\simeq H^i_{(2)}(X,\mathbb{V})$$

gets a polarized Hodge structure of weight $i+m$.

ii) The holomorphic L_2 complex

$$\Omega^\bullet_{(2)}(\overline{X},\mathbb{V})=j_*\left(\Omega^\bullet_{\overline{X}}\otimes_{\mathbb{C}}\mathbb{V}\right)\cap\mathscr{L}^\bullet_{(2)}(\overline{X},\mathbb{V})$$

is also a resolution of $j_*\mathbb{V}$, and the filtration F (induced by (3-8)*) on $H^\bullet(\overline{X},j_*\mathbb{V})$ is the Hodge filtration coming from* (i).

iii) $\Omega^\bullet_{(2)}(\overline{X},\mathbb{V})$ is a subcomplex of

$$\overline{\mathscr{V}}\to\nabla\overline{\mathscr{V}}\ ,$$

and the inclusion is a filtered quasi-isomorphism with respect to F.

Remark 3.13. Given the Hodge norm asymptotics (2.12), the complex $\Omega^\bullet_{(2)}(X,\mathbb{V})$ can be described explicitly in terms of local monodromy weights. On a disc Δ about a point of D, it is given by (see [Z 1: (4.4)])

$$W_m(\overline{\mathscr{V}})\to(dz/z)W_{m-2}(\overline{\mathscr{V}})\ ,$$

where $W_\ell(\overline{\mathscr{V}})=\{\sigma\in\overline{\mathscr{V}}:\sigma(0)\text{ belongs to }W_\ell V(0)\}$.

Briefly, the mixed Hodge theory then proceeds as follows (only $i=1$ is important). The first non-trivial weight level of $H^1(X,\mathbb{V})$ is $H^1(\overline{X},j_*\mathbb{V})$. The rest of the mixed Hodge structure is determined by the Schmid limit mixed Hodge structures at the points of D via residues from $\Omega^\bullet_{\overline{X}}(\log D) \otimes \overline{\mathscr{V}}/\nabla\overline{\mathscr{V}}$ (see [Z 1: § 13] or our Sect. 7). Moreover:

Proposition 3.14 [Z 1: (15.5)]. *If* $\mathbb{V}_{\mathbb{Q}} = R^m f_* \mathbb{Q}$ *for some smooth projective morphism* $f : Z \to X$, *the mixed Hodge structure of* $H^i(X, \mathbb{V}_{\mathbb{Q}})$ *is induced by that of* $H^{i+m}(Z, \mathbb{Q})$.

The proof of Theorem 3.12 is based on the Hodge norm asymptotics that follow from the SL_2-orbit theorem in one variable (2.11). The generalization to higher-dimensional X had to wait until the norm asymptotics were available in several variables (Proposition 2.18). For this purpose, one notes first that X admits (necessarily complete) Kähler metrics with Poincaré singularities along D (see [Z 1: (3.2)]).

Theorem 3.15 [C-K-S 2, K-K 2]. *For* X *with a Kähler metric with Poincaré singularities along* D, $\mathbb{V}$ *metrized by the Hodge metric,* $\mathscr{L}^\bullet_{(2)}(\overline{X}, \mathbb{V})$ *is quasi-isomorphic to the middle perversity intersection cochain complex* $\underline{IC}^\bullet(\overline{X}, \mathbb{V})$.

Corollary 3.16. $IH^i(\overline{X}, \mathbb{V}) \simeq H^i_{(2)}(X, \mathbb{V})$ *thereby gets a Hodge structure of weight* $i+m$.

The complex $\underline{IC}^\bullet(\overline{X}, \mathbb{V})$ occurring in (3.15) is quasi-isomorphic to "Deligne's sheaf" [G-M: II, § 3], which here has the following description. Let

$$D(k) = \{z \in D \mid z \text{ belongs to at least } k \text{ components of } D\} ,$$

$$U(k) = \overline{X} - D(k) , \quad j_k : U(k) \hookrightarrow U(k+1) .$$

Then $\overline{X} = U(n+1)$, where $n = \dim_{\mathbb{C}} X$, $X = U(1)$, and $j = j_n \circ j_{n-1} \circ \ldots \circ j_1$. Deligne's sheaf is

$$\tau_{\leq n-1} \mathbf{R} j_{n*} \ldots \tau_{\leq 1} \mathbf{R} j_{2*} \left[\tau_{\leq 0} \mathbf{R} j_{1*} \mathbb{V}\right] .$$

To prove Theorem 3.15, one must show that on a Poincaré punctured polydisc,

$$H^\bullet_{(2)}\left((\Delta^*)^n, \mathbb{V}\right) \simeq IH^\bullet(\Delta^n, \mathbb{V}) . \tag{3-12}$$

In the case of unipotent local monodromy, the right-hand side admits a combinatorial description in terms of the (commuting) monodromy logarithms N_j, as follows. For any subset $J \subset \{1, \ldots, n\}$, put

$$N_J = \prod_{j \in J} N_j , \quad K_J(V) = N_J V$$

$$K^i(V) = \bigoplus_{|J|=i} K_J(V) .$$

One obtains a finite dimensional complex $K^\bullet(V)$ by taking as differential the sum of

$$\varepsilon_{j,J} N_j : K_{J-\{j\}}(V) \to K_J(V)$$

for all $j \in J$; here $\varepsilon_{j,J}$ is 1 or -1 according to whether j is in odd or even position when the elements of J are listed in increasing order.

Lemma 3.17.

$$IH^\bullet(\Delta^n, \mathbb{V}) \simeq H^\bullet(K^\bullet(V)) .$$

The proof of (3-12) makes use of the Hodge norm estimates (2.18), together with the following consequence of the fact the N_j's make up a nilpotent orbit (Definition 2.3). Let $N = N^n = \sum_{i=1}^n N_j$, and define a filtration Wt of $K^\bullet(V)$ by

$$Wt_\ell(K_J(V)) = K_J(W_{\ell+|J|}V) ,$$

where W denotes the weight filtration of the nilpotent endomorphism N. Then

Proposition 3.18 [C-K-S 2, K-K 2]. *i) $N : K^\bullet(V) \to K^\bullet(NV)$ induces the zero mapping on cohomology.*

ii) ("purity") *Every cohomology class in $H^\bullet(K^\bullet(V))$ has a representative in $Wt_0K^\bullet(V)$.*

The desired isomorphism (3-12) follows from Proposition 3.18(ii) essentially because it is Wt that controls the asymptotics of the integrand occuring in the L_2 seminorm.

We can use the data of $K^\bullet(V)$ to construct a subcomplex $\mathscr{S}^\bullet$ of $\Omega^\bullet_{\overline{X}}(\log D) \otimes \overline{\mathscr{V}}$.

The following makes sense on $\overline{X}$:

(3-13) $$\mathscr{S}^i = \left\{ \varphi \in \Omega^i_{\overline{X}}(\log D) \otimes \overline{\mathscr{V}} : \mathrm{Res}_{\widetilde{D}_J} \varphi \in N_J V \right\} .$$

We suspect that the means are available (cf. [K-K 2: Proposition 3.4.3]) to verify:

Conjecture 3.19. *The complex $\mathscr{S}^\bullet$, together with the filtration F induced from* (3-8), *completes $\underline{IC}^\bullet(\overline{X}, \mathbb{V}_A)$ to a cohomological A-Hodge complex of weight m, such that the Hodge structure it gives on $IH^i(\overline{X}, \mathbb{V})$ coincides with that of* (3.16).

The construction of a mixed Hodge structure on $H^\bullet(X, \mathbb{V})$, such that F is given in the desired manner (3-8), is proven in [Sa 10], based on his machinery of mixed Hodge modules [Sa 5] (see our Sect. 4; also cf. [K-K 3]). An attempt to reprove this without using **D**-modules – a plausible goal – is begun in [E 6].

The isomorphism between L_2-cohomology and intersection cohomology in (3.16) fits into a pattern of similar results and conjectures in other contexts. For instance, there is the following very basic one. Let $\overline{X}$ be any complex projective variety, and X its locus of regular points. Given any projective embedding

$$\overline{X} \hookrightarrow \mathbb{P}^N ,$$

one can restrict the Fubini-Study metric to X, yielding a Kähler metric on X. The class of the metric is intrinsic to $\overline{X}$ in the sense that its asymptotic behavior at any point of $\overline{X}$ depends only on the analytic germ of $\overline{X}$ at that point. On the other hand, the metric is incomplete (unless $X = \overline{X}$), so one cannot be sure that for harmonic forms representing L_2-cohomology classes (see (3-5)) their (p, q)-components do likewise. Still, Cheeger, Goresky and MacPherson have made the following:

Conjecture 3.20 [C-G-M]. *For a projective variety $\overline{X}$ with Fubini-Study metric,*
i) $H^\bullet_{(2)}(X, \mathbb{C}) \simeq IH^\bullet(\overline{X}, \mathbb{C})$
ii) Moreover, a Hodge structure is induced by the decomposition of L_2 harmonic forms.

Here, it is understood that the isomorphism in (i) should hold locally, as in Theorem 3.15. The underlying philosophy was that *something* should induce a pure Hodge structure on the intersection cohomology, for one knew the analogous purity in characteristic p [G] (see also [B-B-D: (5.3)]), and the above seems to be the most natural candidate. Moreover, Cheeger had already established the analogue of (3.20, i) for Riemannian pseudomanifolds whose metric has conical singularities [Ch 1]. This was carried further in [Ch 2], where he verified (3.20, ii) for varieties with (analytically) conical singularities. The assertion for general complex surfaces is treated in the work of Hsiang and Pati [H-P] and Nagase [N 1, N 2].

One way to avert the nasty question of having to prove (ii) for this incomplete metric is to replace the Fubini-Study metric by a suitable complete Kähler metric on X, for which (i) can be shown to hold. This has been carried out in the case where $\overline{X}$ has only isolated singularities by Saper [Sap] and Ohsawa [O].

In some special cases, X possesses a natural complete metric. For instance, when X is a locally symmetric variety (the quotient of an Hermitian symmetric space G/K by an arithmetically defined torsion-free subgroup Γ of G), it carries the Bergman metric. Let $\mathbb{V}$ be the local system corresponding to a finite-dimensional representation of G, metrized by means of a so-called admissible inner product. Let X^* be the Baily-Borel Satake compactification of X, which is a normal projective variety. In [Z 3: §6], one finds (3.16) conjectured for $\overline{X} = X^*$. This has now been proved in two different ways, by Looijenga [Lo] and by Saper and Stern [Sap-S] (see also [Z 6: III]). Though we will not discuss the proofs here, we wish to point out that Looijenga's argument utilizes some of the results about variations of Hodge structure discussed in this article. The local system $\mathbb{V}$ underlies a "locally homogeneous" variation of Hodge structure [Z 7]. The use of the decomposition theorem (see (4.13)) permits the eventual reduction of the problem to the local purity theorem (3.18 (ii)) on a (certain) good resolution of X^*.

In any case, it is appropriate to ask whether the Hodge structure constructed from L_2-cohomology is the "right" one. To explain this, suppose first that $\overline{X}$ has only isolated singularities. Then one has the familiar isomorphisms

$$IH^i(\overline{X}) \simeq \begin{cases} H^i(X) & \text{if } i < n \\ H^i(\overline{X}) & \text{if } i > n \\ \operatorname{im}\{H^n(\overline{X}) \to H^n(X)\} & \text{if } i = n\,. \end{cases} \tag{3-14}$$

Hodge-Deligne theory (Sect. 1) gives a priori *mixed* Hodge structures on the right-hand side, but they are actually pure, and these Hodge structures are patently right. A method for comparing L_2-cohomology to these is given in [Z 5], where the comparison is carried out for some of the above cases. For general varieties $\overline{X}$, the construction given by Morihiko Saito is also clearly right, for it has the desired compatibilities. Methods should be developed to compare L_2-cohomology to that.

4. D-Modules and Hodge Theory

To motivate the recent appearance of **D**-modules in Hodge theory, let us first take another look at the theory of variations of Hodge structure. Recall (Sect. 2) that for a real local system $\mathbb{V}_{\mathbb{R}}$ on a smooth complex manifold X, a variation of Hodge structure of weight m with underlying local system $\mathbb{V}_{\mathbb{R}}$ is a decreasing filtration $\mathscr{F}^p(\mathscr{V})$ of the holomorphic vector bundle $\mathscr{V} = \mathscr{O}_X \otimes \mathbb{V}_{\mathbb{R}}$ by holomorphic vector bundles, which have the following properties:

(1) for each $x \in X$, the real vector space $\mathbb{V}_{\mathbb{R},x}$ (fiber at x), together with the filtration induced by $\mathscr{F}^p(\mathscr{V})$, is a pure Hodge structure of weight m;

(2) The Griffiths' transversality condition is satisfied (see Sect. 2).

Also recall that a *polarization* for a variation of Hodge structures over X is a $(-1)^m$-symmetric bilinear form S on the local system $\mathbb{V}_{\mathbb{R}}$ which induces a polarization on the weight m-Hodge structure on every fiber $(\mathbb{V}_{\mathbb{R}})_x$ (see Sect. 2 for details).

A natural question is to find an extension for this notion, when $\mathbb{V}_{\mathbb{R}}$ is replaced with a real constructible sheaf $K_{\mathbb{R}}$ on X. What one is looking for is something like a "complex of differential operators of order 1" (in the sense of Sect. 1), quasi-isomorphic to $K_{\mathbb{R}} \otimes \mathbb{C}$, together with a filtration. The hypercohomology groups of $K_{\mathbb{R}}$, equipped with this filtration, should be Hodge structures, at least when X is compact. More generally, the objects one is looking for should be functorial under the direct image operation for a projective morphism. One reason such objects should exist is the belief that the notion of *pure complexes*, due to Deligne [D 5: II, § 6.2] must admit some analog in characteristic 0.

For instance, assume that X is a Riemann surface, $S \subset X$ a finite subset, with complement $U := X - S$, and $(\mathbb{V}_{\mathbb{R}}, \mathscr{F}^p(\mathscr{V}))$ is a polarized variation of Hodge structure. Then Zucker's holomorphic L^2-complex [Z 1], which we discussed in Sect. 3, is a filtered complex of sheaves $\Omega^{\bullet}_{(2)}(\mathbb{V})$ which is a resolution of the constructible sheaf $j_*(\mathbb{V}_{\mathbb{C}})$, where $j : U \to X$ is the inclusion. Zucker showed that the hypercohomology groups of $\Omega^{\bullet}_{(2)}(\mathbb{V})$ carry natural Hodge structures (see Sect. 3). Hence this L^2-complex ought to be an example of the sort of object we are looking for.

The reason $j_*(\mathbb{V}_{\mathbb{C}})$ (in the case of curves, with the above notations), is a good sheaf to consider is that it is not only a constructible sheaf, but it is also (up to shift) a *perverse sheaf* (in the sense of [B-B-D: §4.0]). Since we are looking for an analog of Deligne's pure complexes, and since pure complexes are essentially a direct sum of shifted perverse sheaves, it is more natural to consider perverse sheaves than constructible sheaves. In characteristic p, $j_*(\mathbb{V})$ would be pure if $\mathbb{V}$ itself is pure (combine [D 5: Corollaire 1.8.9] with the fact that the operation j_* commutes with Verdier duality); in contrast, the perverse sheaves $Rj_*(\mathbb{V})$ and $j_!\mathbb{V}$ would not be pure, but only mixed.

The necessity to focus on perverse sheaves instead of constructible sheaves is now considered a fact of life in Hodge theory, from the work of Cattani-Kaplan-Schmid on degeneration of Hodge structures over a higher-dimensional base [C-K-S 1, C-K-S 2] (see also Sect. 2 and Sect. 3). Their work involves L^2-cohomology and intersection cohomology, and the intersection complex $IC^{\bullet}_X$ of [G-M: II, § 3], described in Sect. 3 in a special case, is an example (indeed, the most important example) of perverse sheaf. But another fundamental reason has to do with **D**-modules. Indeed, even the theory of variation of Hodge structure brings to mind **D**-modules.

To explain this point, let $\mathbf{D}_X$ denote the sheaf of germs of holomorphic differential operators on X (one may, for instance, define the notion of differential operator of order r, by induction on r, as in [B-B-G]). Write $\mathbf{D}_X(r)$ for the subsheaf of differential operators of order $\leq r$. Hence $\mathbf{D}_X(0) = \mathcal{O}_X$. A (left) $\mathbf{D}_X$-module $\mathcal{M}$ is a sheaf of (left)-modules over $\mathbf{D}_X$.

As usual, one has the notion of coherency for a $\mathbf{D}_X$-module. As $\mathbf{D}_X$ itself is a coherent $\mathbf{D}_X$-module, $\mathcal{M}$ is coherent if and only if it admits locally a finite free presentation: $\mathbf{D}^b_X \to \mathbf{D}^a_X \to \mathcal{M} \to 0$. For such coherent $\mathcal{M}$, there exists, at least locally, a *good filtration* $\mathcal{M} = \cup_{n\in\mathbb{Z}}\mathcal{M}_n$; this means that each $\mathcal{M}_n$ is a coherent $\mathcal{O}_X$-module,

(4-1) $$\mathbf{D}_X(1)\mathcal{M}_n \subset \mathcal{M}_{n+1},$$

and

(4-2) $$\text{for } n \gg 0, \text{the equality holds in (4-1).}$$

Locally on X, such a filtration is easy to construct. Choose a coherent $\mathcal{O}_X$-submodule $\mathcal{M}_0$ of $\mathcal{M}$ which generates $\mathcal{M}_0$ over $\mathbf{D}_X$ and set $\mathcal{M}_j := \mathbf{D}_X(j) \cdot \mathcal{M}_0$ for $j \geq 0$. This gives a good filtration of $\mathcal{M}$. A *filtered* **D**-*module* is a **D**-module equipped with a good filtration.

It is well-known that a $\mathbf{D}_X$-module $\mathcal{M}$ is coherent as an $\mathcal{O}_X$-module if and only if it is the sheaf of germs of sections of a holomorphic vector bundle $\mathcal{V}$, endowed with an integrable connection $\nabla : \mathcal{V} \to \Omega^1_X \otimes \mathcal{V}$; the action of a holomorphic vector field ξ on the $\mathbf{D}_X$-module is then given by ∇_ξ. Now assume that $\mathcal{V}$ is associated with a variation of Hodge structures, and let $\mathcal{F}^p(\mathcal{V})$ be the corresponding Hodge filtration. Let $\mathcal{M}_n$ be the sheaf of germs of sections of $\mathcal{F}^{-n}(\mathcal{V})$. Then this gives a good filtration of $\mathcal{M}$, since condition (4-1) is exactly equivalent to Griffiths' transversality condition.

In view of the fact that filtered $\mathbf{D}_X$-modules arise in this way from variations of Hodge structure, they appear as a natural framework in which to define a notion of "perverse variations of Hodge structures", which is what we are after. To motivate this further, we need to recall the "Riemann-Hilbert correspondence", which gives an equivalence of categories between the category of *holonomic* $\mathbf{D}_X$*-modules with regular singularities* and the category of perverse sheaves on X. First we recall that for a coherent $\mathbf{D}_X$-module $\mathcal{M}$, the characteristic variety $Ch(\mathcal{M})$ is a well-defined conical complex-analytic subvariety of the cotangent bundle T^*X. According to a theorem of Bernstein see [Bo 3: V, Theorem 1.12], unless $\mathcal{M}$ is zero, $Ch(\mathcal{M})$ has dimension at least $n = \dim(X)$.

Definition 4.1. *The coherent* $\mathbf{D}_X$*-module* $\mathcal{M}$ *is holonomic if* $\dim(Ch(\mathcal{M})) \leq n$.

We next recall how one obtains a constructible complex (which will turn out to be a perverse sheaf) by taking the *de Rham complex* of a holonomic $\mathbf{D}_X$-module $\mathcal{M}$. This de Rham complex $\mathrm{DR}(\mathcal{M})$ is the complex of sheaves:

$$\mathcal{M} \xrightarrow{d} \Omega^1_X \otimes \mathcal{M} \xrightarrow{d} \Omega^2_X \otimes \mathcal{M} \ldots \xrightarrow{d} \Omega^n_X \otimes \mathcal{M},$$

where $\Omega^j_X \otimes \mathcal{M}$ is put in degree $j - n$. The differential d is easily described using local coordinates $(z_1, z_2, \ldots, z_n)$. We have:

$$d(\omega \otimes u) = \sum_{i=1}^{i=n} (dz_i \wedge \omega) \otimes \frac{\partial}{\partial z_i} u,$$

for ω a differential form and u a section of $\mathcal{M}$.

For instance, if $\mathcal{M}$ is coherent, $\mathrm{DR}(\mathcal{M})$ is the holomorphic de Rham complex of the corresponding local system, shifted n steps to the left. In general, we have the following:

Theorem 4.2. *For any holonomic* $\mathbf{D}_X$*-module* $\mathcal{M}$*, the complex* $\mathrm{DR}(\mathcal{M})$ *is a perverse sheaf on* X.

This is essentially proven by Kashiwara, around 1974 [K 1], although the notion of "perverse sheaf" was only introduced several years later, by Beilinson, Bernstein and Deligne [B-B-D: § 4.0]. We will refer to their foundational article for that theory.

Definition 4.3. *A complex of sheaves of $\mathbb{C}$-vector spaces $K^\bullet$ on X is perverse if it satisfies:*

(1) the cohomology sheaves $\mathcal{H}^j(K^\bullet)$ are constructible, and their support has dimension at most $-j$;

(2) the cohomology sheaves $\mathcal{H}^j(D(K^\bullet))$ of the Verdier-dual complex

$$D(K^\bullet) := \mathbf{R}\mathcal{H}om_X(K^\bullet, \mathbb{C}_X[2n])$$

also have support of dimension at most $-j$.

Let Perv($\mathbb{C}_X$) *be the category of perverse sheaves of complex vector spaces "up to quasi-isomorphisms" (see* [B-B-D: §4.0] *for a precise definition, in terms of derived categories).*

Every object of the category Perv($\mathbb{C}_X$) has finite length (at least locally on X). The irreducible objects are the intersection complexes $\underline{IC}^\bullet_Z(\mathbb{V})$, for $Z \subset X$ an irreducible closed analytic subvariety, and $\mathbb{V}$ an irreducible local system over a Zariski open subset of Z.

By construction, the category Perv($\mathbb{C}_X$) admits the anti-involution $K^\bullet \to D(K^\bullet)$. The corresponding involution of the category of holonomic $\mathbf{D}_X$-modules is given by

$$\mathcal{M} \to \mathcal{M}^* := Ext^n_{\mathbf{D}_X}(\mathcal{M}, \mathbf{D}_X) \otimes_{\mathcal{O}_X} \Omega^n(X)^{\otimes -1}.$$

We refer to [K 1] and [Bo 3: VI.3] for an explanation of this duality. We simply note here that tensoring with $\Omega^n(X)^{\otimes -1}$ has the effect of transforming a right $\mathbf{D}_X$-module into a left $\mathbf{D}_X$-module.

As is the case already in dimension one, there are many $\mathbf{D}_X$-modules with the same de Rham complex. Therefore to obtain the equivalence of categories, one needs to restrict oneself to a special sort of holonomic $\mathbf{D}_X$-module.

Definition 4.4. *Let $\mathcal{M}$ be a holonomic $\mathbf{D}_X$-module. Let Z be a closed analytic subvariety of X, and let $\mathcal{I}_Z$ be the corresponding sheaf of ideals in $\mathcal{O}_X$. $\mathcal{M}$ is said to have regular singularities along Z if the natural map of complexes of sheaves:*

$$\mathrm{DR}(\mathbf{R}\underline{\Gamma}_{[Z]}(\mathcal{M})) \to \mathbf{R}\underline{\Gamma}_Z(\mathrm{DR}(\mathcal{M}))$$

is an isomorphism, where

$$\mathbf{R}\underline{\Gamma}_{[Z]}(\mathcal{M}) := \varinjlim \mathbf{R}\mathcal{H}om_{\mathcal{O}_X}(\mathcal{O}_X/\mathcal{I}^n_Z, \mathcal{M})$$

is the derived functor of the functor $\underline{\Gamma}_{[Z]}$ of algebraic geometry [SGA 2]. *This is a complex of $\mathbf{D}_X$-modules with holonomic cohomology sheaves, by a theorem of Kashiwara* [K 2]. *$\mathcal{M}$ is said to be a holonomic $\mathbf{D}_X$-module with regular singularities if it has regular singularities along any closed subvariety Z of X.*

This definition is due to Ramis [Ra]. It means, essentially, that the cohomology of the de Rham complex of $\mathcal{M}$, with support in Z, may be computed as the de Rham complex of a complex of algebraic local cohomology of $\mathcal{M}$ along Z.

Theorem 4.5 (Riemann-Hilbert Correspondence). *The functor $\mathcal{M} \mapsto \mathrm{DR}(\mathcal{M})$ realizes an equivalence of categories between the category of holonomic $\mathbf{D}_X$-modules with regular singularities and the category $\mathrm{Perv}(\mathbb{C}_X)$ of perverse sheaves of complex vector spaces over X.*

This theorem is due to Mebkhout [Me 1, Me 2] and to Kashiwara and Kawai [K-K 1]. Actually, these authors prove a stronger theorem, giving an equivalence between derived categories.

There is a precise correspondence between operations on $\mathbf{D}_X$-modules and operations on perverse sheaves. For instance, if $f : X \to Y$ is a proper complex-analytic map between complex-analytic manifolds, for $\mathcal{M}$ a holonomic $\mathbf{D}_X$-module, Kashiwara [K 2] defines a complex $\int_f \mathcal{M}$ of $\mathbf{D}_Y$-modules, whose cohomology sheaves $\mathcal{H}^j(\int_f \mathcal{M})$ are holonomic. This is defined as follows, with $d = \dim(X) - \dim(Y)$:

$$\int_f \mathcal{M} = \mathbf{R}f_* \left(\mathbf{D}_{Y \leftarrow X} \overset{\mathbf{L}}{\underset{\mathbf{D}_X}{\otimes}} \mathcal{M} \right) [d] \, ,$$

where $\mathbf{D}_{Y \leftarrow X}$ is a sheaf of $(f^{-1}(\mathbf{D}_Y), \mathbf{D}_X)$-bimodules (our convention for bimodules is that the first ring acts on the left, the second one on the right). $\mathbf{D}_{Y \leftarrow X}$ is equal to $f^*(\mathbf{D}_Y) \otimes_{\mathcal{O}_X} \omega_{X/Y}$ (here $\omega_{X/Y} := \omega_X \otimes f^*(\omega_Y)^{\otimes -1}$, and tensoring with this line bundle has the effect of turning a $(\mathbf{D}_X, f^{-1}(\mathbf{D}_Y))$-bimodule into a $(f^{-1}(\mathbf{D}_Y), \mathbf{D}_X)$-bimodule).

This construction becomes more concrete when f is a closed immersion (if $(z_1, \ldots z_d)$ are holomorphic functions whose derivatives generate the conormal bundle $\mathcal{N}_{X/Y}$, then $\int_f \mathcal{M} = \mathcal{M}[z_1, \ldots z_d][d]$), and in case f is a smooth map (then $\int_f \mathcal{M}$ is the image under $\mathbf{R}f_*$ of the "relative de Rham complex" $\mathrm{DR}_f(\mathcal{M})$). We refer to [Me 1, KK 1, Bo 3], for the fact that the de Rham complex of $\int_f \mathcal{M}$ is isomorphic to $\mathbf{R}f_*(\mathrm{DR}(\mathcal{M}))$, hence the de Rham complex of $\int_f^j \mathcal{M} := \mathcal{H}^j(\int_f \mathcal{M})$ is isomorphic to the j-th *perverse cohomology sheaf* $\mathcal{H}^j_{\mathbf{p}}(\mathbf{R}f_* \mathrm{DR}(\mathcal{M}))$. Note in particular that, in case Y is a point, the vector space $H^j(\int_f \mathcal{M})$ is the j-th hypercohomology group $H^j(X, \mathrm{DR}(\mathcal{M}))$.

There is also an inverse image operation on $\mathbf{D}_X$-modules, corresponding with inverse image for perverse sheaves; we refer to [Me 2] for a general discussion of the "six operations of Grothendieck" in the context of $\mathbf{D}_X$-modules (see also [Bo 3]). We will later describe the $\mathbf{D}$-module correspondent of the vanishing cycle functor ψ of Grothendieck and Deligne [SGA 7], which we will need to state Saito's construction.

The direct image operations $\mathcal{H}^j \int_f$ may be refined to operations on holonomic $\mathbf{D}$-modules equipped with a good filtration. This was first observed by Brylinski [Bry 2] and then made more precise by Laumon [Lau].

Proposition and Definition 4.6. *Let $f : X \to Y$ be a proper map, and let $\mathcal{M}$ be a holonomic $\mathbf{D}_X$-module, with good filtration $(\mathcal{M}_n)$. Put $d := \dim(X) - \dim(Y)$. Then define the subsheaf $(\int_f^j \mathcal{M})_n$ of $\int_f^j \mathcal{M} = \mathbf{R}^{j+d} f_* \left(\mathbf{D}_{Y \leftarrow X} \overset{\mathbf{L}}{\underset{\mathbf{D}_x}{\otimes}} \mathcal{M} \right)$ as follows:*

(1) If f is a closed immersion, define:

$$\left(\int_f^{-d} \mathcal{M} \right)_n := \sum_{k+l \le n} f_* \left(\mathbf{D}_{Y \leftarrow X}(l) \overset{\mathbf{L}}{\underset{\mathbf{D}_X}{\otimes}} \mathcal{M}_k \right) .$$

(2) If f is a smooth mapping, put:

$$\left(\int_f^{j} \mathcal{M} \right)_n := \sum_{k+l \le n} \mathrm{Image} \left(R^{j+d} f_* (\mathbf{D}_{Y \leftarrow X}(l)) \overset{\mathbf{L}}{\underset{\mathbf{D}_X}{\otimes}} \mathcal{M}_k \right) ,$$

where the image refers to the map induced on $R^{j+d} f_*(\text{——})$ by inclusions of sheaves, and where $\mathbf{D}_{Y \leftarrow X}(l) := f^*(\mathbf{D}_Y(l) \otimes \omega_{X/Y})$.

(3) In general, factor f as a closed immersion followed by a smooth mapping, and define the filtration $(\int_f^j \mathcal{M})_n$ in two steps.

Then the $(\int_f^j \mathcal{M})_n$ give a good filtration of the $\mathbf{D}_Y$-module $\int_f^j \mathcal{M}$.

Note that this process gives a filtration of the derived tensor product $\mathbf{D}_{Y \leftarrow X} \overset{\mathbf{L}}{\underset{\mathbf{D}_X}{\otimes}} \mathcal{M}$ *before* one applies the direct image functor $\mathbf{R} f_*$. In particular, one obtains a filtration of the de Rham complex of $\mathcal{M}$ by the sub-complexes

$$(\mathrm{DR}(\mathcal{M}))_n := \ldots \Omega_X^j \otimes \mathcal{M}_{n-j} \to \Omega_X^{j+1} \otimes \mathcal{M}_{n-j-1} \ldots$$

Upon taking global cohomology, one obtains the filtration of $H^j(X, \mathrm{DR}(\mathcal{M}))$ described in 4.6. See [Bry 2].

Now, the idea of Brylinski was to define, for every complex-analytic manifold X, some category of holonomic $\mathbf{D}_X$-modules with regular singularities, equipped with a good filtration. There should be a direct image operation $\int_f^j$, for f a projective morphism, where $\int_f^j$ is the operation in Definition 4.6. In case of a point, a filtered **D**-module is just a filtered vector space, and the condition will be that the filtration gives rise to a Hodge structure.

Let us discuss how the stability of the category we are looking for under direct images will lead to Hodge structures on hypercohomology groups. Let $(\mathcal{M}, (\mathcal{M}_n))$ be an object in our hypothetical category of filtered $\mathbf{D}_X$-modules. Assume X projective, and let $p : X \to pt$. Then $\int_p^j(\mathcal{M})$ will just be a filtered vector space. Since it is supposed to belong in our category relative to a point, it has to carry a Hodge structure.

Assume for instance that $\mathcal{M}$ has de Rham complex equal to the intersection complex $\underline{IC}_Z^\bullet$ of a subvariety Z of X. Then, according to a conjecture of Cheeger, Goresky and MacPherson [C-G-M] (see Conjecture 3.20), there should be a pure Hodge structure on the global intersection cohomology of Z (if Z is projective). So, in concrete terms, what one wants to find is a good filtration of $\mathcal{M}$ such that the induced filtration on the global cohomology of $\mathrm{DR}(\mathcal{M})$ is the Hodge filtration for the desired Hodge structure.

This sought for stability under direct image for a projective morphism may be used as a guide to find the right kind of filtrations on holonomic **D**-modules. For instance, return to a polarized variation of Hodge structures $(\mathbb{V}_{\mathbb{R}}, \mathscr{F}^p(\mathscr{V}))$. As we discussed earlier in this section, we have a natural good filtration $(\mathcal{M}_n)$ of the sheaf $\mathcal{M}$ of germs of sections of $\mathscr{V}$. If X is projective, Deligne proved that this gives the Hodge filtration for a polarized Hodge structure on the global cohomology of $\mathrm{DR}(\mathcal{M}) \simeq \mathbb{V}_{\mathbb{C}}$ (Proposition 3.7). Hence one can be sure that one has the right notion of good filtration in that ca

This gives the background for the following conjecture of Brylinski.

Conjecture 4.7 [Bry 2]. *There exists a naturally defined filtration on intersection complexes (with twisted coefficients), which induces pure Hodge structures on global cohomology.*

In some sense, this conjecture was established by M. Saito. However, Brylinski gave a precise definition of a good filtration, based on the microlocal theory of pseudodifferential systems with regular singularities, due to Kashiwara and Kawai [K-K 1]. Such a filtration is the right one over a base X of dimension one, in which case the filtered de Rham complex of the holonomic $\mathbf{D}_X$-module with de Rham complex $j_*(\mathbb{V})$, for $\mathbb{V}$ a polarized variation of Hodge structure over an open set $U \xrightarrow{j} X$, is filtered quasi-isomorphic to Zucker's holomorphic L_2 de Rham complex $\Omega^\bullet_{(2)}(\mathbb{V})$.

Brylinski's was shown to have good properties in many examples [Bry 2], but it was very hard to manipulate and to compute. Furthermore, Saito found a counterexample showing that this filtration is not the right one in general! [Sa 1: § 5].

M. Saito followed a different route to define the category $MH(X, n)$ of pure Hodge modules of weight n on a complex manifold X. The bulk of his work is contained in the voluminous article [Sa 3]. A summary of results is given in [Sa 4]. Saito first introduces the category $MF_h(\mathbf{D}_X)$ of filtered $\mathbf{D}_X$-modules $(\mathcal{M}, F)$ (the filtration is then denoted by $F_n(\mathcal{M})_{n\in\mathbb{Z}}$), where $\mathcal{M}$ is holonomic with regular singularities, and the filtration is good. The morphisms in this category are the $\mathbf{D}_X$-linear maps which preserve the filtration. This category is an exact category, if one defines a short exact sequence as an exact sequence of underlying $\mathbf{D}_X$-modules, in which the morphisms are strictly compatible with the filtrations.

Because one wants to generalize variations of Hodge structure, one needs a *real structure* on the perverse sheaf $\mathrm{DR}(\mathcal{M})$. More generally, let A be a subfield of $\mathbb{R}$ (for instance, $\mathbb{Q}$ or $\mathbb{R}$). Then we have the category $\mathrm{Perv}(A_X)$ of A-perverse sheaves over X, which maps to $\mathrm{Perv}(\mathbb{C}_X)$ by "extension of scalars".

Then Saito [Sa 3, Sa 4] defines $MF_h(\mathbf{D}_X, A)$ to be the fiber product of the categories $MF_h(\mathbf{D}_X)$ and $\mathrm{Perv}(A_X)$ over $\mathrm{Perv}(\mathbb{C}_X)$; so an object in $MF_h(\mathbf{D}_X, A)$ is a quadruple $(\mathcal{M}, F, K, \alpha)$, where $(\mathcal{M}, F)$ is a filtered holonomic $\mathbf{D}_X$-module with regular singularities, K is a A-perverse sheaf over X, and α is an isomorphism $\alpha : \mathrm{DR}(\mathcal{M}) \to K \otimes \mathbb{C}$. Usually, α is omitted from the notation. The "n-th Tate twist" $(\mathcal{M}, F, K)(n)$ is defined to be $(\mathcal{M}, F[n], K(n))$, where $F[n]_k := F_{k-n}$, and $K(n) := (2\pi i)^n K$. One may also define the dual of $(\mathcal{M}, F, K)$; to find a filtration on the dual $\mathcal{M}^*$, one uses a free filtered resolution of $\mathcal{M}$, as in [Bry 2].

Inside this category $MF_h(\mathbf{D}_X, A)$, Saito defines a full subcategory $MH(X, n)$, the category of *Hodge modules of weight n*. This is defined by induction on the dimension of X. The crucial inductive assumption involves the "nearby cycle sheaf" $\psi_g(K)$ and the "vanishing cycle sheaf" $\phi_g(K)$, for K a constructible complex of sheaves on X and g a non-constant holomorphic function on X. These are constructible complexes of sheaves on $g^{-1}(0)$. They are introduced and studied in [SGA 7] (see also Definition 2.21).

Proposition 4.8 (Gabber, see [Bry 3]). *Let $g : X \to \mathbb{C}$ be a non-constant holomorphic function. If $K^\bullet$ is a perverse sheaf on X, then $\psi_g(K)[-1]$ and $\phi_g(K)[-1]$ are perverse sheaves on $g^{-1}(0)$.*

We next give the construction of functors ψ and ϕ on the level of holonomic $\mathbf{D}$-modules. This is due to Malgrange [M] in the case of $\mathcal{O}_X$, and to Kashiwara [K 4] for a general holonomic $\mathbf{D}_X$-module $\mathcal{M}$ with regular singularities. Embedding X into $X \times \mathbb{C}$ by $x \to (x, g(x))$, one may replace g with the second projection $X \times \mathbb{C} \to \mathbb{C}$. Hence one may assume g is a smooth map.

If g is smooth, $X_0 := g^{-1}(0)$ is a smooth hypersurface. The *V-filtration* of Malgrange-Kashiwara is defined first on the sheaf $\mathbf{D}_X$; it is an increasing filtration $(V_\alpha \mathbf{D}_X)$, indexed by $\alpha \in \mathbb{Q}$. Let I_{X_0} be the sheaf of ideals of X_0. Then one puts:

$$V_\alpha \mathbf{D}_X := \{P \in \mathbf{D}_X : P(I_{X_0}^j) \subset I_{X_0}^{j-[\alpha]}, \forall j \in \mathbb{Z}\} .$$

Concretely, $V_\alpha \mathbf{D}_X = V_k \mathbf{D}_X$, for $k = [\alpha]$. Locally, we may assume $X = X_0 \times \mathbb{C}$; then $V_k \mathbf{D}_X$ is generated, as a $\mathbf{D}_{X_0}$-module, by products $g^i (\frac{\partial}{\partial g})^j$, for $i - j \geq -k$.

Now if $\mathcal{M}$ is a coherent $\mathbf{D}_X$-module, there is at most one filtration $(V_\alpha(\mathcal{M}))_{\alpha \in \mathbb{Z}}$, indexed by rational numbers, such that:

(1) $\bigcup_{\alpha \in \mathbb{Q}} V_\alpha(M) = M$, and each $V_\alpha(M)$ is a coherent $V_0(\mathbf{D}_X)$-module;

(2) $(V_i \mathbf{D}_X)(V_\alpha(M)) \subset V_{\alpha+i}(M)$, for all $\alpha \in \mathbb{Q}$ and $i \in \mathbb{Z}$; furthermore, the inclusion $g \cdot V_\alpha(M) \subset V_{\alpha-1}(M)$ is an equality for $\alpha < 0$;

(3) for all α, the action of $g \frac{\partial}{\partial g} + \alpha$ on $\mathrm{Gr}_\alpha^V M$ is nilpotent.

Here $\frac{\partial}{\partial g}$ denotes any vector field on X such that $[\frac{\partial}{\partial g}, g] = 1$. It is an easy fact [K 4, Sa 3: Lemma 3.1.2] that this filtration V, if it exists, is unique, and is called the *rational V-filtration.* Kashiwara proved that it does exist if $\mathcal{M}$ is holonomic with regular singularities and the monodromy of $DR(\mathcal{M})$ is quasi-unipotent (see [Me 2]).

Definition 4.9 (Saito [Sa 3: § 3.2]). *Let $(\mathcal{M}, F)$ be a filtered $\mathbf{D}_X$-module. One says that $(\mathcal{M}, F)$ is* quasi-unipotent and regular along X_0 *if the rational V-filtration* *exists and the following compatibility conditions between V and F hold:*

$$(a) \qquad g(F_pV_\alpha \mathcal{M}) = F_pV_{\alpha-1}\mathcal{M}\ , \quad \text{for } \alpha < 1\ ;$$

$$(b) \qquad \frac{\partial}{\partial g}(F_p\mathrm{Gr}^V_\alpha \mathcal{M}) = F_{p+1}\mathrm{Gr}^V_{\alpha+1}\mathcal{M} \cap (\frac{\partial}{\partial g}(\mathrm{Gr}^V_\alpha \mathcal{M}))\ , \quad \text{for } \alpha \geq 0\ .$$

Saito's construction requires a "nearby cycles" functor $(\mathcal{M}, F, K) \mapsto \psi_g(\mathcal{M}, F, K)$ and a "vanishing cycles" functor $(\mathcal{M}, F, K) \mapsto \phi_g(\mathcal{M}, F, K)$ on the level of filtered **D**-modules.

Construction 4.10. *For* $(\mathcal{M}, F, K)$ *as above*

$$\psi_g(\mathcal{M}, F, K) := (\bigoplus_{-1<\alpha\leq 0} \mathrm{Gr}^\alpha_V(\mathcal{M}, F), \psi_g K[-1])$$

$$\phi_{g,1}(\mathcal{M}, F, K) := (\mathrm{Gr}^{-1}_V(\mathcal{M}, F[-1]), \phi_{g,1}K[-1])\ .$$

The map can : $\psi_g \to \phi_{g,1}$ *is induced by* $-\frac{\partial}{\partial g}$, *and the map* Var : $\phi_{g,1} \to \psi_{g,1}(-1)$ *is induced by* g.

This construction gives a refinement (for filtered **D**-modules) of the constructions of Malgrange [M] and Kashiwara [K 4].

Saito proves [Sa 3: Lemma 5.1.4] that the following two conditions are equivalent, for $(\mathcal{M}, F, K^\bullet)$ quasi-unipotent and regular along $X_0 = g^{-1}(0)$.

(I) In the category $MF_h(\mathbf{D}_X, A)$, $\phi_1(\mathcal{M})$ decomposes as the direct sum of $\ker(Var)$ and of $\mathrm{Im}(can)$.

(II) One has a decomposition in $MF_h(\mathbf{D}_X, A)$:

$$(\mathcal{M}, F, K^\bullet) = (\mathcal{M}_1, F_1, K_1^\bullet) \oplus (\mathcal{M}_2, F_2, K_2^\bullet)\ ,$$

where $\mathcal{M}_2$ has support contained in X_0 and $\mathcal{M}_1, F_1, K_1^\bullet$ has no sub-object or quotient object supported in X_0.

We can now (at last) turn to the inductive definition ([Sa 3: § 5.1]) of the full subcategory $MH(X, A, n)$ of $MF_h(\mathbf{D}_X, A)$ of pure Hodge modules of weight n. First let $MF_h(\mathbf{D}_X, A)_{(0)}$ be the full subcategory of $MF_h(\mathbf{D}_X, A)$ whose objects are those which satisfy (I) (or (II)) for any g. Then $MH(X, A, n)$ is the largest full subcategory of $MF_h(\mathbf{D}_X, A)_{(0)}$ satisfying the two conditions:

(1) An object of $MH(X, A, n)$ with support $\{x\}$ is of the form $(\mathcal{M}, F, K) = i_\bullet(H_\mathbb{C}, F, H_A)$, for $i : \{x\} \hookrightarrow X$ the inclusion, where $(H_\mathbb{C}, F, H_A)$ is a pure A-Hodge structure of weight n, with increasing filtration $F_p := F^{-p}$.

(2) If $(\mathcal{M}, F, K)$ belongs to $MH(X, A, n)$, then for every non-constant holomorphic function g, for W the monodromy weight filtration of $\mathcal{M}$ (see Sect. 2), shifted by $n-1$, we have:

$$\mathrm{Gr}^W_i \psi_g(\mathcal{M}, F, K) \in MH(X, A, i)$$

for all $i \in \mathbb{Z}$, and the same condition holds for $\mathrm{Gr}^W_i \phi_{g,1}(\mathcal{M}, F, K)$.

This completes the definition of the category $MH(X, A, n)$. We observe that the filtration F of an object $\mathcal{M}$ of $MH(X, A, n)$ is determined by its restriction to an open set $g \neq 0$ which meets the support of any direct factor of $\mathcal{M}$. Indeed, it is shown in [Sa 3: 3.2.2.1–2] that one has:

$$F_p(\mathcal{M}) = \sum_{i \geq 0} \left(\frac{\partial}{\partial g} \right)^i \left(V_{<0}\mathcal{M} \cap j_* j^* F_{p-i}(\mathcal{M}) \right) ,$$

where $j : X - X_0 \hookrightarrow X$ is the inclusion mapping.

Next, for Z an irreducible closed analytic subvariety of X, one defines the full subcategory $MH_Z(X, A, n)$ of $MF(X, A, n)$ as follows: an object $(\mathcal{M}, F, K)$ is in $MH_Z(X, A, n)$ if and only if it has support Z and admits no sub-object or quotient object with strictly smaller support. One then says that $(\mathcal{M}, F, K)$ has *support strictly equal to* Z. An object of $MH_Z(X, A, n)$ is called a "pure Hodge module of weight n with strict support Z" [Sa 3: 5.1.3].

Proposition 4.11 [Sa 3: Proposition 5.1.6]. *$MH_Z(X, A, n)$ and $MH(X, A, n)$ are abelian categories. One has the decomposition:*

$$MH(X, A, n) := \bigoplus_{Z \subset X} MH_Z(X, A, n) .$$

Every morphism in these categories is strict with respect to the filtrations F. Furthermore, these subcategories of $MF_h(\mathbf{D}_X, A)$ are stable under the operation of taking a direct summand.

As a by-product of the proof of Proposition 4.11, Saito obtains more information about the nearby cycles $\psi_g(\mathcal{M}, F, K)$. The logarithm N of the monodromy induces an isomorphism in the category $MF_h(\mathbf{D}_X, A)$:

$$N^i : \mathrm{Gr}^W_{n-1+i}\psi_g(\mathcal{M}, F, K) \simeq \mathrm{Gr}^W_{n-1-i}\psi_g(\mathcal{M}, F, K)(-i) ,$$

and the variation morphism $\mathrm{Var} : \phi_{g,1}(\mathcal{M}, F, K) \to \psi_{g,1}(\mathcal{M}, F, K)$ is a strict monomorphism.

To obtain results about the stability of these categories under direct image, one needs the notion of "polarization" of a Hodge module $(\mathcal{M}, F, K)$. We may assume that $(\mathcal{M}, F, K)$ belongs to $MH_Z(X, A, n)$, for some irreducible Z. A polarization, according to [Sa 3: 5.2.3] is a pairing $S : K \otimes K \to A_X[2n](-n)$, which satisfies:

(P 1): If ι is the canonical involution of $K \otimes K$, we have: $S \circ \iota = (-1)^n S$ (i.e. S is "$(-1)^n$-symmetric");

(P 2): S, viewed as a morphism from K to the n-th Tate twist of the dual perverse sheaf $D(K) = \mathbf{R}Hom(K, A_X[2n])$, extends to an isomorphism between $(\mathcal{M}, F, K)(n)$ and $D(\mathcal{M}, F, K)$;

(P 3): If $Z = \{x\}$, and $(\mathcal{M}, F, K) = i_*(H_\mathbb{C}, F, H_A)$, for $i : \{x\} \to X$ and $(H_\mathbb{C}, F, H_A)$ a A-Hodge structure, then $S = i_*(S')$, for S' a polarization of this Hodge structure;

(P 4): If $\dim Z > 0$, and g is a germ of holomorphic function at x, with $g(x) = 0$ but g non-identically zero on Z, then for N the logarithm of the monodromy, the duality:

$$\mathrm{Gr}^W_{n-1+i}\psi_g S \circ (id \otimes N^i) : P_N \mathrm{Gr}^W_{n-1+i}\psi_g K[-1] \otimes P_N \mathrm{Gr}^W_{n-1+i}\psi_g K[-1] \\ \rightarrow A_X[2\dim X](\dim X - n - i + 1)$$

is a polarization of $P_N \mathrm{Gr}^W_{n-1+i}\psi_g K[-1]$, for every i. Here P_N denotes the primitive part with respect to N, and one uses the fact that ψ commutes with Verdier duality, and the self-duality of the monodromy filtration W. See [Sa 3: 5.2] for details.

With all these definitions, we may now state the first main theorem of [Sa 3].

Theorem 4.12 (Saito, [Sa 3: Théorème 5.3.2]). *Let $f : X \to Y$ be a projective morphism between smooth complex-analytic varieties, and let l be the first Chern class of a relatively ample line bundle. Let $(\mathcal{M}, F, K) \in MH_Z(X, A, n)$ be endowed with a polarization S. Then:*

(1) the complex $\int_f(\mathcal{M}, F)$ is strict and $\mathcal{H}^j \int_f(\mathcal{M}, F)$ belongs to $MH(Y, A, n+i)$;

(2) the hard Lefschetz theorem holds, i.e.
$l^j : \mathcal{H}^{-j} \int_f(\mathcal{M}, F, K) \simeq \mathcal{H}^j \int_f(\mathcal{M}, F, K)(j)$ *is an isomorphism;*

(3)

$$(-1)^{j(j-1)2} f_* S \circ (id \otimes l^j) : P_l(\mathcal{H}^{-j}K) \otimes P_l(\mathcal{H}^{-j}K) \rightarrow A_Y[2\dim Y](\dim Y - n + j)$$

is a polarization of the primitive part $P_l(\mathcal{H}^{-j}K)$ relative to l.

Implicit in this statement is the fact the direct image operation for f, in the context of filtered **D**-modules (cf. Proposition 4.6) commutes with duality. Note that part (1) of the theorem, in case Y is a point, implies that the spectral sequence

$$E_1^{p,q} = H^{p+q}(X, \mathrm{Gr}^F_{-q}\mathrm{DR}(\mathcal{M})) \Rightarrow H^{p+q}(X, \mathrm{DR}(\mathcal{M}))$$

degenerates at E_1. This should be viewed as a generalization of similar degeneration results of Deligne, which are central to his mixed Hodge theory (see e.g. Theorem 1.5). It means that the differential of the complex is strictly compatible with the filtration F.

The above theorem can be useful only if one can actually construct objects equipped with a polarization in the category $MH_Z(X, A, n)$. In that direction, one has:

Theorem 4.13 [Sa 3: Théorème 5.4.4]. *Let X be a smooth complex-analytic variety of dimension d, let $(\mathbf{V}_A, F, S)$ be a polarized variation of Hodge structures of weight n on X. Then the corresponding filtered $\mathbf{D}_X$-module is a Hodge module of weight n with strict support X.*

Following [Sa 3], let us state a few corollaries:

Theorem 4.14 [Sa 3]. *The decomposition theorem of Beilinson, Bernstein, Deligne and Gabber* [B-B-D: Theorem 6.2.5] *holds for* $\mathbf{R}f_*(K)$, *for* f *a projective morphism, and* K *an object of* $\mathrm{Perv}(A_X)$ *which underlies a Hodge module on* X.

Theorem 4.15 [Sa 3 and Sa 9]. *Let* $\mathbb{V}_A$ *be a polarized variation of Hodge structure of weight* n *over a Zariski open set of a compact Kähler manifold* X. *Then the intersection cohomology groups* $IH^k(X, \mathbb{V})$ *carry polarized Hodge structures. In particular, the polarized Hodge structure on* $IH^k(X, A)$ *is a direct factor of that of* $H^k(\tilde{X}, A)$, *for* $\tilde{X}$ *a resolution of singularities of* X.

The existence of such a Hodge structure was conjectured by Cheeger, Goresky and MacPherson [C-G-M] (Conjecture 3.20).

Saito has generalized the case $\mathcal{M} = \mathcal{O}_X$ of Theorem (4.10) as follows:

Theorem 4.16. *Let* X *be a complex Kähler manifold, let* $f : X \to Y$ *be a proper morphism. Then the conclusion of Theorem 4.13 holds true for* $\mathcal{O}_X$.

There is an ample supply of Hodge modules, according to the following.

Theorem 4.17 [Sa 10; Sa 11: 3.20–3.21]. *Let* $j : U \to X$ *be an open immersion of* d*-dimensional smooth algebraic varieties over* $\mathbb{C}$, *such that* $D := X - U$ *is a divisor with normal crossings. Let* $\mathbb{V}$ *be a polarized variation of Hodge structure of weight* n *on* U, *such that* $\mathbb{V}_A$ *has quasi-unipotent local monodromies. For* $\alpha \in \mathbb{R}$, *let* $\mathcal{V}_X^{>\alpha}$ *be the Deligne extension of* $\mathcal{V}$ *to a vector bundle on* X, *on which* ∇ *has pole of order* ≤ 1, *and its residue along each branch of* D *has spectrum contained in* $(\alpha, \alpha + 1]$.

Let the $\mathbf{D}_X$*-module* $\mathcal{M}$ *be the sub-*$\mathbf{D}_X$*-module of* $j_*\mathcal{V}$ *generated by* $\mathcal{V}_X^{>-1}$. *Then* $\mathrm{DR}(\mathcal{M})$ *is a realization of* $\underline{IC}^\bullet(X, \mathbb{V}_\mathbb{C})$. *Filter* $\mathcal{V}_X^{>-1}$ *by* $F_p\mathcal{V}_X^{>-1} := \mathcal{V}_X^{>-1} \cap j_*F_p(\mathcal{V})$ *and* $\mathcal{M}$ *by* $F_p(\mathcal{M}) := \sum_i F_i(\mathbf{D}_X)F_{p-i}(\mathcal{V}_X^{>-1})$. *Let* $K := \underline{IC}^\bullet(X, \mathbb{V}_A)$. *Then* $(\mathcal{M}, F, K)$ *is a Hodge module on* X *of weight* $d+n$. *Furthermore, the pairing* $S : K \otimes K \to \mathbb{Q}_X(-n)[2d]$ *gives a polarization of* $(\mathcal{M}, F, K)$.

Corollary 4.18 [Sa 10]. *Let* X *be a complex algebraic variety,* $Z \subset X$ *a closed algebraic subvariety. The subcategory of* $MH_Z(X, A, n)$ *consisting of polarizable Hodge modules is equivalent to the category of polarizable variations of Hodge structure on Zariski-dense smooth open subsets of* Z.

Saito has also developed a theory of "mixed Hodge modules" over a complex algebraic subvariety X [Sa 5, Sa 8, Sa 10, Sa 11]. These are objects of a category $MHM(X)$, a full subcategory of the category whose objects are quadruples $(\mathcal{M}, F, K; W)$, with $(\mathcal{M}, F)$ a filtered **D**-module, K is a perverse sheaf over $\mathbb{R}$, $\alpha : \mathrm{DR}(\mathcal{M}) \cong K \otimes \mathbb{C}$ an isomorphism, and W a pair of filtrations on $\mathcal{M}$ and on K, compatible with α. For $\mathcal{M}$ in $MHM(X)$, each $\mathrm{Gr}_j^W(\mathcal{M})$ is a polarizable Hodge module of weight j. All the standard functors $f_*, f^*, f_!, f^!, \psi_g, \phi_{g,1}, \otimes, Hom$ are defined for $MHM(\text{——})$, and also on the derived category of complexes of mixed Hodge modules. In particular, one obtains M.H.S. on hypercohomology. Admissible variations of mixed Hodge structure, in the sense of Definition 7.2, belong to this category.

Saito gives interesting applications of his theory to vanishing cycles. Let $f : X \to \mathbb{C}$ be a holomorphic function on an analytic space X. Assume $X \subset \mathbb{C}^n$, with $0 \in X$, and f extends to a function on a ball of $\mathbb{C}^n$. Let $X(f,x) := B_\varepsilon \cap X_t$ be the Milnor fiber, for B_ε a small ball and $0 < |t| << \varepsilon$. Let $i_x : \{x\} \hookrightarrow X$ be the inclusion. Since $H^j(X(f,x),A) = H^j(i_x^* \psi_f(A_X))$, this cohomology group is endowed with a natural M.H.S. , by the functoriality of $MHM(X)$ under ψ and i_x^*. Similarly $H_c^j(X(f,x),A) = H^j(i_x^! \psi_f(A_X))$ has a M.H.S., and also the homology and Borel-Moore homology of the Milnor fiber carry M.H.S. These are compatible with all the duality pairings. The intersection form

$$H_c^j(X(f,x),A) \otimes H_c^{2n-j}(X(f,x),A) \to A(-n)$$

is a morphism of M.H.S.

Theorem 4.19 (Saito, [Sa 8: Theorem 2.10]). *Assume that X and $X_0 = f^{-1}(0)$ have isolated singularity at x. Let $\underline{IC}_X^\bullet(\mathbb{R})$ denote the intersection complex of X (with real coefficients). Then $\underline{IC}_X^\bullet(\mathbb{R})$ is a cohomological Hodge complex, and we have an isomorphism $\mathrm{Gr}_\bullet^W H^n(X(f,x),\mathbb{R})_1 \cong L_\bullet \oplus L'_\bullet$ of graded Hodge structures with an action of a nilpotent endomorphism N of degree -2, such that N^k induces isomorphisms $N^k : L_{n+k} \cong L_{n-k}(-k)$ and $N^k : L'_{n+1-k} \cong L'_{n+1+k}$.*

Furthermore, using the notation P_N for the primitive part with respect to N, we have:

$$P_N L_{n+k} \cong (\mathrm{Gr}_{n-k}^W H^{-1} i_x^* \underline{IC}_X^\bullet(\mathbb{R}))(-k) \quad \text{for } k \geq 0$$

and

$$P_N L'_{n+k} \cong (\mathrm{Gr}_{n-k}^W (H^{-1} i_x^* \underline{IC}_{X_0}^\bullet(\mathbb{R}) / H^{-2} i_x^* \underline{IC}_X^\bullet(\mathbb{R})))(-k) \quad \text{for } k > 0 \, .$$

Another interesting application of M. Saito's theory concerns the *higher residue pairings* of K. Saito [S]. Consider a cartesian diagram of complex manifolds with base point 0:

$$\begin{array}{ccc} (Z,0) & \xrightarrow{\widehat{\pi}} & (X,0) \\ {\scriptstyle p}\downarrow & & {\scriptstyle q}\downarrow \\ (S,0) & \xrightarrow{\pi} & (T,0) \end{array}$$

with π and q submersions, $\dim(S) = m$, $\dim(T) = m-1$, $\dim(X) = n+m$ and $\dim(Z) = n+m+1$. Let δ_1 (resp. $\widehat{\delta_1}$) be a holomorphic vector field on S (resp. Z) such that by $\pi^{-1}(\mathcal{O}_T) = \{g \in \mathcal{O}_S : \delta_1 g = 0\}$ and $\widehat{\pi}'^{-1}(\mathcal{O}_X) = \{g \in \mathcal{O}_Z : \widehat{\delta}_1 g = 0\}$, and $p_* \widehat{\delta}_1 = \delta_1$.

Let $\mathfrak{g}$ be the sheaf of Lie algebras on T defined by $\mathfrak{g} = \{\delta \in \pi_* Der_S : [\delta_1, \delta] = 0\}$. Then $\mathfrak{g}$ is a $\mathcal{O}_T$-free module of rank m.

K. Saito defines a *hamiltonian system* F as a holomorphic function F on Z such that $F(0) = 0$ and $\widehat{\delta_1} F = 1$. The datum of F is equivalent to that of the section ι of $\widehat{\pi}$ with image $F^{-1}(0)$ or to that of the holomorphic mapping $\phi : (X,0) \to (S,0)$, such that $q = \pi \circ \phi$, given by $\phi = p \circ \iota$. Let $C \subset X$ be the scheme of critical points for ϕ. There is natural $\mathcal{O}_T$-homomorphism $\mathfrak{g} \to q_*\mathcal{O}_C$, sending v to $v \cdot F_{|C}$. If this is bijective, F is said to be a *universal unfolding* of $f := F_{|q^{-1}(0)}$. One assumes that f has an isolated singularity at 0.

Choose a function t_1 on S such that $\delta_1(t_1) = 1$ and let $F_1 := t_1 - F$. Define $\mathcal{O}_S$-modules $\mathcal{H}_F^{(-k)}$ for $k \in \mathbb{N}$ as follows. For $k = 0$,

$$\mathcal{H}_F^{(0)} := \phi_*\Omega_{X/T}^{n+1}/dF_1 \wedge d(\phi_*\Omega_{X/T}^{n-1})$$

is the Brieskorn lattice. Define inductively $\mathcal{H}_F^{(-k-1)} := \{\omega \in \mathcal{H}_F^{(-k)} : \nabla_{\delta_1}\omega \in \mathcal{H}_F^{(-k)}\}$. Let $r^{(0)} : \mathcal{H}_F^{(0)} \to \phi_*\Omega_F := \Omega_{X/S}^{n+1}$ be the natural map. The completion $\widehat{\pi_*\mathcal{H}_F^{(0)}} := \varprojlim_k \pi_*\mathcal{H}^{(0)}/\pi_*\mathcal{H}^{(-k)}$ is in a natural way a $\mathfrak{g}$-module and a $\mathcal{O}_T[[\delta_1^{-1}]]$-module.

Define the first residue mapping J_F be the non-degenerate $\mathcal{O}_T$-bilinear form $J_F : q_*\Omega_F \times q_*\Omega_F \to \mathcal{O}_T$ by

$$J_F(\psi_1(z,t')dz_0\wedge\ldots\wedge dz_n, \psi_2(z,t')dz_0\wedge\ldots\wedge dz_n) = \mathrm{Res}_{X/T}\begin{bmatrix} \psi_1 \cdot \psi_2 dz_0 \wedge \ldots \wedge dz_n \\ \frac{\partial F}{\partial z_0}, \ldots, \frac{\partial F}{\partial z_n} \end{bmatrix},$$

where "Res" denotes the Grothendieck residue. (Here, we introduced coordinates t' on T and (z,t') on Z).

K. Saito's higher residue pairing is a far-reaching extension of J_F. It is defined on the (micro)-localization $\widehat{\pi_*\mathcal{H}_F} := \mathcal{O}_T[[\delta_1^{-1}]][\delta_1] \otimes_{\mathcal{O}_T[[\delta_1^{-1}]]} \widehat{\pi_*\mathcal{H}_F^{(0)}}$.

Theorem 4.20 (K. Saito, [S]). *There is a $\mathcal{O}_T$-bilinear map*

$$K_F : \widehat{\pi_*\mathcal{H}_F} \times \widehat{\pi_*\mathcal{H}_F} \to \mathcal{O}_T[[\delta_1^{-1}]]$$

with the following properties

(1) $K_F(\omega_1, \omega_2) = K_F(\omega_2, \omega_1)^*$, where * denotes the formal adjoint of a pseudo-differential operator.

(2) $K_F(P\omega_1, \omega_2) = K_F(\omega_1, P^*\omega_2) = PK_F(\omega_1, \omega_2)$, for $P \in \mathcal{O}_T[[\delta_1^{-1}]]$.

(3) $\delta K_F(\omega_1, \omega_2) = K_F(\delta\omega_1, \omega_2) + K_F(\omega_1, \delta\omega_2)$, for $\delta \in \mathfrak{g}$.

(4) $\frac{\partial}{\partial\delta_1} K_F(\omega_1, \omega_2) = K_F(t_1\omega_1, \omega_2) - K_F(\omega_1, t_1\omega_2)$.

(5) The restriction of K_F to $\widehat{\pi_*\mathcal{H}_F^{(0)}}$ takes values in $\delta_1^{-n-1}\mathcal{O}_T[[\delta_1^{(-1)}]]$ and is congruent to $J_F(r^{(0)}(—), r^{(0)}(—))$ modulo δ_1^{-n-2}.

The existence of this higher residue pairing is closely related to that of a "primitive form" and to the period mapping, which has also been studied by K. Saito.

M. Saito has related the higher residue pairing to the self-duality on vanishing cycles. Restrict the $\mathcal{O}_T$-modules and the higher residue pairing to the origin. Let $\mathcal{M}$ be the Gauss-Manin $\mathbf{D}$-module for $f := F_{|q^{-1}(0)}$, with filtrations (F, V). Then $F_{-n}(\mathcal{M})$ is the Brieskorn lattice. Furthermore the higher residue pairing induces a perfect pairing

$$\mathrm{Gr}_V^\alpha \mathcal{M} \otimes \mathrm{Gr}_V^\beta \mathcal{M} \to \mathbb{C} \otimes \delta_1^{-i} \quad \text{for } \alpha + \beta = i$$

Up to a sign, this pairing is identified with the natural perfect pairing on $\phi_f \mathbb{C}_X = H^n(X(f,x), \mathbb{C}_X)$ constructed by M. Saito [Sa 8: §2.7]. This gives a topological interpretation for the higher residue pairing.

5. Deligne-Beilinson Cohomology

To motivate the definition of the Deligne cohomology groups $H^q_{\mathcal{D}}(X, \mathbb{Z}(j))$, for X a proper algebraic manifold over $\mathbb{C}$, we follow a line of thought due to Beilinson. We first recall that for $Z \subset X$ an irreducible algebraic subvariety of codimension j in X, the cohomology class c_Z in $H^{2j}(X, \mathbb{C})$ admits two refinements:

(1) a class $c_{Z,H}$ in $F^j(H^{2j}(X, \mathbb{C})) = H^{2j}(X, F^j)$, where $F^j = F^j(\Omega^\bullet_X)$ is the "troncation bête" of the holomorphic de Rham complex $\Omega^\bullet_X$; this just reflects the fact that c_Z is of Hodge filtration $\geq j$.

(2) a class $c_{Z,B}$ in $H^{2j}(X, \mathbb{Z}(j))$ which maps to c_Z under the inclusion $\mathbb{Z}(j) \hookrightarrow \mathbb{C}$; this reflects the integrality of c_Z.

It is therefore natural to consider the fiber product of $H^{2j}(X, F^j)$ and of $H^{2j}(X, \mathbb{Z}(j))$ over $H^{2j}(X, \mathbb{C})$. Deligne performed the analogous construction on the level of complexes of sheaves (see the book [R-S-S]). Recall that a fiber product is the kernel of a morphism, and a kernel of a morphism of complexes coincides with the cone of this morphism, shifted to the right by 1, in case the morphism is surjective. Hence, for $j \in \mathbb{Z}$ and for A a subring of $\mathbb{R}$, Deligne defined a complex of sheaves $A(j)_{\mathcal{D}}$ on X as:

(5-1) $$A(j)_{\mathcal{D}} = Cone(F^j \oplus A(j) \to \Omega^\bullet_X)[-1] \ .$$

For $j \leq 0$, the morphism in question is surjective, hence $A(j)_{\mathcal{D}}$ is equal to the kernel, which is isomorphic to $A(j)$. For $j \geq 0$, the complex $A(j)_{\mathcal{D}}$ is quasi-isomorphic to the following complex of sheaves:

(5-2) $$0 \to A(j) \to \mathcal{O}_X \to \Omega^1_X \cdots \to \Omega^{j-1}_X \ .$$

Definition 5.1 *The* Deligne cohomology *groups* $H^q_{\mathcal{D}}(X, A(j))$ *are the hypercohomology groups of the complex of sheaves* $A(j)_{\mathcal{D}}$. *That is:*

$$H^q_{\mathcal{D}}(X, A(j)) = H^q(X, A(j)_{\mathcal{D}}).$$

Similarly, one defines the Deligne cohomology groups with support in a closed subset Y of X, denoted by $H^q_{Y,\mathcal{D}}(X, A(j))$.

It follows directly from the definitions that the complexes $A(j)_{\mathscr{D}}$, hence the Deligne cohomology groups, are functorial with respect to algebraic mappings.

From its construction, it is clear that the complex of sheaves $A(j)_{\mathscr{D}}$ fits in a distinguished triangle:

$$\ldots A(j)_{\mathscr{D}} \to A(j) \oplus F^j \to \Omega_X^\bullet \to A(j)_{\mathscr{D}}[1] \ldots \tag{5-3}$$

Hence one has a long exact sequence of groups:

(5-4)
$$\ldots H^q_{\mathscr{D}}(X, A(j)) \to H^q(X, A(j)) \oplus F^j H^q(X, \mathbb{C}) \to H^q(X, \mathbb{C}) \to H^{q+1}_{\mathscr{D}}(X, A(j)) \ldots$$

and an analogous exact sequence for cohomology groups with support in a closed subset.

Let us point out that $A(0)_{\mathscr{D}}$ is quasi-isomorphic to A, and that $A(1)_{\mathscr{D}}$ is isomorphic to $\mathcal{O}_X/A(1)[-1]$, hence $\mathbb{Z}(1)_{\mathscr{D}}$ is quasi-isomorphic to $\mathcal{O}_X^*[-1]$.

Proposition 5.2 (Deligne, see [E-Z: §2]). *For Z an algebraic subvariety of X of pure codimension p, there is a canonical cohomology class $c_{\mathscr{D}}(Z)$ in $H^{2p}_{Z,\mathscr{D}}(X, \mathbb{Z}(p))$ which maps to the "Betti" class of Z in $H^{2p}_Z(X, \mathbb{Z}(p))$ and also to the "Hodge" class of Z in $H^{2p}_Z(X, F^p)$.*

Proof. In the above exact sequence the group $H^{2p-1}_Z(X, \mathbb{C})$ is zero for dimension reasons. Hence $H^{2p}_Z(X, \mathbb{Z}(p)_{\mathscr{D}})$ is equal to the fiber product of $H^{2p}_Z(X, \mathbb{Z}(p))$ and $H^{2p}_Z(X, F^p)$ over $H^{2p}_Z(X, \mathbb{C})$. Thus the existence of a class $c_{\mathscr{D}}(Z)$ satisfying the requirements follow from the fact that the Betti class and the Hodge class have the same image in $H^{2p}_Z(X, \mathbb{C})$. □

Remark 5.3. The construction of the Hodge class of Z (in Hodge cohomology with support in Z) may be found in [E 1].

One extends this linearly to a cohomology class for algebraic cycles of codimension p. One obtains a canonical class in $H^{2p}_{\mathscr{D}}(X, \mathbb{Z}(p))$ simply by enlarging the support. In case γ is homologically equivalent to 0, this is related to the class of γ in the Griffiths intermediate Jacobian $J^p = H^{2p-1}(X, \mathbb{C})/H^{2p-1}(X, \mathbb{Z}) + F^p H^{2p-1}(X, \mathbb{C})$ as follows. There is a triangle, in the derived category:

$$\mathbb{Z}(p)[-1] \to \Omega^\bullet/F^p[-1] \to \mathbb{Z}(p)_{\mathscr{D}} \to \mathbb{Z}(p)\ . \tag{5-5}$$

This induces a long exact sequence:

$$\begin{aligned} \ldots H^{q-1}(X, \mathbb{Z}(p)) \to H^{q-1}(X, \mathbb{C})/F^p H^{q-1}(X, \mathbb{C}) &\to H^q_{\mathscr{D}}(X, \mathbb{Z}(p)) \\ &\to H^q(X, \mathbb{Z}(p)) \ldots \end{aligned} \tag{5-6}$$

Therefore the kernel of the map $H^q_{\mathscr{D}}(X, \mathbb{Z}(p)) \to H^q(X, \mathbb{Z}(p))$ is isomorphic to J^p.

Proposition 5.4 (Deligne, see [Bl 1, E-Z and E-V]). *If the codimension p cycle γ is homologically equivalent to 0, the cohomology class in J^p obtained from Deligne cohomology coincides with the class constructed by Griffiths in* [Gr 2] *via the Abel-Jacobi mapping.*

Let us mention an example of application of Deligne cohomology to algebraic cycles, due to El Zein and Zucker [E-Z]. Let $f : X \to \Delta$ be a projective morphism, with X smooth, and f smooth over the punctured disc Δ^*. Let $X^* = f^{-1}(\Delta^*)$. Let Z be a codimension k algebraic cycle on X, such that its class in $H^{2k}(X^*, \mathbb{Z})$ is zero. Then the Griffiths cycle class at each point of Δ^* may be viewed as a section of the sheaf $\mathcal{J}^k_{X^*/\Delta^*} = R^{2k-1}f_*\sigma_k(\Omega^\bullet_{X^*/\Delta^*})/\mathrm{im}\ R^{2k-1}f_*\mathbb{Z}$. of k-th intermediate jacobians, where for $K^\bullet$ a complex $\sigma_k K^\bullet$ denotes the quotient of $K^\bullet$ by its k-th "troncation bête" (see § 1). This sheaf $\mathcal{J}^k_{X^*/\Delta^*}$ admits a natural extension to Δ, namely

$$R^{2k-1}f_*\sigma_k(\Omega^\bullet_{X/\Delta}(\log D))/im\ \ R^{2k-1}f_*\mathbb{Z}\ .$$

Working instead with relative Deligne cohomology, El Zein and Zucker were able to prove that the section of this sheaf over Δ^* actually extends over Δ. This result says that "normal functions" can be extended to the origin.

We now come to the remarkable construction of a product $A(j)_{\mathcal{D}} \otimes A(k)_{\mathcal{D}} \to A(j+k)_{\mathcal{D}}$. The construction below is due to Beilinson [Be 1]. He defines the product in such a way that the natural maps of complexes $A(j)_{\mathcal{D}} \to A(j)$ and $A(j)_{\mathcal{D}} \to F^j$ are compatible with the products (note that the cup-product maps $F^j \otimes F^k$ to F^{j+k}). There is actually a product map $\cup_\alpha : A(j)_{\mathcal{D}} \otimes A(k)_{\mathcal{D}} \to A(j+k)_{\mathcal{D}}$ for each real number α, defined as follows:

$$\begin{aligned}
&f_j \cup_\alpha f_k = f_j \wedge f_k\ , \quad \text{for } f_j \in F^j, f_k \in F^k\\
&a_j \cup_\alpha a_k = a_j a_k \in A(j+k)\ , \quad \text{for } a_j \in A(j), a_k \in F^k\\
&a_j \cup_\alpha f_k = f_j \cup_\alpha a_k = \omega_j \cup \omega_k = 0\\
&f_j \cup_\alpha \omega_k = (-1)^{\deg f_j}\alpha f_j \wedge \omega_k, \quad \omega_j \cup_\alpha f_k = (1-\alpha)\omega_j \wedge f_k\\
&a_j \cup_\alpha \omega_k = (1-\alpha) a_j \omega_k, \quad \omega_j \cup_\alpha a_k = \alpha a_k \omega_j
\end{aligned}$$

for $f_j \in F^j, f_k \in F^k, a_j \in A(j), a_k \in F^k, \omega_j$ and $\omega_k \in \Omega^\bullet$.

Proposition 5.5 [Be 1: § 1]. *(1) Each $\cup_\alpha$ is a morphism of complexes;*
(2) The homotopy class of $\cup_\alpha$ is independent of α, and is denoted by $\cup$;
(3) The multiplication $\cup$ is associative and commutative up to homotopy;
(4) The class $(a, f) = (1, 1) \in A(0)_{\mathcal{D}}$ is a two-sided identity for $\cup$.

Remark 5.6. There is another construction for the product, using the realization 5-2 of $A(j)_{\mathcal{D}}$. This is the original construction of Deligne [Bel, Bl 1].

Bloch [Bl 3] used Deligne cohomology to relate the Chern classes of a vector bundle on a complex projective manifold associated with a representation of the fundamental group with the Cheeger-Simons classes. To state his results, denote by $CH^k(X)_{hom \simeq 0}$ the subgroup of the Chow group of algebraic cycles of codimension k, consisting of cycles homologically equivalent to 0. Let $\Phi^k : CH^k(X)_{hom\simeq 0} \to J^k$ be the cycle map of Griffiths. Note that J^k can be naturally identified with $H^{2k-1}(X,\mathbb{R})/H^{2k-1}(X,\mathbb{Z})$.

Theorem 5.7 (Bloch [Bl 3: Theorem 3.1]). *Let X be a complex projective manifold, let $\rho : \pi_1(X) \to U(n)$ be a unitary representation, and let $\mathscr{E}$ be the corresponding algebraic vector bundle over X. For $k \geq 0$, let $c_k(\mathscr{E}) \in CH^k(X)$ be the Chern class of $\mathscr{E}$ in the Chow group of codimension k. Let N be an integer such that the cycle $N \cdot c_k(\mathscr{E})$ is homologically trivial. Then $\Phi^k(N \cdot c_k(\mathscr{E})) = N \cdot \widehat{c}_k(\rho)$, where $\widehat{c}_k(\rho) \in H^{2k-1}(X,\mathbb{R})/H^{2k-1}(X,\mathbb{Z})$ is the (topological) Chern class of Cheeger-Simons* [C-S] *for local systems.*

For some generalizations of Theorem 5.7, we refer to [So 2] and [Z 8].

Bloch obtains remarkable further results in the case $k = 2$. He uses the (multi-valued) *dilogarithm function* $L_2(x) := \int_0^x \log(1-t)\frac{dt}{t}$ to construct a homomorphism $\varepsilon : \mathscr{A}(\mathbb{C}) \to \mathbb{C} \otimes \mathbb{C}^*$, where for any field F, $\mathscr{A}(F)$ is the kernel of the product map $F^* \times F^* \to K_2(F)$. The map ε is defined by $\varepsilon(a) := [\log(1-a) \otimes a] + [2\pi i \otimes \exp(L_2(a))]$. It is in fact independent of the branches chosen for log and L_2. There is a commutative diagram

$$\begin{array}{ccccccccc}
 & & & & & & 0 & & \\
 & & & & & & \downarrow & & \\
 & & & & & & \mathscr{B}(\mathbb{C}) & & \\
 & & & & & & \downarrow & & \\
 & & & & \mathscr{A}(\mathbb{C}) & \xrightarrow{=} & \mathscr{A}(\mathbb{C}) & & \\
 & & & & {\scriptstyle\varepsilon}\downarrow & & \downarrow & & \\
0 & \longrightarrow & \mathbb{C}^* & \xrightarrow{2\pi i \otimes Id} & \mathbb{C} \otimes \mathbb{C}^* & \longrightarrow & \mathbb{C}^* \otimes \mathbb{C}^* & \longrightarrow & 0 \\
 & & & & & & \downarrow & & \\
 & & & & & & K_2(\mathbb{C}) & & \\
 & & & & & & \downarrow & & \\
 & & & & & & 0 & &
\end{array}$$

From this diagram, one deduces a homomorphism $\Phi : \mathscr{B}(\mathbb{C}) \to \mathbb{C}^*$.

Theorem 5.8 (Bloch, [Bl 3]). *(1) Let B^* be the image of Φ. Then $B^* = \Phi(\mathscr{B}(\overline{\mathbb{Q}}))$, hence is countable.*

(2) Let $\mathscr{E}$ be the algebraic vector bundle associated with a local system on X, and let N be a positive integer such that the algebraic cycle $N \cdot c_2(\mathscr{E})$ is homologically trivial. then $\Phi^2(N \cdot c_2(\mathscr{E}))$ belongs to the image of the composed map $H^3(X,B^*) \to H^3(X,\mathbb{C}^*) \to J^2(X)$.

Bloch first found a regulator map $K_2(X) \to H^1(X,\mathbb{C}^*)$, for X a curve [Bl 3, Bl 4]. He found beautiful applications of it, in particular to zeta-functions of elliptic curves. Deligne found an interpretation of this regulator in terms of line bundles with connections. This was shown by Beilinson to be a special case of his very general regulator. The following theorem also points the way to deep relations between Deligne cohomology and iterated integrals in the sense of Chen.

Theorem 5.9 (Beilinson, Bloch, Deligne) [Be 1: § 1-2]. *Let X be a smooth complex-analytic manifold.*

(1) $H^1_{\mathscr{D}}(X,\mathbb{Z}(1))$ is isomorphic to $\mathscr{O}^*(X)$;

(2) $H^2_{\mathscr{D}}(X,\mathbb{Z}(2))$ is isomorphic to $H^1(X,\mathscr{O}^*_X \xrightarrow{d\log} \Omega^1_X)(1)$ hence, if $i=\sqrt{-1}$ is fixed, to $H^1(X,\mathscr{O}^*_X \xrightarrow{d\log} \Omega^1_X)$;

(3) $H^1(X,\mathscr{O}^*_X \xrightarrow{d\log} \Omega^1_X)$ is isomorphic to the group of isomorphism classes of invertible sheaves on X equipped with a connection;

(4) The pairing $H^1_{\mathscr{D}}(X,\mathbb{Z}(1)) \times H^1_{\mathscr{D}}(X,\mathbb{Z}(1)) \to H^2_{\mathscr{D}}(X,\mathbb{Z}(2))$ may therefore be viewed as associating to invertible holomorphic functions f and g on X a line bundle with connection $(\mathscr{L},\nabla)$;

(5) The curvature of the line bundle in (4) is equal to $d\log(f)\wedge d\log(g)$;

(6) The monodromy of ∇ around a loop γ in X is equal to

$$\exp\left(\frac{1}{2\pi i}\left(\int_\gamma \log(f)d\log(g) - \log g(p)\int_\gamma d\log(f)\right)\right),$$

for p a point of γ, the integration being taken over γ starting at p;

(7) In case X is an open set in a compact Riemann surface $\overline{X}$, and γ is a small loop encircling a point q of $\overline{X}-X$, the monodromy of ∇ around γ is equal to the Tate symbol $(f,g)_q$.

Remark 5.10. (1) The isomorphism from $H^1(\mathscr{O}^*_X \xrightarrow{d\log} \Omega^1_X)$ to the group of isomorphisms classes of line bundle with connection may be made explicit as follows (see [Bl 4]): let $(U_\alpha)_\alpha$ be an open cover of X, and let $h_{\alpha,\beta}\in \mathscr{O}^*(U_\alpha\cap U_\beta), \omega_\alpha\in\Omega^1(U_\alpha)$ be a 1-cocycle of this cover with coefficients in $\mathscr{O}^*_X \xrightarrow{d\log} \Omega^1_X$, i.e. assume that $d\log(h_{\alpha,\beta})=\omega_\alpha-\omega_\beta$ holds on $U_\alpha\cap U_\beta$. Then the line bundle $\mathscr{L}$ is defined as follows: its sections on an open set V consist of families of holomorphic functions f_α on U_α such that $f_\alpha = h_{\alpha,\beta}f_\beta$ on $U_\alpha\cap U_\beta$. The connection ∇ is defined as follows. If σ, a local section of $\mathscr{L}$, is given by the family f_α as above, $\nabla(\sigma)$ is given by the family of 1-forms $df_\alpha+f_\alpha\omega_\alpha$.

(2) It is easy to see that modulo $2\pi i\mathbb{Z}$, the expression $\int_\gamma \log(f)d\log(g) - \log g(p)\int_\gamma d\log(f)$ is independent of the choice of $p\in\gamma$.

(3) The fact that the line bundle with connection associated with $(f,1-f)$ is trivial is closely related to the existence of the dilogarithm function.

Corollary 5.11 [Bl 4]. *Let X be a smooth algebraic curve over $\mathbb{C}$. There is a natural mapping from $K_2(X)$ to $H^2_{\mathscr{D}}(X,\mathbb{Z}(2))$, which maps the symbol $\{f,g\}$ to $f\cup g$.*

We refer to [Ram 1, Ram 2] for Ramakrishnan's interpretation of this regulator in terms of the Heisenberg group.

Deligne cohomology, as defined above for smooth complex-analytic varieties is reasonable in the case of proper algebraic varieties, but not for affine varieties; for instance, $H^1(X, \mathbb{Z}(1)_{\mathcal{D}}) = \mathcal{O}^*(X)$ is a huge group in that case. Beilinson therefore modified the construction of Deligne cohomology, for smooth algebraic varieties, by introducing "growth conditions" – the same ones that enter in Hodge-Deligne theory.

Definition 5.12 [Be 1: E-V]. *Let X be a smooth quasi-projective complex algebraic variety, and let $X \xrightarrow{j} \overline{X}$ be a good compactification, with $D := \overline{X} - X$ as usual. Let $\Omega^\bullet_{\overline{X}}(\log D)$ be the logarithmic de Rham complex of Deligne. Let F^p denote the "troncation bête" of $\Omega^\bullet_{\overline{X}}(\log D)$ (see Sect. 1).*

There are natural morphisms $Rj_*A(p) \to Rj_*\Omega^\bullet_X$ and $F^p \to Rj_*\Omega^\bullet_X$ of complexes of sheaves over $\overline{X}$. Define a complex of sheaves $A(p)_{\mathcal{D}}$ over $\overline{X}$ as

$$A(p)_{\mathcal{D}} = Cone(Rj_*A(p) \oplus F^p \to Rj_*\Omega^\bullet_X)[-1].$$

The Deligne-Beilinson cohomology groups $H^q_{\mathcal{D}}(X, A(p))$ *are the hypercohomology groups $H^q_{\mathcal{D}}(X, A(p)) = H^q(\overline{X}, A(p)_{\mathcal{D}})$. They are independent of the choice of $\overline{X}$, and they are functorial with respect to algebraic mappings.*

The exact sequence (5-4) is easily generalized to the Deligne-Beilinson cohomology groups. To illustrate the effect of the growth conditions, one may observe:

Proposition 5.13 [E-V: Proposition 2.12]. *Let X be a smooth complex algebraic variety.*

(1) $H^1_{\mathcal{D}}(X, A(1)) = \{f \in H^0(\overline{X}, j_*(\mathcal{O}_X/A(1))) : df \in H^0(\overline{X}, \Omega^1_{\overline{X}}(\log D))\}$;

(2) $H^1_{\mathcal{D}}(X, \mathbb{Z}(1))$ is canonically isomorphic to the group of invertible regular functions on X.

The following results are easy to prove.

Proposition 5.14. *(1) If an algebraic morphism between smooth algebraic varieties induces an isomorphism on cohomology with coefficients in A, it also induces an isomorphism on the groups $H^\bullet_{\mathcal{D}}(-, A(*))$.*

(2) Deligne-Beilinson cohomology is homotopy-invariant, in the sense that $H^k_{\mathcal{D}}(X, A(p))$ is isomorphic to $H^k_{\mathcal{D}}(X \times \mathbf{A}^1, A(p))$.

Before discussing Beilinson's construction of Chern characters with values in Deligne-Beilinson cohomology, one needs to further generalize this cohomology to *smooth simplicial schemes* $X_\bullet = (X_n)_{n \in \mathbb{N}}$. For this purpose, one uses a "compactification" $X_\bullet \hookrightarrow \overline{X}_\bullet$, where each $X_n \hookrightarrow \overline{X}_n$ is a good compactification. One may realize the Beilinson complex $A(p)_{\mathcal{D}}$ on $\overline{X}_n$ using compatible injective resolutions of the sheaf $A(p)$ and the complex of sheaves $\Omega^\bullet$ on X_n. One can arrange these complexes to be functorial with respect to face maps of $\overline{X}_\bullet$. Then the Deligne-Beilinson complexes on the $\overline{X}_n$ organize into a simplicial complex of sheaves over the simplicial scheme $\overline{X}$. Taking injective resolutions and global sections, one then obtains a simplicial object in the category of complexes of abelian groups, hence a double complex. Taking cohomology, one obtains the Deligne-Beilinson cohomology $H^\bullet_{\mathscr{D}}(X_\bullet, A(*))$ (they are independent of the choice of the compactification). See [J] for details.

The exact sequence (5-6) generalizes to this situation (this uses mixed Hodge theory for simplicial schemes (Theorem 1.14)). To define Chern classes, one needs to consider the simplicial scheme $B\mathrm{GL}(N, \mathbb{C})$, which is the classifying space of the algebraic group $\mathrm{GL}(N, \mathbb{C})$. This simplicial scheme comes equipped with a principal bundle $E\mathrm{GL}(N, \mathbb{C}) \to B\mathrm{GL}(N, \mathbb{C})$ with structure group $\mathrm{GL}(N, \mathbb{C})$, and total space contractible, hence there are Chern classes $c_p \in H^{2p}(B\mathrm{GL}(N, \mathbb{C}), A(p))$. It is proven in [D 4: III] that the cohomology of $B\mathrm{GL}(N, \mathbb{C})$ is concentrated in even degrees, and that $H^{2p}(B\mathrm{GL}(N, \mathbb{C})) = F^p H^{2p}(B\mathrm{GL}(N, \mathbb{C}))$. It follows that the map from $H^{2p}_{\mathscr{D}}(B\mathrm{GL}(N, \mathbb{C}), A(p))$ to $H^{2p}(B\mathrm{GL}(N, \mathbb{C}), A(p))$ is an isomorphism; hence the Chern class c_p in $H^{2p}(\mathrm{GL}(N, \mathbb{C}), A(p))$ has a unique lifting to $H^{2p}_{\mathscr{D}}(B\mathrm{GL}(N, \mathbb{C}), A(p))$, which we will also denote c_p.

Since $B\mathrm{GL}(n, \mathbb{C})$ classifies principal $\mathrm{GL}(n, \mathbb{C})$-bundles, this gives rise to Chern classes $c_p(E) \in H^{2p}_{\mathscr{D}}(X, A(p))$ for an algebraic vector bundle E over a complex algebraic manifold X. From this, Beilinson deduces a Chern character $ch : K_0(X) \to \oplus_{p \geq 0} H^{2p}_{\mathscr{D}}(X, A(p))$, which is an algebra homomorphism, defined from the c_p by the standard formulae (see Definition 5.15 below). Then he goes on to define a Chern character for higher K-groups. We follow here the presentation of Soulé [So 1], which is entirely analogous to his earlier treatment of the Chern character for l-adic cohomology.

First consider the affine case $X = \mathrm{Spec}\,(R)$, where R is a smooth $\mathbb{C}$-algebra. There is a natural morphism of simplicial schemes $e : X \times B\mathrm{GL}(N, R) \to B\mathrm{GL}(N, \mathbb{C})$; in degree n, e_n is the obvious evaluation map from $X \times \mathrm{GL}(N, R)^n$ to $\mathrm{GL}(N, \mathbb{C})^n$. It should be emphasized that $\mathrm{GL}(N, R)$ is just a discrete scheme, hence $X \times \mathrm{GL}(N, R)^n$ is a disjoint union of varieties, all isomorphic to X. Hence the Deligne-Beilinson cohomology of $X \times B\mathrm{GL}(N, R)$ admits the Künneth decomposition:

$$H^k_{\mathscr{D}}(X \times B\mathrm{GL}(N, R), \mathbb{Z}(p)) = \oplus_{i+j=k} H^i_{\mathscr{D}}(X, \mathbb{Z}(p)) \otimes H^j(B\mathrm{GL}(n, R), \mathbb{Z}).$$

Hence, whenever $m + k = 2p$, the class $e^*(c_p) \in H^{2p}_{\mathscr{D}}(X \times B\mathrm{GL}(N, R), \mathbb{Z}(p))$ induces a homomorphism $H_m(\mathrm{GL}(N, R), \mathbb{Z}) \to H^k_{\mathscr{D}}(X, \mathbb{Z}(p))$. For $m \geq 1$, one denotes by $c_{p,k} : K_m(X) \to H^k_{\mathscr{D}}(X, \mathbb{Z}(p))$ the composition of this homomorphism with the map $K_m(X) \to H_m(\mathrm{GL}(N, R), \mathbb{Z})$ which is given by the Hurewicz map for $B\mathrm{GL}(N, R)^+$ (where "+" denotes Quillen's plus-construction).

To extend the definition of these Chern classes to non-affine smooth varieties, one approach, followed by Beilinson in [Be 1], consists of extending the above construction to the case of smooth affine simplicial schemes over $\mathbb{C}$, and then observing that for any smooth algebraic variety X over $\mathbb{C}$, there exists a smooth affine simplicial scheme $Y_\bullet$ together with an augmentation $Y_0 \to X$, such that the inverse image map gives an isomorphism $H^j(X,\mathbb{Z}) \to H^j(Y_\bullet,\mathbb{Z})$ (one may simply take for $Y_\bullet$ the nerve of an affine open cover of X). Then the inverse image map also gives an isomorphism in Deligne-Beilinson cohomology. One may use instead the construction of a Zariski fiber space $Y \to X$, with fiber an affine space $\mathbf{A}^n$ and total space affine, due to Jouanolou and Karoubi (this approach is detailed in [Schn]).

Definition 5.15 [Be 1]. *Let A be a $\mathbb{Q}$-subalgebra of $\mathbb{R}$. The Chern character $ch : K_m(X) \to \oplus_{p>0} H_{\mathscr{D}}^{2p-m}(X, A(p))$ is defined as follows: $ch = \sum_{p\geq 0} ch_{p,2p-m}$, where:*

(1) For $m \geq 1$, $ch_{p,k} = \frac{(-1)^{p-1}}{(k-1)!} c_{p,k}$.

(2) For $m = 0$ and $p > 0$, $ch_{p,2p}(x) = \frac{N_p((c_{p,2p}(x))}{p!}$ for $x \in K_0(X)$, where N_p is the p-th Newton polynomial.

(3) $ch_{0,0}(x) \in H^0(X,\mathbb{Z})$ is the locally constant function rank (x) *on X.*

Recall [Be 1, So 1] that the groups $K_m(X)$, for X a scheme, admit Adams operations ψ^r, and that if $K_m(X)^{(p)}$ is the subgroup of $K_m(X) \otimes \mathbb{Q}$ where all ψ^r operate by multiplication by r^p, one has the "weight decomposition": $K_m(X) \otimes \mathbb{Q} = \oplus_{p\geq 0} K_m(X)^{(p)}$.

Proposition 5.16 [Be 1]. *Let A be a $\mathbb{Q}$-subalgebra of $\mathbb{R}$. The Chern character of Definition 5.14 has the following properties:*

(1) For given p and k, with $m = 2p-k$, $ch_{p,k}$ vanishes on $K_m(X)^{(j)}$ for $j \neq p$.

(2) $ch(y \cdot z) = ch(y) \cup ch(z)$, for $y \in K_i(X)$, $z \in K_j(X)$, $y \cdot z \in K_{i+j}(X)$.

Beilinson then introduces the *motivic cohomology* $H^k_{\mathscr{M}}(X,\mathbb{Q}(p)) := K_m(X)^{(p)}$, for $m = 2p - k$. Therefore $ch_{p,k}$ may be viewed as a map from $H^k_{\mathscr{M}}(X,\mathbb{Q}(p))$ to $H^k_{\mathscr{D}}(X, A(p))$.

Following [Be 1], we illustrate Beilinson's Chern classes in a few cases.

Examples 5.17 [Be 1, So 1, R-S-S]. *(1) $c_{1,1} : K_1(X) \to H^1_{\mathscr{D}}(X,\mathbb{Z}(1)) = \mathcal{O}(X)^*$ is the classical determinant map.*

(2) Recall that $K_1(X)^{(p)}$ is the $E^{p-1,-p}$-term of the Gersten spectral sequence tensored with $\mathbb{Q}$. Hence $K_1(X)^{(p)}$ is equal to the quotient of the group of finite families (f_α) of non-zero meromorphic functions on irreducible subvarieties Y_α of X of codimension $p-1$, such that the codimension p cycle $\sum_\alpha div(f_\alpha)$ on X is zero, by the subgroup generated by the image of $K_2(\mathbb{C}(Z))$, for Z an arbitrary subvariety of X of codimension $p-2$.

For $A = \mathbb{R}$, the component $ch_{p,2p-1}$ of the Chern character for $K_1(X)$ is a map from $E^{p-1,-p}$ to $H^{2p-1}_{\mathscr{D}}(X,\mathbb{R}(p)) = H^{2p-2}(X,\mathbb{R}(p)) \cap H^{p-1,p-1}(X)$. It associates to the class of the family (f_α) as above, the linear form on differential forms ω

of pure type $(p-1,p-1)$, *with* $(2\pi i)^p\omega$ *real, defined by:*

$$\omega \mapsto \sum_\alpha \int_{Y_\alpha^0} \log|f_\alpha| \wedge \omega,$$

where Y_α^0 *is the smooth locus of* Y_α.

(3) The Chern character $ch_{2,2} : K_2(X) \to H^2_{\mathcal{D}}(X,\mathbb{Z}(2)) = H^1(X,\mathbb{C}^*)(1)$ *is the map described in Corollary 5.11.*

(4) Let $c_p \in H^{2p-1}_{\text{cont}}(\mathrm{GL}(N,\mathbb{C}),\mathbb{R})$ *be the canonical indecomposable element constructed by Borel* [Bo 1]. *The composition of the Hurewicz map with the map* $H_{2p-1}(\mathrm{GL}(N,\mathbb{C})) \to \mathbb{R}$ *induced by* c_p *coincides with the map* $c_{p,1} : K_{2p-1}(\mathbb{C}) \to \mathbb{R}$. [Be 1: § 2, Appendix].

To state Beilinson's conjecture on the relation between motivic cohomology and Deligne-Beilinson cohomology for a projective smooth variety $X_{/\mathbb{Q}}$ over $\mathbb{Q}$, we first need to describe Deligne-Beilinson cohomology groups $H^k_{\mathcal{D}}(X_{/\mathbb{R}},\mathbb{R}(p))$ for a smooth variety $X_{/\mathbb{R}}$ over $\mathbb{R}$. The complex cohomology $H^k(X_{/\mathbb{C}},\mathbb{C})$ of the associated complex manifold $X_\mathbb{C}$ is equal to the algebraic de Rham cohomology [Gro]. This algebraic de Rham cohomology is the complexification of the real vector space $H^k_{DR}(X_{/\mathbb{R}})$, hence it admits a real structure. This real structure is induced by an involution $\imath$ of the ringed space $(X(\mathbb{C}),\Omega^\bullet_X)$, such that $\imath$ is complex conjugation on $X(\mathbb{C})$ and $\imath(\omega)(z) = \overline{\omega}(\overline{z})$, for $\omega \in \Omega^\bullet_X$ and $z \in X(\mathbb{C})$.

One observes that $\imath$ induces complex conjugation on the subsheaf $\mathbb{C}$ and on its subsheaf $\mathbb{R}(p)$. Also $\imath$ preserves the subsheaf F^p. Hence $\imath$ induces an involution of the pair $(X,\mathbb{R}(p)_{\mathcal{D}})$.

Definition 5.18. *For* $X_{/\mathbb{R}}$ *a smooth projective variety over* $\mathbb{R}$, *let* $H^k_{\mathcal{D}}(X_{/\mathbb{R}},\mathbb{R}(p))$ *denote the invariants of* $H^k_{\mathcal{D}}(X_{/\mathbb{C}},\mathbb{R}(p))$ *under the involution* $\imath$.

Proposition and Definition 5.19. *Let* $X_{/\mathbb{R}}$ *be a smooth projective variety over* $\mathbb{R}$. *The Chern character* $H^\bullet_{\mathcal{M}}(X_{/\mathbb{R}},\mathbb{Q}(*)) \to H^\bullet_{\mathcal{D}}(X_{/\mathbb{C}},\mathbb{R}(*))$ *factors through a map from* $H^\bullet_{\mathcal{M}}(X_{/\mathbb{R}},\mathbb{Q}(*))$ *to* $H^\bullet_{\mathcal{D}}(X_{/\mathbb{R}},\mathbb{R}(*))$, *called the regulator, and denoted* reg.

We can now state the following conjecture of Beilinson.

Conjecture 5.20 [Be 1] (see also [So 1, Schn]). *Let* $X_{/\mathbb{Q}}$ *be a smooth projective variety over* $\mathbb{Q}$. *Let* $0 \le i \le 2\dim(X)$, *let* $p \ge 0$ *and let* $m = i+1-p$.

(1) If $m < \frac{i}{2}$, *the regulator map induces an isomorphism*

$$H^{i+1}_{\mathcal{M}}(X,\mathbb{Q}(p)) \otimes \mathbb{R} \xrightarrow{\sim} H^{i+1}_{\mathcal{D}}(X_{/\mathbb{R}},\mathbb{R}(p)).$$

(2) If $m = \frac{i}{2}$ *and* $p = \frac{i}{2}+1$, *let* $N^m(X)$ *be the group of m-codimensional cycles on X (modulo homological equivalence), and let*

$$\overline{z} : N^m(X) \to H^{2m+1}_{\mathcal{D}}(X_{/\mathbb{R}},\mathbb{R}(m+1))$$

be the composition of the cycle map

$$z : N^m(X) \to H^{2m}(X(\mathbb{C}), \mathbb{R}(m))$$

with the connecting homomorphism

$$H^{2m}(X(\mathbb{C}), \mathbb{R}(m)) \to H^{2m+1}_{\mathscr{D}}(X_{/\mathbb{R}}, \mathbb{R}(m+1))$$

(one easily checks that $\bar{z}$ *indeed takes values in* $H^{2m+1}_{\mathscr{D}}(X_{/\mathbb{R}}, \mathbb{R}(m+1))$*). Then the maps* reg *and* $\bar{z}$ *together induce an isomorphism:*

$$(H^{i+1}_{\mathscr{M}}(X, \mathbb{Q}(n)) \otimes \mathbb{R}) \oplus (N^m(X) \otimes \mathbb{R}) \xrightarrow{\sim} H^{i+1}_{\mathscr{D}}(X_{/\mathbb{R}}, \mathbb{R}(n)).$$

As a consequence of these conjectures, the Deligne-Beilinson cohomology groups in question would inherit a natural $\mathbb{Q}$-structure. This $\mathbb{Q}$-structure plays an essential role in the statement of Beilinson's conjectures on special values and, more generally, leading coefficients of L-series associated to cohomology of projective varieties over $\mathbb{Q}$. We will not go into this topic here, since it would take us far astray from our theme, and there are already several excellent expositions [B, So 1, R-S-S]. The evidence for Conjecture 5.20 is mostly due to Beilinson [Be 1: § 2], and concerns elliptic curves with complex multiplication, modular curves and products of two modular curves. More recently, Ramakrishnan treated the case of Hilbert-Blumenthal surfaces [Ram 4].

We will present another conjecture of Beilinson, the "Hodge-$\mathscr{D}$" conjecture.

Conjecture 5.21. *Let X be a smooth projective variety over $\mathbb{C}$. Let $p \geq 0$. Then the vector space $H^{2p}(X, \mathbb{R}(p))$ is generated by the classes of cycles of codimension p and by the classes of the currents $\sum_\alpha \log|f_\alpha|$ considered in Example* 5.17(2).

We now turn to the notion of "absolute Hodge cohomology" due to Beilinson [Be 2] and to its relation with Deligne-Beilinson cohomology. Another reference on this topic is [J: § 2]. As motivation, we first recall the description of extensions of mixed Hodge structures.

Proposition 5.22 [Mo: Proposition 8.1]. *Let A be a subfield of $\mathbb{R}$, and let E and F be mixed Hodge structures over A. The group* $\mathrm{Ext}^1_{A\text{-}MH}(E, F)$ *of equivalence classes of extensions* $0 \to F \to H \to E \to 0$ *of A-mixed Hodge structures is canonically isomorphic to*

$$\mathrm{Hom}^W(E \otimes_A \mathbb{C}, F \otimes_A \mathbb{C}) / \mathrm{Hom}^W_F(E \otimes_A \mathbb{C}, F \otimes_A \mathbb{C}) + \mathrm{Hom}^W(E, F)\ ,$$

hence to

$$W^0(\mathrm{Hom}(E, F) \otimes_A \mathbb{C}) / (W^0 \cap F^0)(\mathrm{Hom}(E, F) \otimes_A \mathbb{C}) + W^0\mathrm{Hom}(E, F).$$

As a corollary, one obtains: $\mathrm{Ext}^j_{A\text{-}MH}(E,F) = 0$ for $j \geq 2$. The relevance of Proposition 5.22 for Deligne-Beilinson cohomology is that it allows one to rewrite the exact sequence (5-5) as:

$$\ldots \mathrm{Ext}^1_{A\text{-}MH}(A, H^{q-1}(X, A(p))) \to H^q_{\mathscr{D}}(X, A(p)) \to \mathrm{Ext}^0_{A\text{-}MH}(A, H^q(X, A(p))) \ldots$$

This suggests that $H^q_{\mathscr{D}}(X, A(p))$ should be interpreted as $\mathrm{Ext}^q(A, R\Gamma(X, A(p))$ in a suitable derived category of mixed Hodge structures. This was done by Beilinson in [Be 2]. To explain his result, we recall the notion of *mixed A-Hodge complex* due to Deligne (Definition 1.8). Then:

Definition 5.23. *Let A be a subring of* $\mathbb{R}$, *and let the A-mixed Hodge complex* $K^\bullet$ *be given by the diagram (in the derived category):*

$$K^\bullet_A \xrightarrow{\alpha} (K^\bullet_{A\otimes Q}, W) \xrightarrow{\beta} (K^\bullet_{\mathbb{C}}, W, F).$$

The absolute Hodge cohomology $H^l_{\mathscr{H}}K^\bullet$ *is the l-th cohomology group of the complex*

$$R\Gamma_{\mathscr{H}}(K^\bullet) := \mathrm{Cone}\left(K^\bullet_A \oplus \widehat{W}_0 K^\bullet_{A\otimes\mathbb{Q}} \oplus (\widehat{W}_0 \cap F^0) K^\bullet_{\mathbb{C}} \xrightarrow{(\alpha,\beta)} {'K}^\bullet_{A\otimes\mathbb{Q}} \oplus \widehat{W}'_0 K^\bullet_{\mathbb{C}} \right)[-1],$$

where the complexes ${'K}^\bullet_{A\otimes\mathbb{Q}}$ *and* $({'K}^\bullet_{\mathbb{C}}, W)$ *are those introduced to express the morphisms* α *and* β *(in the derived category) in terms of actual morphisms of complexes; more precisely* α *is the composition of* $\alpha_1 : K^\bullet_A \to {'K}^\bullet_{A\otimes\mathbb{Q}}$ *and of the "inverse" of* $\alpha_2 : (K^\bullet_{A\otimes\mathbb{Q}}, W) \to {'K}^\bullet_{A\otimes\mathbb{Q}}$, *and* β *is the composition of* β_1 : $(K^\bullet_{A\otimes\mathbb{Q}}, W) \to ({'K}^\bullet_{\mathbb{C}}, W)$ *and of the "inverse" of* $\beta_2 : (K^\bullet_{\mathbb{C}}, W, F) \to ({'K}^\bullet_{\mathbb{C}}, W)$.

The maps α *and* β *in the construction of the complex* $R\Gamma_{\mathscr{H}}(K^\bullet)$ *are* $\alpha = (\alpha_1, -\alpha_2, 0)$ *and* $\beta = (0, \beta_1, -\beta_2)$. *The notation* $\widehat{W}$ *refers to the "filtration décalée" DecW of Deligne, which we mentioned in Sect. 1.*

According to Deligne (see Theorem 1.9), there is a mixed A-Hodge complex associated to a smooth complex algebraic variety (and more generally to a smooth simplicial algebraic variety). This mixed Hodge complex is denoted $R\Gamma(X, A)$. To compute *Ext* groups in the derived category of A-mixed Hodge complexes, one needs the following result.

Proposition 5.24 (Beilinson, [Be 2: § 3.11]). *The natural functor from the derived category of A-mixed polarizable Hodge structures to the subcategory of the derived category of A-mixed polarizable Hodge complexes, consisting of bounded complexes, is an equivalence of triangulated categories.*

Corollary 5.25. *There is a natural hyperext spectral sequence, for* $K^\bullet$ *an A-mixed Hodge complex:*

$$E_2^{p,q} = \mathrm{Ext}^p_{A\text{-}MH}(A, H^q(K^\bullet_A)) \Rightarrow H^{p+q}_{\mathscr{H}}(K^\bullet).$$

Corollary 5.26. *For $X_\bullet$ a smooth simplicial scheme, the Deligne-Beilinson cohomology group $H^q_{\mathscr{D}}(X_\bullet, A(p))$ is canonically isomorphic to* $\mathrm{Ext}^q_{A\text{-}MH}(A, R\Gamma(X_\bullet, A(p))$.

This also gives a means to construct Deligne-Beilinson homology (with closed supports) for complex algebraic varieties. We refer to [J] for this construction. A more direct definition of this homology theory is given in [Be 1: § 1]. It is proven in [J] that the pair of a cohomology and of a homology theory thus obtained satisfy the axioms of Bloch and Ogus [Bl-O].

Beilinson [Be 3: § 3] has used Deligne-Beilinson cohomology to define an archimedean height pairing in the following situation: X is a proper smooth algebraic variety over $\mathbb{C}$ of dimension n, Z_1 and Z_2 are algebraic cycles of respective codimension d_1 and d_2, with $d_1 + d_2 = n - 1$, with disjoint supports. One assumes that Z_1 is homologically trivial; then the class c_{Z_1} of Z_1 in $H^{2d_1}_{\mathscr{D}}(X, \mathbb{R}(d_1)) \simeq H^{2d_1}(X, \mathbb{R}(d_1))$ is zero; hence the class α_1 of Z_1 in Deligne-Beilinson cohomology with support in $|Z_1|$ is the image, under the connecting homomorphism, of a class, say $\tilde{\alpha}_1$, in $H^{2d_1-1}_{\mathscr{D}}(W_1, \mathbb{R}(d_1))$, for $W_1 := X - |Z_1|$.

The cup product of $\tilde{\alpha}_1$ with the class $\alpha_2 \in H^{2d_2}_{|Z_2|,\mathscr{D}}(X, \mathbb{R}(d_2))$ gives a class in $H^{2n+1}_{|Z_2|,\mathscr{D}}(X, \mathbb{R}(n+1))$. There is a natural trace map $Tr : H^{2n+1}_{\mathscr{D}}(X, \mathbb{R}(n+1)) \to \mathbb{R}$, which is actually an isomorphism. Beilinson then defines the height pairing to be:

$$\langle Z_1, Z_2 \rangle = Tr(\tilde{\alpha}_1 \cup \alpha_2).$$

For X a curve, one recovers the local archimedean Néron-Tate height pairing.

We will finally mention important recent work of Beilinson, MacPherson and Schechtman [B-M-S] and of Hain and MacPherson [H-M] relating algebraic K-theory, polylogarithms and Deligne-Beilinson cohomology. The main new geometric object is a smooth truncated strict simplicial algebraic manifold $G^p_\bullet$ defined for an integer $p \geq 1$. "Truncated" means that the simplicial manifold is only defined in degrees k where $p \leq k \leq 2p$. "Strict" means that there are face maps, but no degeneracy maps. For $q \geq 1$, G^p_q is defined as the open set of the grassmannian of codimension p linear subspaces of $\mathbb{C}^{p+q}$ which do not meet any of the coordinate planes of dimension $p-1$. There are $p+q+1$ natural face maps $A_j : G^p_q \to G^p_{q-1}$. The truncated simplicial manifold $G^p_\bullet$ is equal to G^p_{k-p} in degree k.

The existence of a natural vector bundle E of rank p over $G^p_\bullet$ determines a class c_p in $H^{2p}_{\mathscr{D}}(G^p_\bullet, \mathbb{Z}(p))$. The point of view on the p-logarithm adopted in [B-M-S] and [H-M] is that it should be a multivalued holomorphic function on G^p_{p-1}. More precisely, these authors introduce a multi-valued Deligne-Beilinson cohomology $H^\bullet_{\mathscr{M}\mathscr{D}}(X_\bullet, \mathbb{Q}(p))$ for suitable simplicial algebraic manifolds $X_\bullet$, which maps to Deligne-Beilinson cohomology. They conjecture that c_p has a natural lift to $H^{2p}_{\mathscr{M}\mathscr{D}}(G^p_\bullet, \mathbb{Q}(p))$. Concretely, such a lift would consist of a multivalued function p-log on G^p_{p-1}, a 1-form on G^p_{p-2}, etc., which give a cocycle in a double complex. This implies that the function p-log satisfies $\sum_{j=0}^{2p}(-1)^j A_j^*(p\text{-log}) = \text{constant}$.

The $(p-1)$-form ω_{p-1} on $G_0^p = (\mathbb{C})^{*p}$ is normalized by $d\omega_{p-1} = \frac{dz_1}{z_1} \wedge \ldots \wedge \frac{dz_p}{z_p}$. Such a class is constructed in [H-M] for $p = 1, 2, 3$; it is also proven there that the groups $H^{2p}_{\mathcal{M}\mathcal{D}}(G^p_\bullet, \mathbb{Q}(p))$, $H^{2p}_{\mathcal{D}}(G^p_\bullet, \mathbb{Q}(p))$ and $\Omega^p(G^p_0)$ are all isomorphic in that case. In general, there are many fascinating open questions on the topolog the G^p_q, for which we refer to [H-M].

6. Mixed Hodge Structures on Homotopy Groups

Deligne, Griffiths, Morgan and Sullivan [D-G-M-S] combined the Hodge theory of Deligne and the rational homotopy theory of Sullivan to obtain deep results on the rational homotopy theory of complex projective manifolds (and more generally, of compact Moishezon spaces). Then Morgan [Mo] proved the existence of natural mixed Hodge structures on the homotopy groups of smooth algebraic varieties. His method is again based on Sullivan's theory of minimal models [Su].

We begin by recalling Sullivan's theory of minimal models for differential graded algebras (D.G.A. for short) over a fixed ground field L. These are graded in non-negative degrees, and the differential raises degrees by 1. A D.G.A. $A^\bullet = \oplus_{k \geq 0} A^k = A^0 \oplus A^+$ is said to be *connected* if $A^0 = L$. For such $A^\bullet$, the space $I(A^\bullet) := A^\bullet / A^+ \cdot A^+$ of *indecomposable elements* is defined as a quotient of $A^\bullet$. The significance of $I(A^\bullet)$ is that it is the space spanned by a minimal set of generators of $A^\bullet$ as an algebra.

First we introduce the notion of *Hirsch extension* of degree k of D.G.A. $A^\bullet \hookrightarrow B^\bullet$. To construct $B^\bullet$, one needs to know $A^\bullet$ and the datum of a vector space V and a linear map $\phi : V \to A^{k+1}$. Then $B^\bullet := A^\bullet \otimes S(V_k)$ as an algebra, where $S(V_k)$ is the symmetric algebra on V (placed in degree k) if k is even, and the exterior algebra if k is even. The differential of $B^\bullet$ agrees with the given one on $A^\bullet$, and $d(x) = \phi(x) \in A^{k+1}$ for x in V. The Hirsch extension $A^\bullet \hookrightarrow B^\bullet$ is said to be *of finite dimension* if V is. If $B^\bullet := A^\bullet \otimes S(V)$ is a Hirsch extension, then $I(B^\bullet) = I(A^\bullet) \oplus V$.

Definition 6.1 [Su]. *A D.G.A. $M^\bullet$ is a* minimal algebra *if*

(a) it is connected;

(b) it is the increasing union of sub-D.G.A.'s

$$L = M_0^\bullet \subset M_1^\bullet \subset M_2^\bullet \subset \ldots$$

where $M_j^\bullet \subset M_{j+1}^\bullet$ is a Hirsch extension;

(c) the map from $I(M^\bullet)$ to itself induced by d is 0, or equivalently $dM^\bullet \subset M^+ \cdot M^+$.

The sequence of subalgebras $M_j^\bullet$ in (b) is called a series *for $M^\bullet$. It is called a finite-dimensional series if each extension is of finite dimension.*

We observe that for $M^\bullet$ minimal, when $I^1(M^\bullet) = 0$, the series $M_j^\bullet$ may be made canonical by requiring that the Hirsch extension $M_j^\bullet \subset M_{j+1}^\bullet$ be of degree $j+1$, as $M_j^\bullet$ is then the subalgebra generated by the indecomposables of degree $\leq j$. We come now to the concept of minimal model of a D.G.A.

Definition 6.2. *Let $A^\bullet$ be a D.G.A. A* minimal model *for $A^\bullet$ is a map $\phi : M^\bullet \to A^\bullet$ of D.G.A's such that*
(a) $M^\bullet$ is minimal;
(b) $\phi_* : H(M^\bullet) \to H(A^\bullet)$ is an isomorphism.

Sullivan developed the homotopy theory of D.G.A's, in which Hirsch extensions play the role of fibrations with $K(\pi, k)$ as fiber. We will need the notion of homotopy between two maps $f_0, f_1 : A^\bullet \to B^\bullet$. A homotopy H from f_0 to f_1 is a map of D.G.A's $H : A^\bullet \to B^\bullet \otimes L[t, dt]$ such that $f_0 = ev_0 \circ H$ and $f_1 = ev_1 \circ H$, where ev_0 (resp. ev_1) denotes evaluation at 0 (resp. 1). The following result is central to this theory.

Proposition 6.3. *If $A^\bullet$ is a D.G.A., there exists a minimal model for $A^\bullet$. If $M^\bullet \xrightarrow{\phi} A^\bullet$ and $M^{\bullet\prime} \xrightarrow{\phi'} A^\bullet$ are two minimal models, there exists an isomorphism $I : M^\bullet \to M^{\bullet\prime}$ such that ϕ and $\phi' \circ I$ are homotopic maps from $M^\bullet$ to $A^\bullet$. Furthermore, I itself is unique up to homotopy.*

To apply the homotopy theory of D.G.A to topological spaces, one needs a D.G.A. which computes the cohomology of a space. For a smooth manifold M, one has the de Rham algebra, i.e. the real D.G.A. $A^\bullet(M)$ of smooth differential forms. Sullivan [Su: § 7] shows that for a simplicial complex X, the $\mathbb{Q}$-algebra $E^\bullet(X)$ of compatible families of polynomial differential forms on the simplices of X computes the rational cohomology of X. The minimal model of such D.G.A's associated to X captures the "rational nilpotent homotopy type" of X, in the sense of Kan [B-K]. The result is especially neat in the simply-connected case.

Proposition 6.4 [Su]. *Let X be a simply-connected polyhedron, such that $H^j(X, \mathbb{Q})$ is finite-dimensional for every j. Let $E^\bullet(X)$ be the D.G.A. of $\mathbb{Q}$-polynomial forms on X and $M^\bullet$ a minimal model for it, and let $M_0^\bullet = \mathbb{Q} \subset M_1^\bullet \subset M_2^\bullet \subset \ldots$ be the (canonical) series of $M^\bullet$. Then*
(1) $M_1^\bullet = \mathbb{Q}$ (one says that the D.G.A. $M^\bullet$ is simply-connected);
(2) for $j \geq 2$, $M_j^\bullet$ is the minimal algebra for the j-th stage X_j of the rational Postnikov tower of X;
(3) $I^k(M^\bullet)$ is canonically isomorphic to the dual of $\pi_k(X) \otimes \mathbb{Q}$.

A similar result holds for the de Rham algebra of a smooth manifold. In the non-simply connected case, the minimal model $M^\bullet$ has indecomposable elements of degree 1. $M^\bullet$ contains the "1-minimal model" $M_1^\bullet$, which admits a canonical series $A_1^\bullet \subset A_2^\bullet \subset \ldots$, with each $A_l^\bullet$ generated by elements of degree 1. By taking the dual spaces one obtains a tower of Lie algebras

$\ldots \to \mathfrak{L}_2 \to \mathfrak{L}_1 \to 0$. Clearly $\mathfrak{L}_1$ is abelian, $\mathfrak{L}_2$ is a central extension of $\mathscr{L}_1$, etc.

Proposition 6.5. *The tower of Lie algebras* $\ldots \to \mathfrak{L}_2 \to \mathfrak{L}_1 \to 0$ *is (non-canonically) isomorphic to the tower of nilpotent Lie algebras associated to the nilpotent completion of* $\pi_1(X)$.

Morgan's method to put mixed Hodge structures on homotopy groups of algebraic varieties is to show that bigraduations on de Rham complexes can be transferred, under certain assumptions, to bigraduations on their minimal models.

Definition 6.6. *A* mixed Hodge diagram *(over* $\mathbb{R}$*) consists of a filtered differential* $\mathbb{R}$*-algebra* $(A, W_\bullet)$, *and two bi-filtered differential* $\mathbb{C}$*-algebras* $(E^\bullet, W_\bullet, F^\bullet)$, $(\overline{E}^\bullet, W_\bullet, \overline{F}^\bullet)$, *and a diagram of morphisms of differential algebras*

$$E^\bullet \overset{\phi}{\leftarrow} A \otimes_{\mathbb{R}} \mathbb{C} \overset{\overline{\phi}}{\to} \overline{E}^\bullet$$

such that:

(1) ϕ *and* $\overline{\phi}$ *are compatible with the filtrations* $W_\bullet$ *and induce isomorphisms on the* E_1 *terms for the spectral sequences associated with* $W_\bullet$*; hence the spectral sequences* ${}_W E_r^{p,q}$ *for* $E^\bullet$ *and* $\overline{E}^\bullet$ *are equipped with a real structure;*

(2) the spectral sequences for the $W_\bullet$ *filtrations degenerate at* E_2*;*

(3) on each term ${}_W E_r^{p,q}(E^\bullet)$ *(resp.* ${}_W E_r^{p,q}(\overline{E}^\bullet)$*), viewed as a subquotient of* $E^\bullet$ *(resp.* $\overline{E}^\bullet$*), for* $r \geq 1$*, the filtration induced by* $F^\bullet$ *(resp.* $\overline{F}^\bullet$*) is q-opposed to its complex-conjugate;*

(4) the filtration on ${}_W E_\infty^{p,q}(E^\bullet)$ *considered in (3) coincides with the induced filtration on* $\mathrm{Gr}_p^W H(E^\bullet)$*; and the corresponding condition holds for* ${}_W E_\infty^{p,q}(\overline{E}^\bullet)$*;*

(5) for all $r \geq 0$*, the differential* d_r *of the W-spectral sequence is strictly compatible with the filtration induced by* $F^\bullet$ *as in (3);*

(6) the filtration induced by $F^\bullet$ *on* $H^*(A \otimes_{\mathbb{R}} \mathbb{C})$ *is the complex conjugate of the filtration induced by* $\overline{F}^\bullet$.

To illustrate this concept, let X be a smooth complex algebraic variety, and let $X \hookrightarrow \overline{X}$ be a good compactification, so that $D = \overline{X} - X$ is a divisor with normal crossings. Let $D = \cup_j D_j$, where the D_j are smooth divisors. Choose a C^∞ tubular neighborhood N_j of each divisor D_j and a closed smooth 2-form ω_j on $\overline{X}$, with support in N_j, such that the class of ω_j in $H^2(N_j, \partial N_j)$ is the Thom class of the normal bundle of $D_j \hookrightarrow \overline{X}$. Let A be the D.G.A. over $\mathbb{R}$ generated by $C^\infty(\overline{X})$ and by symbols θ_j of degree 1, with $d\theta_j = \omega_j$. The filtration W counts the number of symbols θ_j needed to write an element of A. The filtration $F^\bullet$ is the "filtration bête".

Let $E^\bullet(\log D)$ be the subalgebra of the complex de Rham algebra $A^\bullet(X) \otimes \mathbb{C}$ consisting of those differential forms which locally near a point of D which belongs to components $D_1, D_2, \ldots, D_r$, may be written in the form $\sum_{I \subset \{1,\ldots r\}} \alpha_I \wedge$

$\prod_{j\in I}(\frac{dz_j}{z_j})$, where z_j is a local equation of D_j and α_I is a smooth differential form. In other words, $E^\bullet(\log D)$ is the image of the space of global sections of the tensor product sheaf $\Omega^\bullet_{\overline{X}}(\log D) \otimes_{\mathcal{O}_{\overline{X}}} A^{0,\bullet}_{\overline{X}}$, where $A^{0,\bullet}_{\overline{X}}$ is equipped with its usual bigrading.

The filtration W counts the cardinality of subsets I in local expressions as above. The Hodge filtration $F^\bullet$ is the "filtration bête" of Deligne. Let $E^\bullet(\overline{\log D})$ be the complex-conjugate sub-algebra, where the local expression at a point of D now involves $\frac{d\overline{z}_j}{\overline{z}_j}$ instead of $\frac{dz_j}{z_j}$.

To construct a map $\phi : A \to E^\bullet(\log D)$ of D.G.A's, one needs only to find a 1-form β_j in $E^\bullet(\log D)$ such that $d\beta_j = \omega_j$. As shown in [Mo: Lemma 3.2], one can find such a β_j which is locally of the form $\frac{-1}{2\pi i}(\frac{dz_j}{z_j}) + \alpha_j$, with α_j smooth. One then puts $\phi(\theta_j) = \beta_j$. One next defines $\overline{\phi} : A \to E^\bullet(\overline{\log D})$ simply as the complex-conjugate of ϕ.

Proposition 6.7 [Mo: Proposition 3.5]. *The diagram*

$$E^\bullet(\log D) \overset{\phi}{\leftarrow} A \otimes_{\mathbb{R}} \mathbb{C} \overset{\overline{\phi}}{\to} \overline{E}^\bullet(\log D)$$

is a mixed Hodge diagram.

This is essentially a restatement of Deligne's mixed Hodge theory for smooth complex varieties (see Sect. 1).

For a mixed Hodge diagram

$$E^\bullet \overset{\phi}{\leftarrow} A \otimes_{\mathbb{R}} \mathbb{C} \overset{\overline{\phi}}{\to} \overline{E}^\bullet$$

there is a refined notion of minimal models [Mo: § 6]. A bigrading on a D.G.A. $M^\bullet$ is a decomposition $M^\bullet = \sum_{r,s\geq 0}(M^\bullet)^{r,s}$, where $(M^\bullet)^{0,0} = L$, and d and the product map are of bi-degree $(0,0)$. A morphism from a bigraded D.G.A. $M^\bullet = \sum_{r,s\geq 0}(M^\bullet)^{r,s}$ to a mixed Hodge diagram $E^\bullet \overset{\phi}{\leftarrow} A \otimes_{\mathbb{R}} \mathbb{C} \overset{\overline{\phi}}{\to} \overline{E}^\bullet$ consists of maps of D.G.A's $\psi : M^\bullet \to E^\bullet$, $\rho : M^\bullet \to A \otimes \mathbb{C}$ and $\psi' : M^\bullet \to \overline{E}^\bullet$, together with homotopies $H : M^\bullet \to E^\bullet \otimes L[t, dt]$ from $\phi\rho$ to ψ and $H' : M^\bullet \to \overline{E}^\bullet \otimes L[t, dt]$ from $\overline{\phi}\rho$ to ψ', such that:

(1) $\rho((M^\bullet)^{r,s}) \subset W_{-(r+s)}(A \otimes \mathbb{C})$, $\psi((M^\bullet)^{r,s}) \subset W_{-(r+s)}(E^\bullet) \cap F^r(E^\bullet)$ and $\psi'((M^\bullet)^{r,s}) \subset W_{-(r+s)}(\overline{E}^\bullet) \cap \overline{F}^s(E^\bullet) + \sum_{i\geq 2} W_{i-r-s}(\overline{E}^\bullet) \cap \overline{F}^{s-i+1}(\overline{E}^\bullet)$

(2) $H((M^\bullet)^{r,s}) \subset W_{-r-s}(E^\bullet) \otimes L[t, dt]$ and $H'((M^\bullet)^{r,s}) \subset W_{-r-s}(\overline{E}^\bullet) \otimes L[t, dt]$.

If in addition $\rho : M^\bullet \to A \otimes \mathbb{C}$ is a minimal model, then $(M^\bullet, \psi, \rho, \psi', H, H')$ is said to be a *bigraded minimal model* for the mixed Hodge diagram.

Proposition 6.8 [Mo: Theorem 6.6]. *Any mixed Hodge diagram admits a bigraded minimal model. The induced bigrading on the cohomology of the diagram by its mixed Hodge structure agrees with the bigrading of the cohomology of the minimal model.*

The proof uses a "principle of two types" for mixed Hodge diagrams. Morgan also proves that the bigraded minimal model is unique up to bigraded isomorphisms [Mo: Corollary 6.9]. He also proves a naturality property of the bigraded minimal model. In the case of an algebraic variety, he obtains the following.

Proposition 6.9 *(1) Let X be a smooth complex algebraic variety and let $N_X^\bullet$ be the minimal model for the de Rham complex of X. Then $N_X^\bullet$ has a bigrading which is unique up to isomorphism.*

(2) Let $f : X_1 \to X_2$ be an algebraic map. Then there exist good compactifications $X_1 \hookrightarrow \overline{X}_1$ and $X_2 \hookrightarrow \overline{X}_2$ such that f extends to an algebraic map $f : \overline{X}_1 \to \overline{X}_2$. Then there is an induced map of D.G.A's $f^* : N_{X_2}^\bullet \to N_{X_1}^\bullet$ unique up to homotopy. This map can be chosen to preserve the bigradings.

The next question is to define a weight filtration $W_\bullet$ on the minimal model. If $M^\bullet$ is a minimal D.G.A., a filtration $W_\bullet(M^\bullet)$ is called a *minimal filtration* if

(a) $W_\bullet$ is compatible with the product structure;

(b) d maps W_l to W_{l-1};

(c) $dM^\bullet \cap W_{l-1} = dW_l$.

Proposition 6.10 [Mo: Definition 7.4]. *Let $A^\bullet$ be a D.G.A. with a multiplicative filtration $W_\bullet(A^\bullet)$ such that the associated spectral sequence degenerates at E_2. One says that the filtration W* passes to the minimal model *if there is a minimal model $\rho : M^\bullet \to A^\bullet$ and a minimal filtration $W_\bullet(M^\bullet)$ such that:*

(1) ρ is compatible with the filtrations;

(2) the map $H(M^\bullet) \to H(A^\bullet)$ induced by ρ is an isomorphism of filtered vector spaces.

$(M^\bullet, W_\bullet)$ is then called a *filtered minimal model* for $(A^\bullet, W_\bullet)$. Such a filtered minimal model is unique up to isomorphism.

Proposition 6.11 [Mo: Theorem 7.7]. *For X a smooth algebraic manifold, the weight filtration of the D.G.A. of $\mathbb{Q}$-polynomial forms on some triangulation passes to the minimal model M_X.*

Hence for a smooth algebraic manifold X, one has a filtration $W_\bullet$ on the $\mathbb{Q}$-minimal model $M_X^\bullet$, as well as a bigrading on the complex minimal model $N_X^\bullet$. Setting $\tilde{W}_l(N_X^\bullet) = \oplus_{p+q \le l}(N^\bullet)^{p,q}$, one defines an algebra filtration $W_\bullet$ by setting $W_l = \tilde{W}_{n+l}$ in degree n. It is a minimal filtration, and there is a filtered isomorphism $I : N_X^\bullet \to M_X^\bullet \otimes \mathbb{C}$. I is unique up to a homotopy preserving the filtrations.

Theorem 6.12. *Any such filtered isomorphism $I : N_X^\bullet \to M_X^\bullet \otimes \mathbb{C}$ defines a mixed Hodge complex. The induced mixed Hodge structure on $H(M_X^\bullet)$ coincides with Deligne's mixed Hodge structure on $H^\bullet(X)$.*

Since the rational homotopy groups of X are dual to the spaces of indecomposables of $M_X^\bullet$ for X simply-connected (Proposition 6.4), one deduces mixed Hodge structures on homotopy groups.

Theorem 6.13 [Mo: Theorem 9.1]. *Let X be a simply-connected smooth algebraic manifold.*

(1) The homotopy groups $\pi_n(X) \otimes \mathbb{Q}$ carry natural mixed Hodge structures. The Whitehead product $\pi_n(X) \otimes \pi_m(X) \to \pi_{n+m-1}(X)$ is a morphism of mixed Hodge structures.

(2) The cohomology ring of the j-th stage X_j of the rational Postnikov tower for X also carries natural mixed Hodge structures. The maps $X_j \to X_{j-1}$ and $X \to X_j$ induce morphisms of mixed Hodge structures.

In the case of projective manifolds, one has more precise results.

Theorem 6.14 (Deligne, Griffiths, Morgan and Sullivan) [D-G-M-S: § 6]. *Let X be a compact complex manifold for which the dd^c-lemma holds (for instance, a Kähler manifold, or a Moishezon manifold). Then*

(1) The minimal model is isomorphic to that of $H^\bullet(X)$, where $H^\bullet(X)$ is viewed as a D.G.A. with 0 differential (one says that the real, hence the rational, homotopy type of X is "formal").

(2) If X is simply-connected, the real homotopy groups $\pi_k(X) \otimes \mathbb{R}$ depend only on the cohomology algebra $H^\bullet(X, \mathbb{R})$. Moreover, all Massey products of all orders are rationally zero.

(3) The real form of the canonical tower of nilpotent quotients of the fundamental group of X is determined by $H^1(X, \mathbb{R})$ and by the cup-product mapping $H^1(X, \mathbb{R}) \otimes H^1(X, \mathbb{R}) \to H^2(X, \mathbb{R})$.

The vanishing of Massey triple products was apparently first observed by Atiyah, who deduced that the group of upper-triangular (3, 3)-matrices with integer coefficients cannot be the fundamental group of a compact Kähler manifold.

Because of the absence of base points in Morgan's constructions, his results on the fundamental group describe the Hodge filtration up to inner conjugation.

Theorem 6.15 [Mo: Theorem 9.2]. *The tower of rational Lie algebras associated with the rational nilpotent completion of $\pi_1(X)$ is a tower of Lie algebras in the category of mixed Hodge structures. The weight filtration is a filtration by subalgebras. It is unique and functorial with respect to algebraic mappings. The Hodge filtration is only unique up to inner automorphisms.*

Morgan uses this to obtain a description of the structure of this tower of Lie algebras.

Theorem 6.16 [Mo: Theorem 9.4]. *Let X be a smooth algebraic variety, with good compactification $X \hookrightarrow \overline{X}$, and let $A = H_1(\overline{X}, \mathbb{C}) \oplus \mathrm{Coker}\{H_2(\overline{X}) \to \oplus_j H_0(D_j, \mathbb{C})(1)\}$. This space, non-canonically isomorphic to $H_1(X, \mathbb{C})$, admits a bigrading, where the bi-degrees are $(-1,0)$, $(0,-1)$ and $(-1,-1)$.*

There exists some bigraded Lie algebra ideal I in the free Lie algebra $\mathscr{F}(A)$ such that the tower of nilpotent complex Lie algebras associated with $\pi_1(X)$ is isomorphic to the tower of nilpotent quotients of $\mathscr{F}(A)/I$. The isomorphisms of Lie algebras respect the bigradings. The ideal I has generators of bi-degrees $(-1,-1)$, $(-1,-2)$, $(-2,-1)$ and $(-2,-2)$ only.

This implies that many finitely-generated groups cannot be the fundamental group of a smooth complex algebraic variety. This result was extended to normal algebraic varieties by Deligne.

Although, according to Theorem 6.14, for X a simply-connected complex projective manifold, the $\pi_n(X) \otimes \mathbb{Q}$ may be computed from the cohomology ring $H^\bullet(X, \mathbb{Q})$, Carlson, Clemens and Griffiths [C-C-M] proved that the mixed Hodge structure on the $\pi_n(X)$ is not determined by the ring $H^\bullet(X, \mathbb{Q})$ together with its Hodge structure, already for $n = 3$. They also relate $\pi_3(X)$ to the Griffiths intermediate Jacobian $J_2(X)$ as follows. Let $\mathscr{N} = NS(X)$ be the Néron-Severi group of X , let $\alpha : S^2(\mathscr{N}) \to H_{2N-4}(X, \mathbb{Z})$ be the natural pairing. Then the cycle map of Griffiths from homologically trivial cycles (see Sect. 5) induces a homomorphism: $\Phi : \ker(\alpha) \to J_2(X)$. One may view Φ as an element of

$$\mathrm{Hom}(\ker(\alpha) \otimes \mathbb{C}, H^3(X, \mathbb{C}))/\mathrm{Hom}_F(\ker(\alpha) \otimes \mathbb{C}, H^3(X, \mathbb{C})) + \mathrm{Hom}(\ker(\alpha), H^3(X, \mathbb{Z})) .$$

On the other hand, we have an exact sequence $0 \to H^3(X, \mathbb{Z}) \to \pi_3(X)^\bullet \to K \to 0$ in the category of mixed Hodge structures, where $K = \ker(S^2 H^2(X, \mathbb{Z}) \to H^4(X, \mathbb{Z}))$. The class of this extension determines an element u in $\mathrm{Hom}(K \otimes \mathbb{C}, H^3(X, \mathbb{C}))/\mathrm{Hom}_F(K \otimes \mathbb{C}, H^3(X, \mathbb{C})) + \mathrm{Hom}(K, H^3(X, \mathbb{Z}))$. Then Carlson, Clemens and Morgan prove that under the map

$$\mathrm{Hom}(K \otimes \mathbb{C}, H^3(X, \mathbb{C})) \to \mathrm{Hom}(\ker(\alpha) \otimes \mathbb{C}, H^3(X, \mathbb{C}))$$

induced by the inclusion $\ker(\alpha) \subset K$, u maps to Φ.

We now turn to an entirely different approach to mixed Hodge structures on homotopy groups, due to Hain [H 1, H 2], which is based on the *bar construction*. This approach has several advantages; it is very natural from the topological viewpoint, it incorporates base points, and gives very precise results on the fundamental group. It is also closely related to iterated integrals. The classical bar construction is an algebraic version of the construction of the pointed loop space of a space as a cosimplicial space. Let $A^\bullet$ be a (not necessarily commutative) augmented D.G.A. over a ring L, with augmentation ideal $IA^\bullet$. Define

$$B^{-s,t}(A^\bullet) = \left[\otimes^s IA^\bullet\right]^t .$$

We use the traditional notation $[a_1|\dots|a_s]$ for $a_1 \otimes \dots \otimes a_s$. Define the involution τ on any graded vector space V by $\tau(v) = (-1)^{\deg v} v$. One defines two differentials on B, the *internal differential* d_I (extension of d to the tensor algebra) and the *combinatorial differential* d_C. They are as follows:

$$d_I[a_1|\dots|a_s] = \sum_{i=1}^{s}(-1)^i[\tau a_1|\dots|\tau a_{i-1}|da_i|a_{i+1}|\dots|a_s]$$

$$d_C[a_1|\dots|a_s] = \sum_{i=1}^{s}(-1)^{i+1}[\tau a_1|\dots|\tau a_{i-1}|\tau a_i \wedge a_{i+1}|a_{i+2}|\dots|a_s]\ .$$

One thus obtains a double complex. The *bar construction on* $A^\bullet$ is the associated total complex. It is denoted by $B(A^\bullet)$. Define the diagonal map $\Delta : B(A^\bullet) \to B(A^\bullet) \otimes B(A^\bullet)$ by $\Delta[a_1|\dots|a_s] = \sum_{i=0}^{s}[a_1|\dots|a_i] \otimes [a_{i+1}|\dots|a_s]$. Then $B(A^\bullet)$ becomes a D.G. coalgebra. If $A^\bullet$ is commutative, then $B(A^\bullet)$, equipped with the shuffle-product, is a D.G. Hopf algebra.

The *bar filtration* $0 \subset L = \mathbf{B}^0 \subset \mathbf{B}^{-1} \subset \mathbf{B}^{-2} \subset \dots B(A^\bullet)$ is defined by $\mathbf{B}^{-s} = \sum_{u \le s} B^{-u,v}$ leads to the so-called *Eilenberg-Moore* spectral sequence.

Theorem 6.17 [A]. *Let X be a simply-connected space. Let $A^\bullet(X)$ denote the D.G.A. of simplicial polynomial differential forms on the simplicial complex $Simp_\bullet(X)$. Then the cohomology of the pointed loop space $P_{x,x}(X)$ is isomorphic, as a Hopf algebra, to the cohomology of the bar construction for $A^\bullet(X)$.*

Geometrically, this means that $P_{x,x}(X)$ is homotopically equivalent to the homotopy limit of the cosimplicial space which in degree n is X^n.

Chen [Chen 1] has defined a variant of the bar construction, which involves a D.G. right $A^\bullet$-module $M^\bullet$ and a D.G. left $A^\bullet$-module $N^\bullet$. Define $T^{-s,t}(M^\bullet, A^\bullet, N^\bullet) = [M^\bullet \otimes (\otimes^s A^+) \otimes N^\bullet]^t$. We use the notation $m[a_1|\dots|a_s]n$ for elements of $T^{-s,t}(M^\bullet, A^\bullet, N^\bullet)$. We refer to [Chen 1] and to [H 1: § 1.2] for the precise formula for the differentials d_C and d_I in this context. One introduces the graded subspace R of $T^{-s,t}(M^\bullet, A^\bullet, N^\bullet)$ generated by elements of the form

(*a*) $$m[df|a_1|\dots|a_s]n + mf[a_1|\dots|a_s]n - m[fa_1|\dots|a_s]n\ ,$$

(*b*) $$m[a_1|\dots|a_s|df]n + m[a_1|\dots|fa_s]n - m[a_1|\dots|a_s]fn\ ,$$

(*c*) $$m[a_1|\dots|a_{i-1}|df|a_i|\dots|a_s]n + m[a_1|\dots|fa_{i-1}|a_i|\dots|a_s]n - m[a_1|\dots|a_{i-1}|fa_i|\dots|a_s]n\ ,$$

where $f \in A^0$.

One checks that R is actually a subcomplex. The quotient complex $\overline{B}(M^\bullet, A^\bullet, N^\bullet) = T(M^\bullet, A^\bullet, N^\bullet)/R$ is called the *reduced bar construction.* Again there is a bar filtration of $\overline{B}$ and an Eilenberg-Moore spectral sequence. Note that d_C becomes 0 on $\mathrm{Gr}_{\mathbf{B}}$.

Proposition 6.18 [Chen 1: § 32]. *The E_1-term of the spectral sequence for $\overline{B}(M^\bullet, A^\bullet, N^\bullet)$ is isomorphic to $\overline{B}(H(M^\bullet), H(A^\bullet), H(N^\bullet))$.*

If $A^\bullet$ is commutative and $M^\bullet$ and $N^\bullet$ are D.G.A's, then $\overline{B}(M^\bullet, A^\bullet, N^\bullet)$ becomes a D.G.A. with the shuffle-product. To recover the usual bar construction, one assumes that $A^\bullet$ is augmented and one takes $M^\bullet = N^\bullet = L$. One denotes $\overline{B}(L, A^\bullet, L)$ by $\overline{B}(A^\bullet)$. There is an obvious map $B(A^\bullet) \to \overline{B}(A^\bullet)$. It is a quasi-isomorphism if $H(A^\bullet)$ is connected.

For X a space and $A^\bullet$ the associated D.G.A., the reduced bar construction $\overline{B}(A^\bullet, A^\bullet, A^\bullet)$ is a model for the de Rham algebra of the path space PX. More precisely, Chen's *iterated integral* construction defines a map $\int$ from $A^{\bullet\,\otimes r}$ to $A^\bullet(PX)$, which may be described as follows. Let $w_1, w_2, \ldots, w_r$ be elements of $A^\bullet(PX)$. Let $\sigma : \Delta^n \to PX$ be a n-simplex for PX, and let $\Phi : [0,1] \times \Delta^n \to X$ be the corresponding map. Let t denote the first coordinate on the product $[0,1] \times \Delta^n$ and $\frac{\partial}{\partial t}$ the corresponding vector field. Then the differential form $\sigma^*(\int w_1 w_2 \ldots w_r)$ is described as the integration $\int \ldots \int_{0 \le t_1 \le t_2 \ldots \le t_r \le 1} w_1'(t_1) \wedge w_2'(t_2) \ldots \wedge w_r'(t_r) dt_1 \ldots dt_r$, where $w_j'(t) := \frac{\partial}{\partial t} \rfloor \Phi^*(w_j)$.

Let $p_t : PX \to X$ denote evaluation of a path at $t \in [0,1]$.

Proposition 6.19 (Chen, [Chen 1]). *If X is a simply-connected space, then the map from $\overline{B}(A^\bullet, A^\bullet, A^\bullet)$ to $A^\bullet(PX)$ defined by*

$$w'[w_1|\ldots|w_r]w'' \mapsto p_0^* w' \wedge (\int w_1 \ldots w_r) \wedge p_1^* w''$$

is an isomorphism of D.G.A.'s.

The reduced bar construction $\overline{B}(A^\bullet(Y), A^\bullet(X), A^\bullet(Z))$ is the algebraic machine designed to compute the cohomology of the pull-back of the fibration $PX \to X \times X$ by a map $f = (f_0, f_1) : Y \times Z \to X \times X$.

Proposition 6.20. *Let $f_0 : Y \to X$ and $f_1 : Z \to X$ be continuous maps, let $f = (f_0, f_1) : Y \times Z \to X \times X$, and define P_f by the cartesian diagram:*

$$\begin{array}{ccc} P_f & \xrightarrow{F} & PX \\ \downarrow & & \downarrow \\ Y \times Z & \xrightarrow{f} & X \times X \end{array}$$

Assume X is simply-connected. Then the map $\overline{B}(A^\bullet(Y), A^\bullet(X), A^\bullet(Z)) \to A^\bullet(P_f)$ is a D.G.A. quasi-isomorphism.

This applies in particular to the homotopy fiber $E_f(y)$ of a map $f : X \to Y$ at a point $y \in Y$, since it is the pull-back of the path fibration by the inclusion $X \times \{y\} \hookrightarrow X \times Y$.

Concerning the pointed loop space $P_{x,x}$ for X non-simply-connected, we have an integration pairing

$$B(A^\bullet(X)) \otimes S_\bullet(P_{x,x}(X)) \to L,$$

where $S_\bullet(-)$ denotes the singular chain complex. Since $H_0(P_{x,x}(X), L) = L[\pi_1(X,x)]$, taking degree 0-homology, one obtains a pairing

$$H^0B(A^\bullet(X)) \otimes L[\pi_1(X,x)] \to L.$$

Theorem 6.21 (Chen, [Chen 1: § 2.6]). *If X is path-connected, then under the pairing*

$$H^0B(A^\bullet(X)) \otimes L[\pi_1(X,x)] \to L,$$

the bar filtration

$$0 \subset \mathbf{B}_0 \subset \mathbf{B}_1 \subset \ldots H^0B(A^\bullet(X))$$

is dual to the filtration

$$L[\pi_1(X,x)] \supset J \supset J^2 \supset \ldots$$

by the powers of the augmentation ideal J of the group algebra $L[\pi_1(X,x)]$. Hence the integration pairing gives an isomorphism $\mathbf{B}_s H^0B(A^\bullet X) \xrightarrow{\sim} \mathrm{Hom}_L(L[\pi_1(X,x)]/J^{s+1}, L)$.

It is useful to observe that $\mathrm{Gr}^{\mathbf{B}}_s H^0B(A^\bullet(X))$ is naturally a subspace of $\otimes^s H^1(X)$.

To apply the bar construction to mixed Hodge theory, one needs to show that the bar construction of a D.G.A. which is a mixed Hodge complex, is again a mixed Hodge complex. For this purpose, Hain introduces the notion of *multiplicative L-mixed Hodge complex*, which is a mixed Hodge complex

$$(A^\bullet_L, W) \xrightarrow{\alpha} (A^\bullet_{\mathbb{C}}, W, F),$$

where $(A^\bullet_L, W)$ and $(A^\bullet_{\mathbb{C}}, W, F)$ are filtered D.G.A's and α induces a quasi-isomorphism of filtered D.G.A.'s over $\mathbb{C}$. He also similarly defines the notion of an L-MHC which is a module over a multiplicative MHC.

He then obtains the fundamental technical result

Proposition 6.22 [H 1: Theorem 3.2.1]. *Let A be a positively graded multiplicative L-MHC whose weight filtration is bounded below. Assume that $H^0(A) = L$. If the MHC M is a right A-module and the MHC N is a left A-module, then $\overline{B}(M,A,N) = ((\overline{B}(M_L, A_L, N_L), W_\bullet), (\overline{B}(M_{\mathbb{C}}, A_{\mathbb{C}}, N_{\mathbb{C}}), W_\bullet, F^\bullet))$, where the filtrations $W_\bullet$ and $F^\bullet$ are defined by the natural diagonal process, is an L-MHC.*

Hain [H 1: Theorem 5.6.3] proved that there is a natural L-MHC associated to any smooth simplicial manifold $X_\bullet$ and a good compactification $X_\bullet \hookrightarrow \overline{X}_\bullet$. Similar results were also proven by Deligne (unpublished notes) and by Navarro Aznar [NA 1].

Together with Proposition 6.20 above, this gives mixed Hodge structures on the cohomology of bar-constructions, hence in particular on the cohomology of $P_{x,x}$. Since this cohomology is a Hopf algebra, one also gets mixed Hodge structures on the primitive part of the cohomology, hence on the homotopy groups of X.

The constructions of [H 1] and [NA 1] yield:

Theorem 6.23. *Suppose that X is a complex algebraic variety and x a point of X. Then*

(1) If (X, x) is nilpotent (in the sense of [B-K]*), the higher homotopy groups of X have a mixed Hodge structure for which the Whitehead products are morphisms of MHS.*

(2) The Hurewicz map for (X, x) is a morphism of mixed Hodge structures.

(3) The completion $\mathbb{Q}[\pi_1(X, x)]^\wedge$ of the group ring $\mathbb{Q}[\pi_1(X, x)]$ with respect to the augmentation ideal J has a mixed Hodge structure in the category of Hopf algebras. If X is proper and smooth, the filtration by powers of J is the weight filtration.

All these mixed Hodge structures are functorial with respect to algebraic maps $f : (X, x) \to (Y, y)$.

There is also a version of this theorem for relative homotopy groups (see [H 1: Theorem 4.2.2]). The following result on homotopy fibers and the corresponding monodromy representations is of particular interest to geometers.

Theorem 6.24 [H 1: Theorem 4.3.1]. *Let $f : (X, x) \to (Y, y)$ be an algebraic map. Suppose that Y is connected, and that the action of $\pi_1(Y, y)$ on the cohomology $H^k(E_f(y))$ of the homotopy fiber is unipotent for all k. Then there are natural MHS on the cohomology groups and (if $E_f(y)$ is nilpotent) on the homotopy groups of the homotopy fiber $E_f(y)$. The monodromy representation $\mathbb{Q}\pi_1(Y, y) \to$* End $H^k(E_f(y))$ *is a morphism of MHS.*

Applying this theorem to the situation $(Y, y) = (\mathbb{C}^*, t)$, with f a fibration in a neighborhood of t, Hain [H1] obtains mixed Hodge structures on the cohomology and homotopy of the homotopy fiber, which coincide with the limit M.H.S. on the cohomology or homotopy of the general fiber $f^{-1}(t)$ (in the sense of § 2).

The mixed Hodge structure on the completion of the fundamental group of an algebraic curve has interesting connections with the harmonic volumes of B. Harris [Ha] and with the image of cycles of dimension two in its Jacobian variety in the appropriate Griffiths intermediate Jacobian [Pu]. We will mention the following Torelli type theorem for pointed curves.

Theorem 6.25 (Hain [H 2], Pulte [Pu]). *Let V and W be smooth complex projective curves. Let x be a point of V, y a point of W. If there is a ring homomor-*

phism between $\mathbb{Z}\pi_1(V,x)/J^3$ *and* $\mathbb{Z}\pi_1(W,y)/J^3$*, then there is an isomorphism* $f : V \to W$*. Furthermore, except if* $x = a$ *or* $x = b$*, for two specific points* $a, b \in V$*,* f *can be chosen to map* x *to* y*.*

Applications of the MHS on quotients of the group algebra of π_1 to unipotent variations of mixed Hodge structures are discussed in Sect. 7.

7. Variation of Mixed Hodge Structure

The notion of a variation of Hodge structure (as an abstract concept) was presented in Sect. 2. We have seen that there is a complete local theory of their degeneration (Theorems 2.4 and 2.16). Moreover, this feeds into the corresponding Hodge theory for cohomology with coefficients in the underlying local system (Theorem 3.15). Recall also that algebraic geometry produces variations of Hodge structure over $\mathbb{Z}$ from proper smooth morphisms $f : Z \to X$, with local systems $\mathbb{V} = R^m f_* \mathbb{C}$.

If we drop the smoothness and/or properness conditions on $f : Z \to X$, and replace them with a sharp notion of topological local constancy – one that is automatic in the proper smooth case, and is true over a Zariski-open subset of X for any morphism of varieties (see [Ve 1]) – then f gives rise to a family of mixed Hodge structures on $H^m(Y_x, \mathbb{Z})$ as x varies over X, subject to certain rules. This leads one to the abstract notion of a *(graded-polarized) variation of mixed Hodge structure*, which we now define:

Definition 7.1. *Let A be a subring of $\mathbb{R}$, and X a complex manifold. A graded-polarized variation of A-mixed Hodge structure on X consists of a local system $\mathbb{V}_A$ on X of A-modules of finite rank, together with the following data:*

(1) An increasing filtration $\ldots \subset \mathbb{W}_k \subset \mathbb{W}_{k+1} \subset \ldots$ *of* $\mathbb{V}_A \otimes_{\mathbb{Z}} \mathbb{Q}$,
(2) A decreasing filtration $\ldots \subset \mathcal{F}^{p+1} \subset \mathcal{F}^p \subset \ldots$ *of* $\mathcal{V} = \mathcal{O}_X \otimes_A \mathbb{V}_A$,
(3) A sequence of flat bilinear forms

$$S_k : \mathrm{Gr}_k^W (\mathbb{V}_A \otimes \mathbb{Q}) \times \mathrm{Gr}_k^W (\mathbb{V}_A \otimes \mathbb{Q}) \to A(-k) \otimes \mathbb{Q}$$

subject to the conditions
(a) $\nabla \mathcal{F}^p \subset \Omega_X^1 \otimes \mathcal{F}^{p-1}$,
(b) the data $\mathrm{Gr}_k^W (\mathbb{V}_A \otimes \mathbb{Q})$*, the filtration induced by* $\mathcal{F}$ *on* $\mathcal{O}_X \otimes \mathrm{Gr}_k^W \mathbb{V}_A$*, and* S_k *give a polarized variation of Hodge structure* (2.1) *of weight* k*.*

As in the case of pure Hodge structure, a variation of mixed Hodge structure can be described in terms of the appropriate classifying space of mixed Hodge structures. Let V be a real vector space, with increasing filtration W, and fix bilinear forms S_k on $\mathrm{Gr}_k^W V$ as in 7.1 (3). We wish to describe the set M of all decreasing filtrations F on $V_{\mathbb{C}}$ that induce Hodge structures with given Hodge numbers on each $\mathrm{Gr}_k^W V$, polarized by the form S_k. If we denote by M_k the classifying space for the latter (see §2), then M fibers over $\prod_k M_k$, with fiber homogeneous under a unipotent complex Lie group. Roughly, this unipotent group contains the extension data in the mixed Hodge structures. Indeed, the nature of extensions in the setting of mixed Hodge theory has been studied in several places ([Ca 1, Be 2, Ca-H, Mo]).

The space M can be approached from the point of view of equivalence classes of splittings, which have been mentioned before (2.6). Consider specifying a splitting

$$V_{\mathbb{C}} = \oplus V^{p,q}$$

such that $W_\ell = \bigoplus_{p+q\le\ell} V^{p,q}$, $V^{p,q} \equiv \overline{V^{q,p}} \left(\mathrm{mod}\, W_{p+q-1}\right)$. The splitting is real when $V^{p,q} = \overline{V^{q,p}}$.

Let $W_0\mathrm{GL}(V)$ denote the group of automorphisms of V that preserve the filtration W of V. Let $W_{-1}\mathrm{GL}(V)$ denote the kernel of the canonical mapping

$$W_0\mathrm{GL}(V) \xrightarrow{\pi} \prod_k \mathrm{GL}\left(\mathrm{Gr}_k^W V\right) .$$

Let $G_k \subset \mathrm{GL}\left(\mathrm{Gr}_k^W V\right)$ be the subgroup that preserves S_k, and write

$$\widetilde{G} = \pi^{-1}\left(\prod_k G_k\right) .$$

The set of real splittings is a principal homogeneous space under $\widetilde{G}_{\mathbb{R}}$, while the set of all splittings is principal homogeneous under

$$\widetilde{H} = \widetilde{G}_{\mathbb{R}} \cdot W_{-1}\mathrm{GL}(V)_{\mathbb{C}} .$$

We select a real one (which thus determines the Hodge filtration F of a mixed Hodge structure that is split over $\mathbb{R}$) as basepoint. Let $\widetilde{B}$ denote the subgroup $F^0\widetilde{H}$ of $\widetilde{H}$ that preserves F; then $M \simeq \widetilde{H}/\widetilde{B}$, while $\widetilde{G}_{\mathbb{R}}/\left(\widetilde{B} \cap \widetilde{G}_{\mathbb{R}}\right)$ gives the subspace of mixed Hodge structures split over $\mathbb{R}$. The horizontal subbundle can be defined as before, and a variation of mixed Hodge structure gives rise to a horizontal holomorphic mapping $\widetilde{\Phi} : \widetilde{X} \to M$, equivariant for the action of the monodromy group.

In marked contrast to the pure case, there is no analogue of the Nilpotent Orbit Theorem for variations of mixed Hodge structure. There is a clue as to the reason in our Definition 7.1 itself. With the exception of the conditions on the total Hodge filtration F, viz. (2) and (a), one is talking only about the sequence of pure subquotients. In [St-Z: (3.16)] very simple examples are provided of graded-polarized variation of mixed Hodge structure having either no limit Hodge filtration, or one which behaves poorly with respect to Gr^W. Though the pure variations are "internally controlled", some conditions must be imposed to control the extension data.

The following may seem artificial, but it is apparently the right definition:

Definition 7.2 [St-Z, K 6]. *i) A graded-polarizable variation of* $\mathbb{Q}$*-mixed Hodge structure over* Δ^* *is said to be* admissible *if it satisfies:*

(a) The limit Hodge filtration (in the sense of (2.10)*) exists, and induces that of Schmid for each* $\mathrm{Gr}_k^W \mathbb{V}$.

(b) The nilpotent part of the monodromy logarithm (which necessarily respects W*) admits a weight filtration relative to* W *(see* 2.14*).*

ii) A graded-polarizable variation of mixed Hodge structure on X *is* admissible *if its restriction to every meromorphically embedded* Δ^* *is admissible in the sense of (i).*

The viability of this notion of admissibility comes from the following two results:

Theorem 7.3. *Every geometric variation of mixed Hodge structure is admissible.*

Theorem 7.4. *For admissible variations of mixed Hodge structure* $\mathbb{V}$*, on a quasi-projective manifold* X*, there is a functorial mixed Hodge structure on* $H^\bullet(X, \mathbb{V})$.

We will outline their proofs.

One proves Theorem 7.3 by generalizing Steenbrink's construction in [St 1] (see our Theorem 2.22) beyond the proper smooth case. This was carried out in the early 1980's, with some overlap of priority, by Du Bois [Du 2], El Zein [E 7], Guillén, Navarro Aznar and Puerta [G-NA-PG-P], and Steenbrink and Zucker [St-Z]. By using simplicial hypercoverings (cf. 1.14), one sees that the smooth (but not necessarily proper) case is the key one, and we summarize that here.

Let $f : X \to \Delta$ be a proper morphism with f smooth over Δ^*, such that $D = f^{-1}(0)$ is a divisor with normal crossings. Let $Y \subset X$ be a divisor with normal crossings, with the property that the mapping induced by f of each $\widetilde{Y}^n$ (notation as in 1.3) to Δ is smooth over Δ^*, and moreover $D \cup Y$ is also with normal crossings. Using the nearby cycle construction (cf. 2.21) for the pair (X, Y), one eventually obtains as in Proposition 2.19:

Proposition 7.5. *For each* $m \geq 0$*, the higher direct image sheaf*

$$\mathscr{V}^m = R^m f_* \left(\Omega^\bullet_{X/\Delta}(\log Y \cup D)\right)$$

is free on Δ*, and for all* $s \in \Delta$*, the canonical mapping*

$$\mathscr{V}^m \otimes_{\mathcal{O}_\Delta} \kappa(s) \to H^m \left(X_s, \Omega^\bullet_{X/\Delta}(\log Y \cup D) \otimes_{\mathcal{O}_X} \mathcal{O}_{X_s}\right)$$

is an isomorphism.

Remark 7.6. For $s \in \Delta^*$, the right-hand side is just $H^m(X_s - Y_s, \mathbb{C})$, whereas for $s = 0$ one gets canonically the cohomology of nearby cycles for $X - Y$, isomorphic to $H^m(\widetilde{X}^* - \widetilde{Y}^*, \mathbb{C})$.

We assume without irrecoverable loss of generality, as in Sect. 2, that the monodromy automorphism of $H^m(X_s - Y_s, \mathbb{Q})$ is unipotent. A cohomological mixed Hodge complex is constructed parallel to that in [St 1] – which is really the current situation for $Y = \phi$; the construction reduces to Steenbrink's in that case – as follows.

We have a diagram:

$$\text{(7-1)}\qquad \begin{array}{ccccccc} X - Y - D & \xrightarrow{j} & X - Y & \xrightarrow{g} & X & \xleftarrow{i} & D \\ \downarrow & & \downarrow & \swarrow_f & & & \downarrow \\ \Delta^* & \longrightarrow & \Delta & & \longleftarrow & & \{0\} \end{array}$$

Let $A^{\bullet}_{\mathbb{Z}} = \psi_{f,1}(Rg_*\mathbb{Z}_{X-Y})$. Put

$$\text{(7-2)}\qquad H^q_{\mathbb{Q}} = i^*Rg_*Rj_*\mathbb{Q}_{X-Y-D}(q+1)[q+1]/i^*Rg_*\left(\tau_q Rj_*\mathbb{Q}_{X-Y-D}\cdots\right).$$

and define $\theta : H^q_{\mathbb{Q}} \to H^{q+1}_{\mathbb{Q}}$ as before. Take for $(A^{\bullet}_{\mathbb{Q}}, W)$ the resulting simple complex, with filtration determined by

$$\text{(7-3)}\qquad W_k H^q_{\mathbb{Q}} = \text{image of } i^*\tau_{\leq k+2q+1}R(gj)_*\mathbb{Q}_{X-Y-D}(k+1)[k+1].$$

Finally, $A^{\bullet}_{\mathbb{C}}$ is the simple complex of the double complex with terms

$$\text{(7-4)}\qquad A^{p,q} = \Omega^{p+q+1}_X(\log Y \cup D)/W(D)_q\Omega^{p+q+1}_X(\log Y \cup D),$$

and differentials induced by d and $\theta = ds/s$; its filtrations (W, F) are as before:

$$\text{(7-5)}\qquad \begin{aligned} W_k A^{p,q} &= \text{image } W_{k+2q+1}\Omega^{p+q+1}_X(\log Y \cup D) \\ F^p A^{\bullet}_{\mathbb{C}} &= \bigoplus_{r \geq p} A^{r,s}. \end{aligned}$$

In (7-4), $W(D)$ denotes the weight filtration determined by the divisor D alone (i.e., poles along Y do not count). Of course, the total weight filtration W of $\Omega^{\bullet}_X(\log Y \cup D)$ is "comprised of" $W(D)$ and $W(Y)$, indeed is the *convolution* (or *amalgamation*) of these two filtrations:

$$\text{(7-6)}\qquad W_k = \sum_{l+m=k} W(D)_\ell \cap W(Y)_m.$$

It is also useful to consider a third filtration $W(Y)$ of $A^{\bullet}_{\mathbb{C}}$, defined in the most naive manner:

$$W(Y)_\ell A^{p,q} = \text{image of } W(Y)_\ell\Omega^{p+q+1}_X(\log Y \cup D).$$

As was the case when $Y = \phi$, the complex $A^{\bullet}_{\mathbb{C}}$ admits an endomorphism ν, given up to signs by the canonical projections

$$\nu : A^{p,q} \to A^{p-1,q+1} \ .$$

Clearly, ν respects the filtration $W(Y)$. On hypercohomology, ν induces the monodromy logarithm N.

Proposition 7.7. *(1) A filtration $W(Y)$ can be placed on a complex quasi-isomorphic to $A^\bullet_{\mathbb{Q}}$ so that $\mathrm{Gr}_\ell^{W(Y)} A^\bullet$, together with the endomorphism induced by ν, is isomorphic to Steenbrink's $\mathbb{Q}$-cohomological mixed Hodge complex for the proper smooth situation $\widetilde{Y}^{(\ell)} \to \Delta$, with its endomorphism, shifted by $[-\ell]$.*

(2) The mapping

$$\Omega^\bullet_{X/\Delta}(\log(Y \cup D)) \otimes \mathcal{O}_{D\,\mathrm{red}} \to A^\bullet_{\mathbb{C}}$$

is a bifiltered quasi-isomorphism with respect to $W(Y)$ and F.

Before proceeding, we need to introduce a concept due to El Zein [E 2]. This will be presented in a simplified version that incorporates an observation from [St-Z: § 6] (see also [Z 4: § 3]), which seems to be sufficient for all applications.

Definition 7.8. *A $\mathbb{Q}$-cohomological mixed Hodge complex $(K^\bullet_{\mathbb{Q}}, W)$, $(K^\bullet_{\mathbb{C}}, W, F)$ is said to be filtered by the increasing filtrations P of $K^\bullet_{\mathbb{Q}}$ and $K^\bullet_{\mathbb{C}}$ if*

(a) $(P_j K^\bullet_{\mathbb{Q}}/P_k K^\bullet_{\mathbb{Q}}, W)$, $(P_j K^\bullet_{\mathbb{C}}/P_k K^\bullet_{\mathbb{C}}, W, F)$ is a cohomological mixed Hodge complex whenever $j > k$;

(b) $\mathrm{Gr}_\ell^W P_j K^\bullet_{\mathbb{C}} \simeq \bigoplus_{k \le j} \mathrm{Gr}_\ell^W \mathrm{Gr}_k^P K^\bullet_{\mathbb{C}}$, i.e., P is split on $\mathrm{Gr}_\ell^W K^\bullet_{\mathbb{C}}$.

Proposition 7.9 [E 2, E 7]. *For a $\mathbb{Q}$-cohomological mixed Hodge complex filtered by P (as before),*

(1) The two direct filtrations and the recursive filtration determined by W on ${}_PE_r$ all coincide, and for $r = \infty$ they also coincide with what the filtration W of $H^\bullet(K^\bullet_{\mathbb{C}})$ induces on $\mathrm{Gr}^P H^\bullet(K^\bullet_{\mathbb{C}})$; the same holds when W is replaced by F.

(2) The filtrations W and F determine a mixed Hodge structure on ${}_PE_r$, and d_r is a morphism of mixed Hodge structures. On $E_\infty^{-j,n+j} \simeq \mathrm{Gr}_j^P H^n(K^\bullet_{\mathbb{C}})$, the mixed Hodge structure is the quotient of the mixed Hodge structure on

$$\mathrm{im}\,\{H^n(P_j K^\bullet_{\mathbb{C}}) \to H^n(K^\bullet_{\mathbb{C}})\} \ .$$

Remark 7.10. The role of condition (b) in Definition 7.8 is only to insure that the constructions "Dec W" and "P_j/P_k" on $R\Gamma K^\bullet_{\mathbb{C}}$ commute. The condition is satisfied whenever W can be written as the convolution of P with some other filtration of $K^\bullet_{\mathbb{C}}$ [Z 2: (A.1)].

We can now assert:

Theorem 7.11. *The pair $(A^\bullet_{\mathbb{Q}}, W)$, $(A^\bullet_{\mathbb{C}}, W, F)$ is a $\mathbb{Q}$-cohomological mixed Hodge complex filtered by $W(Y)$. On $H^m(D, A^\bullet_{\mathbb{C}}) \simeq \mathcal{V}_0^m$,*

(1) $W(Y)$ induces Deligne's weight filtration of the cohomology group

$$H^m(\widetilde{X}^* - \widetilde{Y}^*, \mathbb{C}) \simeq H^m(X_t - Y_t, \mathbb{C}) \quad (t \in \Delta^*) ,$$

(2) F induces the limit Hodge filtration,

(3) W induces the weight filtration of the local monodromy logarithm N relative to $W(Y)$.

The first assertion above follows from 7.7, 7.8 and 7.10, as $W(Y)$ is a convolutant of W. One can actually compute directly that there is a quasi-isomorphism (compare 2–4):

$$\mathrm{Gr}_r^W A_{\mathbb{C}}^{\bullet} \simeq \bigoplus_{k \geq 0, -r} \bigoplus_{\ell \leq k+r} (a_{r+2k+1})_* \, \Omega^{\bullet}_{\widetilde{Y}^{(\ell)} \cap \widetilde{D}^{(r+2k+1-\ell)}}(-r-k)[-r-2k] .$$

From 7.5, 7.7 (2) and 7.9, one sees that the limit Hodge filtration is induced by F, and that it behaves well under passage to $\mathrm{Gr}^{W(Y)}$. Finally, 7.6 (1) comes close to giving 7.11 (3) (recall 2.14); precisely, it gives the analogous assertion for

$$_{W(Y)}E_1^{-\ell,\bullet} \simeq H^{\bullet}\left(\widetilde{Y}^{\ell}, \mathbb{C}\right)[\ell] .$$

To get the same for $_{W(Y)}E_2 \simeq \mathrm{Gr}^{W(Y)} H^{\bullet}(A_{\mathbb{C}}^{\bullet})$ requires a strictness argument (cf. [E 7: (3.30)]; see also [St-Z: (5.7)]).

With that, we complete the verification of Theorem 7.3 in the smooth case, as we had set out to do. We turn now to a brief discussion of Theorem 7.4.

First, we wish to state an important consequence of (7.4), namely the *rigidity* of admissible variations of mixed Hodge structure. (This does *not* assert that there are no non-trivial deformations, but rather that in any deformation "*everything* must change".) If $\mathbb{V}$ and $\mathbb{V}'$ underlie admissible variations, then $\mathrm{Hom}(\mathbb{V}, \mathbb{V}')$ is also admissible (see [St-Z: ((A.10)]), so then

$$H^0(X, \mathrm{Hom}(\mathbb{V}, \mathbb{V}')) \hookrightarrow \mathrm{Hom}(V(x), V'(x)) \tag{7-7}$$

is a morphism of mixed Hodge structures for any $x \in X$. When $\mathbb{V} = \mathbb{V}'$ as local systems filtered by W, there is a tautological element

$$\mathrm{Id} \in W_0 H^0\big(X, \mathrm{Hom}(\mathbb{V}, \mathbb{V}')\big) .$$

Then 7-7 asserts that the identity element is in F^0 (i.e., is an isomorphism of variations of mixed Hodge structure) whenever the same holds at any one point of X. Thus, we have:

Theorem 7.12 (Rigidity). *An admissible variation of mixed Hodge structure over a compactifiable Kähler manifold is determined completely by its monodromy representation and its value at one point. In other words, given a filtered local system $(\mathbb{V}, W)$ on X, a point $x \in X$, and a filtration $F(x)$ of $V(x)$ such that $(V(x), W(x), F(x))$ determine a mixed Hodge structure, there is at most one extension of this data to an admissible variation of mixed Hodge structure.*

Remark 7.13. For variations of (pure) Hodge structure, this was proved first in the compact case by Griffiths [Gr 1: III, § 7], in the geometric case by Deligne [D 4: II, (4.1)], and in general – without the Kähler hypothesis – by Schmid [Sc: (7.24)].

The construction of the mixed Hodge structure for $H^\bullet(X, \mathbb{V})$, where $\mathbb{V}$ underlies an admissible variation of mixed Hodge structure over the compactifiable Kähler manifold X, in full generality, was done by Morihiko Saito [Sa 10], using his theory of mixed Hodge modules [Sa 5, Sa 11].

In the case where X is a curve, the construction of the mixed Hodge structure can be carried out "by hand", as was done in [St-Z: § 4]. Assume for simplicity that the local monodromy transformations around $x \in D$ are unipotent, and let $N(x)$ denote the nilpotent logarithms. We take

$$K^\bullet_{\mathbb{C}} = \Omega^\bullet_{\bar{X}}(\log D) \otimes \overline{\mathscr{V}} ,$$

with its usual filtration F (as in 3-8). The weight filtration Wt of $K^\bullet_{\mathbb{C}}$ is comprised of subcomplexes of the form

(7-8) $$\overline{\mathscr{A}} \to \mathscr{B} ,$$

where $\overline{\mathscr{A}}$ is the canonical extension associated to a local subsystem $\mathbb{A}$ of $\mathbb{V}$, and

(7-9) $$\mathscr{B} = \left\{ \varphi \in \Omega^1_{\bar{X}}(\log D) \otimes \overline{\mathscr{A}} : \mathrm{Res}_x(\varphi) \in B(x) \quad \text{for all } x \in D \right\} ;$$

here, $B(x)$ must satisfy

$$N(x)A(x) \subset B(x) \subset A(x) .$$

We write $\{\mathbb{A}, B\}$ for (7-8). Let

$$B_k(x) = N(x)W_k(x) + M_{k-1}(x)W_k(x) ,$$

where $M(x)$ denotes the weight filtration of $N(x)$ relative to W, and define $\mathscr{B}_k$ accordingly (as in (7-9)). Then $Wt_k K^\bullet_{\mathbb{C}}$ is the complex $\{\mathbb{W}_k, B_k\}$, i.e.

$$\overline{\mathscr{W}}_k \to \mathscr{B}_k .$$

One computes an isomorphism (filtered by F):
(7-10)

$$\mathrm{Gr}^{Wt}_k K^\bullet_{\mathbb{C}} \simeq \{\mathrm{Gr}^W_k \mathscr{V}, N\mathrm{Gr}^W_k V\} \oplus \left(\bigoplus_{x \in D} \mathrm{Gr}^{M(x)}_{k-1} \left(W_{k-1}(x)/N(x)W_{k-1}(x)\right) (-1)[-1] \right) .$$

Remark 7.14. Formula (7-9) includes the omitted description of the weight filtration in the pure case (before 3.14).

Theorem 7.15. *Given an admissible variation of A-mixed Hodge structure on the smooth algebraic curve X, with underlying local system* $\mathbb{V}$, $(K^\bullet_{\mathbb{C}}, Wt, F)$ *is the complex part of a cohomological A-mixed Hodge complex for which*

$$H^\bullet(K^\bullet_{\mathbb{C}}) \simeq H^\bullet(X, \mathbb{V}) .$$

There is another context in which Theorem 7.4 can be verified directly (see [H-Z 1: § 8]), and which is also of independent interest:

Definition 7.16. *A variation of mixed Hodge structure with underlying local system* $\mathbb{V}$ *and weight filtration W, is said to be* unipotent *if the variation of Hodge structure on the pure subquotients* $\mathrm{Gr}_k^W \mathbb{V}$ *are constant.*

The classifying mapping $\widetilde{\Phi}$ of a unipotent variation maps into $W_{-1}\mathrm{GL}(V)_{\mathbb{C}}/(\widetilde{B} \cap W_{-1}\mathrm{GL}(V)_{\mathbb{C}})$. The reason for the terminology lies in:

Proposition 7.17. *Each of the following conditions on the monodromy representation*

$$\rho : \pi_1(X, x) \to W_0\mathrm{GL}(V)$$

of a variation of mixed Hodge structure is equivalent to "unipotence" (as defined above):

(1) ρ *is unipotent;*

(2) the linear extension of ρ *to the group ring,* $\overline{\rho} : \mathbb{C}\pi_1(X, x) \to W_0\mathfrak{gl}(V)$, *is trivial on some power* J^ℓ *of the augmentation ideal;*

(3) $\overline{\rho}(J) \subset W_{-1}\mathfrak{gl}(V)$.

We give some examples of unipotent variations coming from geometry.

Examples 7.17. (1) Let X be a smooth variety. For any $s \geq 0$, the family $\mathbb{Z}[\pi_1(X, x)]/J^{s+1}$ as x varies over X determines a unipotent variation of mixed Hodge structure. This is because for the dual variation, $\mathbf{B}_s H^0(H(A^\bullet X))$, – recall 6.19 and 6.20 – $\mathrm{Gr}^{\mathbf{B}}$ is given completely in terms of cohomology, so its Hodge structure is independent of x. We call the above the s-th *tautological variation.*

(2) Let X be a smooth, complete curve. Fix a point $y \in X$, and consider $V(x) = H^1(X, \{x, y\})$ (which can be identified with H^1 of the singular quotient curve where x and y are identified), as x varies over $X - \{y\}$. From the exact sequence

$$0 \to \mathbb{C}([x] - [y]) \to H^1(X, \{x, y\}) \to H^1(X) \to 0 ,$$

one can see that the resulting variation of mixed Hodge structure is unipotent. It is a cousin of the first example (1), for it is dual to

$$H^1(X, \{x, y\})(1) \simeq H_0(P_{x,y}X)/J^2$$

(see [H-Z 1: (5.39)]).

(3) Let V be a vector space of dimension 2, and consider the set of mixed Hodge structures on V with

$$h^{p,q} = \begin{cases} 1 & \text{if } (p,q) = (0,0), (-1,-1), \\ 0 & \text{otherwise.} \end{cases}$$

By what we said earlier, the set of such is parametrized by $W_{-1}\mathrm{GL}(V)_{\mathbb{C}} \simeq \mathbb{C}$. A unipotent variation of $\mathbb{Z}$-mixed Hodge structure is thereby defined over

$$\mathbb{C}^* \simeq \mathbb{C}/\mathbb{Z}(1) .$$

It can be seen to come from geometry (see [St-Z: (2.13)]).

From 7.17 (1), we have that all local monodromy transformations in a unipotent variation are, of course, unipotent, and their nilpotent logarithms N satisfy

$$NW_\ell(V) \subset W_{\ell-1}(V) \quad \text{for all } \ell . \tag{7-11}$$

Suppose further that the variation of mixed Hodge structure is also admissible. Then condition 7.2 (b) is equivalent to the following improvement upon (7-11):

$$NW_\ell(V) \subset W_{\ell-2}(V) \quad \text{for all } \ell ,$$

and then the weight filtration of N relative to W coincides with W itself.

For admissible unipotent variations, there is a classification theorem, which provides a complete complement to Theorem 7.12 in this case, given in terms of a construction from the monodromy representation.

Definition 7.18. *An* A-mixed Hodge theoretic unipotent representation *of* $\pi_1(X,x)$ *is an A-algebra homomorphism,*

$$\rho : A[\pi_1(X,x)]/J^{s+1} \to \mathfrak{gl}(V_A)$$

for some $s \geq 0$, *which is a morphism of mixed Hodge structures; here* V_A *underlies some A-mixed Hodge structure.*

Theorem 7.19 [H-Z 1: (1.6)]. *Admissible unipotent variations are in natural one-to-one correspondence with mixed Hodge theoretic unipotent representations of* $\pi_1(X,x)$.

The correspondence is easy to describe, and goes as follows. Inside $\mathbb{C}[\pi_1(X,x)]/J^{s+1}$ is the s-th *Malcev group* G_s (the subset of group-like elements). Given ρ as above, it restricts to a group homomorphism

$$G_s \to W_{-1}\mathrm{GL}(V) ;$$

furthermore, because ρ is mixed Hodge theoretic, it defines a mapping:

$$G_{s,\mathbb{C}}/F^0G_{s,\mathbb{C}} \xrightarrow{\widehat{\rho}} W_{-1}\mathrm{GL}(V)_{\mathbb{C}}/F^0W_{-1}\mathrm{GL}(V)_{\mathbb{C}} .$$

Note that the target of $\widehat{\rho}$ is the space of all mixed Hodge structures whose pure subquotients coincide with those of the given mixed Hodge structure on V. We might have started with $V = \mathbb{C}[\pi_1(X,x)]/J^{s+1}$ itself, and ρ the adjoint representation, in which case one must observe

Proposition 7.20 [H-Z 1: (5.20)]. *The classifying mapping of the s-th tautological variation factors through* $\widehat{\rho}$.

We write $\alpha_s : \widetilde{X} \to G_{s,\mathbb{C}}/F^0G_{s,\mathbb{C}}$ for the resulting mapping. (One checks that α_1 gives the usual Albanese mapping [H-Z 1: (5.40)].) Then given any ρ (and corresponding $\widehat{\rho}$), one gets a unipotent variation of mixed Hodge structure by taking $\widehat{\rho} \circ \alpha_s$ as classifying mapping. The tautological variations, being geometric, are admissible, and it is easy to see that the ones constructed as above are likewise [H-Z 1: p. 115]. Conversely, the monodromy representation

$$\mathbb{C}[\pi_1(X,x)]/J^{s+1} \to W_0\mathrm{gl}(V)$$

of an admissible unipotent variation is always a morphism of mixed Hodge structures [H-Z 1: § 7]. That we get a one-to-one correspondence follows from rigidity (7.12).

8. Monodromy Groups

In Sect. 7, we have seen that there is a phenomenon of rigidity for admissible variations of mixed Hodge structure (Theorem 7.12): if one specifies a filtered local system of A-modules on a variety X, and a mixed Hodge structure on the fiber at one point $x \in X$ (with the given filtration as weight filtration), the number of admissible variations of A-mixed Hodge structure extending this data is either zero or one. In this section we discuss recent results in the direction of whether a given local system admits any variations of (mixed) Hodge structure at all. When $A = \mathbb{Z}$, this is a broadening of the question: what are the global monodromy groups for families of algebraic varieties? (Recall that giving the local system $\mathbb{V}$ is equivalent to specifying its monodromy representation

$$\rho : \pi_1(X,x) \to \mathrm{GL}(V_x)\ .)$$

For unipotent representations, the answer is given by Theorem 7.18. The set of admissible unipotent variations on $\mathbb{V}$ is parametrized by the set of mixed Hodge structures on V_x compatible with that of the nilpotent completion of $\pi_1(X,x)$. Deligne has informed us that there is a generalization to the case where the pure subquotients of $\mathbb{V}$ have finite monodromy. Results in general are just beginning to emerge.

We will restrict ourselves to the pure case from here on. First, it is useful to drop the reality condition ($\overline{V^{p,q}} = V^{q,p}$) from Definition 2.1:

Definition 8.1 (see [D 8: (1.1), Z 7: (4.6)]). *Let A be a subring of $\mathbb{C}$ that is stable under complex conjugation. A complex variation of A-Hodge structure of weight m on X consists of a local system $\mathbb{V}_A$ of A-modules of finite rank, together with the following data:*

(1) a collection of C^∞ sub-bundles $\mathscr{V}^{p,q}$ of $\mathscr{V} = \mathscr{O}_X \otimes_A \mathbb{V}_A$, for pairs of integers (p, q) with $p+q = m$, such that $\mathscr{V}$ is C^∞-isomorphic to their direct sum;

(2) a flat $(-1)^m$-hermitian form $S : \mathbb{V}_A \times \mathbb{V}_A \to A(-m)$,

which satisfy

(a) for any r, the bundle $\bigoplus_{p \geq r} \mathscr{V}^{p,q}$ defines a holomorphic sub-bundle $\mathscr{F}^p$ of $\mathscr{V}$, and $\bigoplus_{q \geq s} \mathscr{V}^{p,q}$ defines an anti-holomorphic sub-bundle $\overline{\mathscr{F}}^q$.

(b) $\nabla(\mathscr{F}^p) \subset \Omega^1_X \otimes_{\mathscr{O}_X} \mathscr{F}^{p-1}$, and the same for the anti-holomorphic objects:

$$\nabla(\overline{\mathscr{F}}^q) \subset \overline{\Omega}^1_X \bigotimes_{\overline{\mathscr{O}}_X} \overline{\mathscr{F}}^{q-1} .$$

(c) for every $x \in X$, the form S_x on the A-module $\mathscr{V}_{A,x}$ satisfies: $(2\pi i)^m S_x(w, C_x w) > 0$ for $w \neq 0$; here C_x is the Weil operator, the direct sum of i^{p-q} on $V^{p,q}_x = F^p_x \cap \overline{F}^q_x$.

Remark 8.2. The *conjugate* of a complex variation $\mathscr{V}$ is defined by the complex conjugate, i.e. dual (by 8.1 (2)), local system $\mathbb{V}$, with summands $\overline{\mathscr{V}}^{p,q} = \overline{\mathscr{V}^{p,q}}$. A polarized variation of $\mathbb{R}$-Hodge structure (2.1) is just a complex variation that is isomorphic to its conjugate. Moreover, given any complex variation $\mathscr{V}$, $\mathscr{V} \oplus \overline{\mathscr{V}}$ is a real one.

The following theorem of Deligne shows that underlying a variation of Hodge structure is a strong condition on local systems:

Theorem 8.3 [D 8: Theorem 0.5]. *Fix a variety X and a positive integer d. Then there are only finitely many isomorphism classes of monodromy representations for local systems of rank d underlying complex variations of $\mathbb{Z}$-Hodge structure on X.*

The main point in the proof is to realize that the existence of a variation implies (elementwise) bounds on the trace of the monodromy. Sufficiently many traces determine the representation up to isomorphism.

Remark 8.4. The case of variations of Hodge structure of weight one and Hodge types $(1,0)$, $(0,1)$ (families of abelian varieties) is implicit in [F 1].

Recent work of Simpson [Si 1–Si 4] sheds much light on the original question. Suppose that a local system $\mathbb{V}$ on X underlies a complex variation of Hodge structure. Then the vector bundle $\mathscr{V}$ is, of course, C^∞-isomorphic to

$$\mathscr{E} = \bigoplus_p \left(\mathscr{F}^p / \mathscr{F}^{p+1}\right) .$$

Moreover, the latter admits an endomorphism-valued 1-form θ, given as the direct sum of "the linear parts of the flat connection", i.e.

$$\left(\mathscr{F}^p / \mathscr{F}^{p+1}\right) \longrightarrow \Omega^1_X \otimes \left(\mathscr{F}^{p-1} / \mathscr{F}^p\right)$$

(cf. (3-9)). This provides an example of the following:

Definition 8.5. *A* Higgs bundle *on X is a pair $(\mathscr{E}, \theta)$, consisting of a holomorphic vector bundle $\mathscr{E}$ on X and a section θ of $\Omega^1_X \otimes \mathrm{End}(\mathscr{E})$ such that $\theta \wedge \theta = 0$.*

An isomorphism of Higgs bundles is defined in the obvious way. There is an action of $\mathbb{C}^*$ on the set of Higgs bundles, given by the simple rule

$$t \cdot (\mathscr{E}, \theta) = (\mathscr{E}, t\theta) .$$

Lemma 8.6. *The following statements about $(\mathscr{E}, \theta)$ are equivalent:*
(1) $\mathscr{E}$ admits a decomposition $\oplus_p \mathscr{E}^p$ such that $\theta \mathscr{E}^p \subset \Omega^1_X \otimes \mathscr{E}^{p-1}$;
(2) $t \cdot (\mathscr{E}, \theta) \simeq (\mathscr{E}, \theta)$ for some t of infinite order in $\mathbb{C}^*$;
(3) the isomorphism class of $(\mathscr{E}, \theta)$ is fixed by the $\mathbb{C}^*$-action.

Thus, a complex variation of Hodge structure gives rise to a Higgs bundle that is invariant under the $\mathbb{C}^*$-action. But actually, any local system $\mathbb{V}$ on a compact Kähler manifold gives rise to a Higgs bundle, by the following construction. Let $\nabla = \nabla' + \nabla''$ be the flat connection on C^∞ sections of $\mathscr{V}$, where ∇' is its (1,0)-component and $\nabla'' = \overline{\nabla'}$ the (0,1)-component. For any hermitian metric h on $\mathscr{V}$, one has the (holomorphic) metric connection $D_h + \nabla''$. Put

$$\bar{\partial} = \frac{1}{2}\left(\overline{D}_h + \nabla''\right) , \quad \theta = \frac{1}{2}\left(\nabla' - D_h\right) . \tag{8-1}$$

If the metric h is *harmonic* (i.e., classified by a harmonic mapping into $\mathrm{GL}(d, \mathbb{C})/U(d)$) – one must argue the existence and uniqueness of the harmonic metric – the operators in (8-1) will satisfy

$$\left(\bar{\partial} + \theta\right)^2 = 0 , \tag{8-2}$$

from which it follows that the data $(\mathscr{E}, \theta)$, where $\mathscr{E}$ denotes the underlying C^∞ bundle of $\mathscr{V}$ but with complex structure determined by $\bar{\partial}$ above, defines a Higgs bundle. Of course, $c(\mathscr{E}) = 1$ in $H^\bullet(X, \mathbb{Q})$. By the way, if one starts with a complex variation, the harmonic metric will be just the Hodge metric.

Conversely, one produces a local system from a Higgs bundle $(\mathscr{E}, \theta)$ with vanishing Chern classes as follows. One seeks to construct a flat connection on the underlying C^∞ bundle of $\mathscr{E}$. Let h be an hermitian metric on $\mathscr{E}$, and ∂_h denote the (1,0)-component of the metric connection. Also, let $\overline{\theta}_h$ denote the operator of type (0, 1) that is adjoint to θ with respect to h. Use $\bar{\partial}+\theta$ to define a new complex structure on $\mathscr{E}$, and consider the connection

$$\nabla_h = (\partial_h + \overline{\theta}_h) + (\bar{\partial} + \theta) . \tag{8-3}$$

If h satisfies the hermitian Yang-Mills equation [Si 1: p. 878], the Chern class condition will imply that it is also harmonic, and then ∇_h will be flat.

In order to proceed, we must recall the notion of stability for vector bundles. Let $\mathscr{C}$ be a class of vector bundles, perhaps with extra structure.

Definition 8.7. *A vector bundle $\mathscr{E} \in \mathscr{C}$ is said to be a* semi-stable *(resp.* stable*) $\mathscr{C}$-bundle if the function* $(\deg \mathscr{E}')/(\operatorname{rk} \mathscr{E}')$ *on non-zero $\mathscr{C}$-sub-bundles $\mathscr{E}'$ of $\mathscr{E}$ is maximized (resp. strictly maximized) by $\mathscr{E}' = \mathscr{E}$.*

Here, the *degree* of $\mathscr{E}'$ means $c_1(\mathscr{E}') \cdot \omega^{n-1}$, where ω is the Kähler class of X.

Remark 8.8. Since the degree of a flat bundle is always zero, stability for flat bundles means irreducibility (i.e., having no flat sub-bundles), and all flat bundles are semi-stable.

The main result is:

Theorem 8.9 [Si 3: § 2]. *Let X be a compact Kähler manifold. Then the two constructions:*

$$\nabla \mapsto \theta_h \ : \ \text{(8-1) } \textit{for a harmonic metric}$$

$$\theta \mapsto \nabla_h \ : \ \text{(8-3) } \textit{for an hermitian Yang-Mills metric}$$

are mutually inverse, and set up a one-to-one correspondence between isomorphism classes of irreducible flat bundles and stable Higgs bundles with vanishing Chern classes.

This theorem is based on work of Corlette [Co], Hitchin [Hit], and others (see [Si 2: § 1, Si 3: § 1]). It admits a rather direct extension to semi-simple flat bundles [Si 3: (2.3)]. An immediate consequence is the complement to Lemma (8.6):

Corollary 8.10 [Si 3: (4.2)]. *A (semi-simple) local system underlies a complex variation of Hodge structure if and only if the corresponding Higgs bundle is invariant under the $\mathbb{C}^*$-action.*

Note that both the correspondence of Theorem 8.9 and the $\mathbb{C}^*$-action on Higgs bundles are functorial for holomorphic mappings. From this, we get the following striking result:

Theorem 8.11 [Si 3: (4.3)]. *Let X and Y be compact Kähler manifolds, and $f : Y \to X$ a holomorphic mapping. Assume that*

$$f_* : \pi_1(Y, y) \to \pi_1(X, f(y))$$

is surjective. Then a local system $\mathbb{V}$ on X underlies a complex variation of Hodge structure if (and only if) $f^*\mathbb{V}$ does.

It is easy to identify when a Higgs bundle comes from a *real* representation of $\pi_1(X, x)$:

Proposition 8.12 [Si 3: (3.20)]. *The Higgs bundle $(\mathscr{E}, \theta)$ corresponds to a real local system if and only if there is a symmetric non-degenerate pairing*

$$S : \mathscr{E} \otimes_{\mathcal{O}_X} \mathscr{E} \to \mathcal{O}_X$$

such that $S(\theta e, e') = S(e, \theta e')$.

On the other hand, it can be shown, via the discussion in [D 4: II, § 4] and [D 8], that if an irreducible real local system underlies a complex variation of Hodge structure, then it also underlies a real one. Thus, the assertion in Theorem 8.11 (and also Theorem 8.16 below) is valid over $\mathbb{R}$.

Next, the correspondence in Theorem 8.9 is made topological via the introduction of coarse moduli spaces. In the case of flat bundles, the construction is rather direct, for it depends only on the finitely-generated group $\pi_1(X, x)$. Let $\{\gamma_1, \ldots, \gamma_\ell\}$ be a set of generators for $\pi_1(X, x)$. A representation ρ of dimension d comes down to assigning ℓ invertible $d \times d$ matrices $\rho(\gamma_j)$, subject only to the algebraic equations determined by the relations among the generators. Isomorphism classes of representations are just the orbits under the ℓ-fold adjoint action of $\mathrm{GL}(d, \mathbb{C})$. A coarse moduli space of semi-simple isomorphism classes M_{flat} of flat vector bundles can be constructed by geometric invariant theory. (Here, a reducible bundle must be equated with the sum of its composition series factors.) M_{flat} is an affine variety, and it has only quadratic singularities [G-Mi, Si 3: (3.4)].

For the purpose of obtaining the moduli space of semi-stable Higgs bundles with vanishing Chern classes, one adds the hypothesis that X is projective. Then a typical method of constructing moduli spaces in algebraic geometry can be used to produce M_{Higgs} [Si 2: § 3]. We consider these moduli spaces with their classical (complex) topologies. Then

Proposition 8.13 [Si 2: § 5]. *The correspondence given in Theorem 8.9 induces a homeomorphism between M_{flat} and M_{Higgs}.*

Remark 8.14. i) It is expected that the homeomorphism is actually real-analytic.

ii) It would be nice to find a construction of M_{Higgs} that works for arbitrary compact Kähler manifolds.

Although M_{Higgs} is non-compact, the action of $\mathbb{C}^*$ extends to $\mathbb{C}$, i.e., for any $(E,\theta) \in M_{\text{Higgs}}$, $\lim_{t\to 0} t \cdot (E,\theta)$ exists in M_{Higgs} [Si 3: §4]. But this limit is clearly $\mathbb{C}^*$-invariant. Thus one gets from Corollary 8.10 and Theorem 8.13:

Theorem 8.15. *Let X be a smooth projective variety. Then every connected component of* M_{flat} *contains a flat bundle that underlies a complex variation of Hodge structure.*

This has the following remarkable corollary:

Theorem 8.16. *Every rigid representation of* $\pi_1(X,x)$ *is the monodromy representation of a complex variation of Hodge structure.*

From this, Simpson concludes that certain discrete groups cannot be fundamental groups of compact Kähler manifolds.

It is natural to try to extend this theory to the case of non-compact smooth varieties. This has been carried out in the case of curves by Simpson in [Si 4]. The objects that intervene in the generalization of Theorem 8.9 and 8.12 must be given filtrations at infinity.

References

[A] Adams, J.: On the cobar construction. In: Colloque de topologie algébrique (Louvain, 1956). Masson 1957, pp. 81–87

[An] Anderson, G.: Cyclotomy and an extension of the Taniyama group. Comp. Math. **57** (1986) 153–217

[A-V] Andreotti, A., Vesentini, E.: Carleman estimates for the Laplace-Beltrami equation on complex manifolds. Publ. Math. I.H.E.S. **25** (1966) 81–130

[Ar] Arapura, D.: Hodge theory with local coefficients and fundamental groups of varieties. Bull. Am. Math. Soc. **20** (1989) 169–172

[Ba 1] Barlet, D.: Filtration de Hodge asymptotique et poles de $\int |f|^{2\lambda}$. Prépubl. Inst. Henri Cartan 88/7

[Ba 2] Barlet, D.: Symétrie de Hodge pour le polynôme de Bernstein-Sato. (Lecture Notes in Mathematics, vol. 1295). Springer 1987, pp. 1–10

[Be 1] Beilinson, A.: Higher regulators and values of *L*-functions. J. Soviet Math. **30** (1985) 2036–2070

[Be 2] Beilinson, A.: Notes on absolute Hodge cohomology. (Contemporary Mathematics, vol. 55, part I). Amer. Math. Soc. 1986, pp. 35–68

[Be 3] Beilinson, A.: Height pairing between algebraic cycles. In: K. Ribet (ed.) Current trends in arithmetical algebraic geometry (Contemporary Mathematics, vol. 67). American Mathematical Society 1987, pp. 1–24

[Be 4] Beilinson, A.: On the derived category of perverse sheaves. In: *K*-Theory, Arithmetic and Geometry, Yu. Manin, ed. (Lecture Notes in Mathematics, vol. 1289) Springer, pp. 27–41

[B-B-D] Beilinson,A, Bernstein, I.N., Deligne, P.: Faisceaux Pervers. (Astérisque, vol. 100). Soc. Math. Fr., 1982

[B-G] Beilinson, A., Ginsburg, V.: Mixed categories, Ext-duality and representations (results and conjectures). Preprint 1988

[B-G-G] Bernstein, I.N., Gel'fand, I.M., Gel'fand, S.I.: Differential operators on the base affine space and a study of g-modules. In: Lie Groups and their Representation. Halsted 1975, pp. 21–64

[B-M-S] Beilinson, A., MacPherson, R., Schechtman, V.: Notes on motivic cohomology. Duke Math. J. **54** (1987) 679–710

[Bl 1] Bloch, S.: Deligne groups info. Manuscript (unpubl.) c. 1972

[Bl 2] Bloch, S.: Algebraic K-theory and zeta-functions of elliptic curves. In: Proc. Intern. Congress Helsinki, 1978, pp. 511–515

[Bl 3] Bloch, S.: Applications of the dilogarithm function in algebraic K-theory and algebraic geometry. In: Proceedings of the International Symposium on Algebraic Geometry 1977. Kinokuniya Book Store, 1978, pp. 103–114

[Bl 4] Bloch, S.: The dilogarithm and extensions of Lie algebras. (Lecture Notes in Mathematics, vol. 854). Springer, 1981, pp. 1–23

[Bl 5] Bloch, S.: Height pairing between algebraic cycles.

[Bl-O] Bloch, S., Ogus, A.: Gersten's conjecture and the homology of schemes. Ann. Sci. Ec. Norm. Sup. **7** (1984) 181–202

[B-H] Bloom, T., Herrera, M.: The de Rham cohomology of an analytic space. Invent. math. **7** (1969) 275–296

[Bo 1] Borel, A.: Cohomologie de SL_n et valeurs de fonctions zêta. Ann. Sc. Norm. Sup. **7** (1974) 613–636

[Bo 2] Borel, A., et al.: Intersection Cohomology. (Progress in Mathematics, vol. 50). Birkhäuser, 1984

[Bo 3] Borel, A., et al.: Algebraic **D**-modules. (Perspective in Mathematics, vol. 2). Academic Press, 1987

[B-K] Bousfield, A., Kan, D.: Homotopy limits, completions and localizations. (Lectures Notes in Mathematics, vol. 304). Springer, 1972

[Br] Brieskorn, E.: Die Monodromie der isolierten Singularitäten von Hyperflächen. Manuscr. Math. **2** (1970) 103–161

[Bry 1] Brylinski, J.-L.: Cohomologie d'intersection et faisceaux pervers. (Séminaire Bourbaki Exposé 585, February 1982, in: Astérisque, vol. 92-93). Soc. Math. Fr., 1982, pp. 129–157

[Bry 2] Brylinski, J.-L.: Modules holonomes à singularités régulières et filtration de Hodge. In: Algebraic Geometry. Proceedings, la Rabida 1981. (Lecture Notes in Math., vol. 961). Springer, 1982, pp. 1–21. II (Astérisque, vol. 101-102). Soc. Math. Fr., 1983, pp. 75–117

[Bry 3] Brylinski, J.-L.: Transformations canoniques, dualité projective, théorie de Lefschetz, transformation de Fourier et sommes trigonométriques. In: Géométrie et analyse Microlocales. (Astérisque, vol. 140-141). Soc. Math. Fr., 1986, pp. 3–134

[Ca 1] Carlson, J.: Extensions of mixed Hodge structures. In: Journées de Géométrie Algébrique d'Angers 1979. Sijthoff and Nordhoff, 1980, pp. 107–127

[Ca 2] Carlson, J.: The geometry of the extension class of a mixed Hodge structure. In: Algebraic Geometry, Bowdoin 1985, Proc. Symp. Pure Math. 46, Part 2. Amer. Math. Soc., pp. 199–222

[C-C-M] Carlson, J., Clemens, H., Morgan, J.: On the mixed Hodge structure associated to π_3 of a simply-connected projective manifold. Ann. Sci. Ec. Norm. Sup. **14** (1981) 323–338

[Ca-H] Carlson, J., Hain, R.: Extensions of variations of mixed Hodge structure. In: Proc. of the Luminy Conference on Hodge Theory, 1987, Astérisque. Soc. Math. Fr. (to appear)

[C-K 1] Cattani, E., Kaplan, A.: On the $SL(2)$-orbits in Hodge theory (with an appendix by P. Deligne). I.H.E.S., preprint M/82/58 (1982)

[C-K 2] Cattani, E., Kaplan, A.: Polarized mixed Hodge structures and the monodromy of a variation of Hodge structure. Invent. math. **67** (1982) 101–115

[C-K 3] Cattani, E., Kaplan, A.: Degenerating variations of Hodge structures. In: Proc. of the Luminy Conference on Hodge Theory, 1987, Astérisque. Soc. Math. Fr. (to appear)

[C-K-S 1] Cattani, E., Kaplan, A., Schmid, W.: Degeneration of Hodge structures. Ann. Math. **123** (1986) 457–535

[C-K-S 2] Cattani, E., Kaplan, A., Schmid, W.: L^2 and intersection cohomologies for a polarizable variation of Hodge structure. Invent. math. **87** (1987) 217–252

[Ch 1] Cheeger, J.: On the Hodge theory of Riemannian pseudomanifolds. In: Geometry of the Laplace operator. Proc. Symp. Pure Math., vol. 36, 1980, pp.91–146

[Ch 2] Cheeger, J.: Hodge theory of complex cones. In: Analyse et Topologie sur les Espaces Singuliers. (Astérisque, vol. 101-102). Soc. Math. Fr., 1983, pp. 118–134

[C-G-M] Cheeger, J., Goresky, M., MacPherson, R.: L^2-cohomology and intersection homology of singular algebraic varieties. In: Ann. Math. studies, vol. 102. Princeton Univ. Press, 1982, pp. 303–340

[C-S] Cheeger, J., Simons, J.: Differential characters and geometric invariants. In: Geometry and Topology Lecture Notes in Math., vol. 1167. Springer, 1985, pp. 50–80

[Chen 1] Chen, K.-T.: Iterated path integrals. Bull. Am. Math. Soc. **83** (1977) 831–879

[Chen 2] Chen, K.-T.: Reduced bar constructions on de Rham complexes. In: Algebra, Topology and Category Theory (ed. A. Heller and M. Tierny). Academic Press, 1976, pp. 19–32

[Chen 3] Chen, K.-T.: Circular bar construction. J. Alg. **57** (1979) 466–483

[Cl 1] Clemens, C.H.: Picard-Lefschetz theorem for families of nonsingular algebraic varieties acquiring ordinary singularities. Trans. Am. Soc. **136** (1969) 93–108

[Cl 2] Clemens, C.H.: Degeneration of Kähler manifolds. Duke Math. J. **44** (1977) 215–290

[Co] Corlette, K.: Flat G-bundles with canonical metrics. J. Diff. Geom. **28** (1988) 361–382

[D 1] Deligne, P.: Théorème de Lefschetz et critères de dégénérescence de suites spectrales. Publ. Math. I.H.E.S. **35** (1968) 107–126

[D 2] Deligne, P.: Equations différentielles à points singuliers réguliers. (Lecture Notes in Mathematics, vol. 163). Springer, 1970

[D 3] Deligne, P.: Travaux de Griffiths. Séminaire Bourbaki, Exposé 376, June 1970. (Lecture Notes in Mathematics, vol. 180). Springer, 1971, pp. 213–237

[D 4] Deligne, P.: Théorie de Hodge I. Actes, Congrès Intern. Math. Nice 1970, pp. 425–430. II, Publ. Math. I.H.E.S. **40** (1971) 5–58. III, Publ. Math. I.H.E.S. **44** (1974) 5–77

[D 5] Deligne, P.: La conjecture de Weil I. Publ. Math. I.H.E.S. **43** (1974) 273–308. II, Publ. Math. I.H.E.S. **52** (1980) 137–252

[D 6] Deligne, P.: Structures de Hodge mixtes réelles. Appendix to: On the $SL(2)$-orbits in Hodge theory, by E. Cattani and A. Kaplan

[D 7] Deligne, P.: Théorie de Hodge irrégulière. Handwritten notes 1984

[D 8] Deligne, P.: Un théorème de finitude pour la monodromie. In: Discrete Groups in Geometry and Analysis (Papers in the Honor of G. D. Mostow on His Sixtieth Birthday). Progress in Mathematics, vol. 67. Birkhäuser, 1987, pp. 1–19

[D 9] Deligne, P.: Valeurs de fonctions L et périodes d'intégrales. In: Automorphic Forms, Representations and L-functions. Proc. Symp. Pure Math., vol. 33, Part 2. Amer. Math. Soc. 1979, pp. 313–346

[D-G-M-S] Deligne, P., Griffiths, P., Morgan, J., Sullivan, D.: Real homotopy theory of Kähler manifolds. Invent. math. **29** (1975) 245–274

[D-M-O-S] Deligne, P., Milne, J.-S., Ogus, A., Shih, K.-Y.: Hodge cycles, motives and Shimura varieties. (Lecture Notes in Mathematics, vol. 900). Springer, 1982

[Du 1] Du Bois, P.: Complexe de de Rham filtré d'une variété singulière. Bull. Soc. Math. Fr. **109** (1981) 41–81

[Du 2] Du Bois, P.: Structure de Hodge mixte sur la cohomologie évanescente. In: Ann. Inst. Fourier **35** (1985) 191–213

[Dur] Durfee, A.: A naive guide to mixed Hodge theory. In: Proc. Symp. Pure Math., vol. 40, part I. Amer. Math. Soc. 1983, pp. 313–320

[D-H] Durfee, A., Hain, R.: Mixed Hodge structures on the homotopy of links. Math. Ann. **280** (1988) 69–83

[D-S] Durfee, A., Saito, M.: Mixed Hodge structures on intersection cohomology of links. Preprint 1988

[E 1] El Zein, F.: Complexe dualisant et applications à la classe fondamentale d'un cycle. Mémoire Bull. Soc. Math. Fr. **58** (1978) 5–66

[E 2] El Zein, F.: Complexe de Hodge mixte filtré. CRAS Paris **295** (1982) 669–672

[E 3] El Zein, F.: Mixed Hodge structures. Proc. Symp. Pure Math., vol. 40, 1983, 345–352

[E 4] El Zein, F.: Dégénérescence diagonale I. C.R.A.S. Paris **296** (1983) 51–54. II, C.R.A.S. Paris **296** (1983) 199–202

[E 5] El Zein, F.: Suites spectrales de structures de Hodge mixtes. (Astérisque, vol. 130). Soc. Math. Fr. 1985, pp. 308–329

[E 6] El Zein, F.: Théorie de Hodge à coefficients: étude locale. C.R.A.S. Paris **307** (1988) 593–598

[E 7] El Zein, F.: Théorie de Hodge des cycles évanescents. Ann. Sci. Ec. Norm. Sup. **19** (1986) 107–194

[E 8] El Zein, F.: Mixed Hodge structures. Trans. Amer. Math. Soc. **275** (1983) 71–106

[E-Z] El Zein, F., Zucker, S.: Extendability of normal functions associated to algebraic cycles. In: Topics in Transcendental Algebraic Geometry, P. Griffiths, ed. Princeton University Press, vol. 106, 1986, pp. 269–288

[E-V] Esnault, H., Viehweg, E.: Deligne-Beilinson cohomology. In: Beilinson's Conjectures on Special Values of L-functions. (Perspectives in Mathematics, vol. 4). Academic Press, pp. 43–91

[F 1] Faltings, G.: Arakelov's theorem for abelian varieties. Invent. math. **73** (1983) 337–347

[F 2] Faltings, E.: p-adic Hodge theory. J. Amer. Math. Soc. **1** (1988) 255–299

[Fr] Friedman, S.: Global smoothings of varieties with normal crossings. Ann. Math. **118** (1983) 75–114

[G] Gabber, O.: Pureté de la cohomologie de Goresky-MacPherson. Prépublication I.H.E.S. 1981

[Gaf 1] Gaffney, M.: A special Stokes' theorem for complete Riemannian manifolds. Ann. Math. **60** (1954) 140–145

[Gaf 2] Gaffney, M.: Hilbert space methods in the theory of harmonic integrals. Trans. Amer. Math. Soc. **78** (1955) 426–444

[Gi] Gillet, H.: Deligne homology and Abel-Jacobi maps. Bull. Amer. Math. Soc. **10** (1984) 284–288

[Go] Godement, R.: Topologie Algébrique et Théorie des Faisceaux. Hermann, 1958

[G-Mi] Goldman, W., Millson, J.: The deformation theory of representations of fundamental groups of compact Kähler manifolds. J. Diff. Geom. **67** (1988) 43–96

[G-M] Goresky, M., MacPherson, R.: Intersection homology. Topology **19** (1980) 77–129. II, Invent. math. **72** (1983)

[Gr 1] Griffiths, P.: Periods of integrals on algebraic manifolds I. Amer. J. Math. **90** (1968) 568–626. II, Amer. J. Math. **90** (1968) 805–865. III, Publ. Math. I.H.E.S. (1970) 228–296

[Gr 2] Griffiths, P.: Algebraic cycles on algebraic manifolds. In: Algebraic Geometry. (Papers presented at the Bombay Colloquium 1969). Oxford University Press, 1969, pp. 93–191

[Gr 3] Griffiths, P.: On the periods of certain rational integrals. Ann. Math. **90** (1969) 460–541

[Gr 4] Griffiths, P.: Periods of integrals on algebraic manifolds: Summary of main results and discussion of open problems. Bull. Amer. Math. Soc. **76** (1970), 228–296

[Gr-S] Griffiths, S., Schmid, W.: Recent developments in Hodge theory: A discussion of techniques and results. In: Proc. Bombay Colloq. on Discrete Subgroups of Lie Groups (Bombay, 1973). Oxford University Press, 1975, pp. 31–127

[Gro] Grothendieck, A.: On the de Rham cohomology of algebraic varieties. Publ. Math. I.H.E.S. **29** (1966) 96–103

[G-NA-P] Guillén, F., Navarro Aznar, V., Puerta, F.: Théorie de Hodge via schémas cubiques. Preprint 1982

[G-P] Guillén, F., Puerta, F.: Hyperrésolutions cubiques et applications à la théorie de Hodge-Deligne. In: Hodge Theory Proceedings, Sant Cugat, 1985. (Lecture Notes in Mathematics, vol. 1246). Springer, 1987

[G-NA-PG-P] Guillén, F., Navarro Aznar, V., Pascual-Gainza, P., Puerta, F.: Hyperrésolutions cubiques et descente cohomologique. (Lecture Notes in Mathematics, vol. 1335). Springer, 1988

[H 1] Hain, R.: The de Rham homotopy theory of complex algebraic varieties I. *K*-Theory **1** (1987) 271–324. II, *K*-Theory

[H 2] Hain, R.: The geometry of the mixed Hodge structure on the fundamental group. In: Proc. Symp. Pure Math. Amer. Math. Soc. 1987, pp. 247–282

[H-M] Hain, R., MacPherson, R.: Higher logarithms. To appear in a volume in the honor of Chen, *K*-Theory

[H-Z 1] Hain, R., Zucker, S.: Unipotent variations of mixed Hodge structure. Invent. math. **88** (1987) 83–124

[H-Z 2] Hain, R., Zucker, S.: A guide to unipotent variations of mixed Hodge structures. In: Hodge Theory Proceedings, Sant Cugat, 1985. (Lecture Notes in Mathematics, vol. 1246). Springer, 1987, pp. 92–106

[Ha] Harris, B.: Harmonic volumes. Acta Math. **150** (1983) 91–123

[Hi] Hironaka, H.: Resolution of singularities of an algebraic variety over a field of characteristic zero. Ann. Math. **79** (1964) 109–326

[Hit] Hitchin, N.: The self-duality equation over a Riemann surface. Proc. Lond. Math. Soc. **55** (1987) 59–126

[Ho] Hodge, W.V.D.: The Theory and Applications of Harmonic Integrals. Cambridge University Press, 1941

[H-P] Hsiang, W.-C., Pati, V.: L^2-cohomology of normal algebraic surfaces. Invent. math. **81** (1985) 395–412

[J] Janssen, U.: Deligne homology, Hodge-*D* conjecture and motives. In: Beilinson's Conjectures on Special Values of *L*-functions. (Perspectives in Mathematics, vol. 4). Acad. Press, 1988, pp. 305–372

[K 1] Kashiwara, M.: On the maximally overdetermined system of linear differential equations I. Publ. Res. Inst. Math. Kyoto University **10** (1974/75) 563–579

[K 2] Kashiwara, M.: *B*-functions and holonomic systems, rationality of roots of *b*-functions. Invent. math. **38** (1976) 33–53

[K 3] Kashiwara, M.: On the holonomic systems of differential equations II. Invent. math. **49** (1978) 121–135

[K 4] Kashiwara, M.: Vanishing cycles and holonomic systems of differential systems of differential equations. In: Algebraic Geometry, M. Raynaud and T. Shioda, eds. (Lecture Notes in Mathematics vol. 1016). Springer, 1983, pp. 134–142 1983

[K 5] Kashiwara, M.: The asymptotic behavior of a variation of polarized Hodge structure. Publ. RIMS. Kyoto University **21** (1985) 853–875

[K 6] Kashiwara, M.: A study of a variation of mixed Hodge structure. Publ. RIMS. Kyoto University **22** (1986) 991–1024

[K-K 1] Kashiwara, M., Kawai, T.: On the holonomic systems of microdifferential equations III. Systems with regular singularities. Publ. R.I.M.S. Kyoto University **17** (1981) 813–979

[K-K 2] Kashiwara, M., Kawai, T.: The Poincaré lemma for variations of polarized Hodge structures. Publ. R.I.M.S. Kyoto University **23** (1987) 345–407

[K-K 3] Kashiwara, M., Kawai, T.: Hodge structures and holonomic systems. Proc. Japan Acad. (A) **62** (1986) 1–4

[Ka] Katz, N.: The regularity theorem in algebraic geometry. In: Actes, Congrès Intern. Math. Nice 1970, **1** (1970) 437–443

[La] Landman, A.: On the Picard-Lefschetz transformation for algebraic manifolds acquiring general singularities. Trans. Amer. Math. Soc. **181** (1973) 89–126

[Lau] Laumon, S.: Sur la catégorie dérivée des **D**-modules filtrés. In: Algebraic geometry, M. Raynaud and T. Shioda, eds. (Lecture Notes in Mathematics, vol. 1016). Springer, 1983, pp. 151–237

[Le] Lê, D.-T.: Sur les noeuds algébriques. Comp. Math. **25** (1972) 281–321

[Lef] Lefschetz, S.: L'analysis Situs et la Géométrie Algébrique. Gauthier-Villars, 1924

[Li] Lieberman, D.: Higher Picard varieties. Amer. J. Math. **90** (1968) 1165–1199

[Loe] Loeser, F.: Evaluation d'intégrales et théorie de Hodge. In: Hodge Theory Proceedings, Sant Cugat, 1985. (Lecture Notes in Mathematics, vol. 1246). Springer, 1987, pp. 125–142

[Lo] Looijenga, E.: L^2-cohomology of locally symmetric varieties. Comp. Math. **67** (1988) 3–20

[M] Malgrange, B.: Polynôme de Bernstein-Sato et cohomologie évanescente. (Astérisque, vol. 101-102). Soc. Math. Fr. 1983, pp. 243–267

[M-V] MacPherson, R., Vilonen, K.: Elementary construction of perverse sheaves. Invent. math. **84** (1986) 403–435

[Me 1] Mebkhout, Z.: Une équivalence de catégories et une autre équivalence de catégories. Ark. Math. **50** (1984) 51–88

[Me 2] Mebkhout, Z.: Le formalisme des six opérations de Grothendieck pour les **D**-modules cohérents. (Travaux en cours, vol. 35). Hermann, 1989

[Mi] Milne, J.S.: Canonical models of (mixed) Shimura varieties and automorphic vector bundles. In: Automorphic Forms, Shimura Varieties and L-Functions, vol. I. Academic Press, 1990, pp. 283–414

[Mo] Morgan, J.: The algebraic topology of smooth algebraic varieties. Publ. Math. I.H.E.S. **48** (1978) 137–204. Correction, Publ. Math. I.H.E.S. **64** (1986)

[N 1] Nagase, M.: Remarks on the L^2-cohomology of singular algebraic surfaces. J. Math. Soc. Japan **41** (1989) 97–116

[N 2] Nagase, M.: Pure Hodge structure on the harmonic L^2-forms on singular algebraic surfaces. Publ. RIMS. Kyoto University **24** (1988) 1005–1023

[NA 1] Navarro Aznar, V.: Sur la théorie de Hodge-Deligne. Invent. math. **90** (1987) 11–76

[NA 2] Navarro Aznar, V.: Sur la théorie de Hodge des variétés algébriques à singularités isoleés. (Astérisque, vol. 130). Soc. Math. Fr. 1985, pp. 272–307

[NA 3] Navarro Aznar, V.: Sur les structures de Hodge mixtes associées aux cycles évanescents. (Lecture Notes in Mathematics, vol. 1246). Springer, 1987, pp. 143–153

[O] Ohsawa, T.: On extension of Hodge theory to Kähler spaces with isolated singularities of restricted type. Publ. RIMS. Kyoto University **24** (1988) 253–263

[P] Pham, F.: Structures de Hodge mixtes associées à un germe de fonction à point singulier isolé. (Astérisque, vol. 101-102). 1983, pp. 268–285

[Pu] Pulte, M.: The fundamental group of a Riemann surface: Mixed Hodge structures and algebraic cycles. Duke J. Math. **57** (1988) 721–760

[Ra] Ramis, J.-P.: Variations sur le thème GAGA. In: P. Lelong, H. Skoda (eds.) Seminaire Pierre Lelong – Henri Skoda (Analyse) Année 1976/77. (Lecture Notes in Mathematics, vol. 694). Springer, 1978, pp. 228–289

[Ram 1] Ramakrishnan, D.: A regulator for curves via the Heisenberg group. Bull. Amer. Math. Soc. **5** (1981) 191–195

[Ram 2] Ramakrishnan, D.: Regulators, algebraic cycles and values of L-functions. In: Algebraic K-Theory and algebraic number theory. (Contemporary Mathematical). Amer. Math. Soc. 1989, pp. 183–310

[Ram 3] Ramakrishnan, D.: Valeurs de fonctions L des surfaces d'Hilbert-Blumenthal en $s = 1$. C.R.A.S. Paris **301** (1985) 809–812

[Ram 4] Ramakrishnan, D.: Periods of integrals arising from K_1 of Hilbert-Blumenthal surfaces. Preprint 1984

[Ram 5] Ramakrishnan, D.: Arithmetic of Hilbert-Blumenthal surfaces. In: Number theory (Montréal 1985) CMS Conf. Proc. Amer. Math. Soc. 1987, pp. 285–370

[Ram 6] Ramakrishnan, D.: Analogs of the Bloch-Wigner function for higher polylogarithms. In: Number theory (Montréal 1985) CMS conf. Proc. Amer. Math. Soc. 1987, pp. 371–376

[R-S-S] Rapoport, M., Schappacher, N., Schneider, P.: Beilinson's Conjectures on Special Values of L-functions. (Perspectives in Math., vol. 4). Academic Press, 1988

[R-Z] Rapoport, M., Zink, T.: Über die lokale Zetafunktionen von Shimuravarietäten. Monodromie filtration und verschwindende Zyklen in ungleicher Charakteristik. Invent. math. **68** (1982) 21–101

[S] Saito, K.: Period mapping associated to a primitive form. Publ. RIMS. Kyoto University **19** (1983) 1231–1264

[Sa 1] Saito, M.: Gauss-Manin systems and mixed Hodge structure. Proc. Japan Acad. Sci. (A) **58** (1982) 29–32. Supplement (1983) 320–331. (Astérisque, vol. 101-102)

[Sa 2] Saito, M.: Hodge filtrations on Gauss-Manin systems. J. Fac. Sci. Univ. Tokyo, Sect. IA (Math.) **30** (1984) 489–498. II, Proc. Japan Acad. (A) **59** (1983)

[Sa 3] Saito, M.: Modules de Hodge polarisables. Publ. RIMS. Kyoto University **24** (1988) 849–995

[Sa 4] Saito, M.: Hodge structure via filtered **D**-modules. In: Astérisque, vol. 130, 1985, pp. 342–351

[Sa 5] Saito, M.: Mixed Hodge modules. Proc. Japan Acad. (A) **62** (1986) 360–363

[Sa 6] Saito, M.: On the derived category of mixed Hodge modules. Proc. Japan Acad. (A) **62** (1986) 364–366

[Sa 7] Saito, M.: On the structure of Brieskorn lattice. Ann. Inst. Fourier Grenoble **39** (1989) 27–72

[Sa 8] Saito, M.: Vanishing cycles and mixed Hodge modules. Preprint 1988

[Sa 9] Saito, M.: Decomposition theorem for proper Kähler morphisms. Preprint I.H.E.S./M/88/34

[Sa 10] Saito, M.: Mixed Hodge modules and admissible variations. CRAS Paris **1989** (1989) 351–356

[Sa 11] Saito, M.: Mixed Hodge modules. RIMS-585, preprint 1987

[Sa 12] Saito, M.: Introduction to mixed Hodge modules. RIMS-605, preprint 1987

[Sa-Z] Saito, M., Zucker, S.: The kernel spectral sequence of vanishing cycles. Duke Math. J. (to appear)

[Sap] Saper, L.: L^2-cohomology and intersection cohomology of certain algebraic varieties with isolated singularities. Invent. math. **82** (1985) 207–255

[Sap-S] Saper, L., Stern, M.: L_2-cohomology of arithmetic varieties. Proc. Nat. Acad. Sci. USA **84** (1987) 5516–5519

[Sch] Scherk, J.: A note on two local Hodge filtrations. In: Proc. Symp. Pure Math. 46, Part 2. Amer. Math. Soc. 1983, pp. 473–477

[Sch-St] Scherk, J., Steenbrink, J.: On the mixed Hodge structure of the Milnor fiber. Math. Ann. **271** (1985) 641–665

[Sc] Schmid, W.: Variation of Hodge structure: The singularities of the period mapping. Invent. math. **22** (1973) 211–319

[Schn] Schneider, P.: Introduction to the Beilinson conjectures. In: Beilinson's Conjectures on Special Values of *L*-functions. (Perspectives in Math., vol. 4). Academic Press, 1988, pp. 1–35

[Se] Serre, J.P.: Géométrie algébrique et géométrie analytique. Ann. Inst. Fourier **6** (1955-56) 1–42

[SGA 2] Grothendieck, A.: Séminaire de Géométrie Algébrique du Bois-Marie. Cohomologie Locale des Faisceaux Cohérents et Théorèmes de Lefschetz Locaux et Globaux. North-Holland, 1968

[SGA 4] Artin, M., Grothendieck, A., Verdier, J.-L.: Séminaire de Géométrie Algébrique du Bois-Marie. Théorie des topos et cohomologie étale des schémas (SGA 4). (Lecture Notes in Mathematics, vol. 269). Springer, 1972. II, (Lecture Notes in Mathematics, vol. 270). 1972. III, (Lecture Notes in Mathematics, vol. 305), 1973

[SGA 7] Grothendieck, A.: Séminaire de Géométrie Algébrique du Bois-Marie. Groupes de Monodromie en Géométrie Algébrique (SGA 7 I). (Lecture Notes in Mathematics, vol. 288). Springer, 1972. Groupes de Monodromie en Géométrie algébrique (SGA 7 II), (Lecture Notes in Mathematics, vol. 340), Springer, 1973

[Si 1] Simpson, C.: Constructing variations of Hodge structures using Yang-Mills theory and application to uniformization. J. Amer. Math. Soc. **1** (1988) 867–918

[Si 2] Simpson, C.: Moduli of representations of the fundamental group of a smooth projective variety. Preprint 1989

[Si 3] Simpson, C.: Higgs bundles and local systems. Preprint 1989

[Si 4] Simpson, C.: Harmonic bundles on non-compact curves. Preprint 1989

[So 1] Soulé, C.: Régulateurs. Exposé Séminaire Bourbaki 644, in: Séminaire Bourbaki 84/85. (Astérisque, vol. 133-134). Soc. Math. Fr. 1966, pp. 237–253

[So 2] Soulé, C.: Connexions et classes caractéristiques de Beilinson. In: Algebraic K-Theory and algebraic number theory. (Contemporary Mathematics, vol. 83). Amer. Math. Soc. 1989, pp. 349–376

[St 1] Steenbrink, J.: Limits of Hodge structures. Invent. math. **31** (1976) 229–257

[St 2] Steenbrink, J.: Mixed Hodge structures on the vanishing cohomology. In: Real and Complex Singularities. Sijthoff and Noordhoff, 1977, pp. 525–563

[St 3] Steenbrink, J.: Mixed Hodge structures associated with isolated singularities. In: Proc. Symp. Proc. Pure Math. Amer. Math. Soc. 1983, pp. 513–536

[St 4] Steenbrink, J.: The spectrum of hypersurface singularities. In: Proc. of the Luminy Conference on Hodge Theory, 1987. Astérisque. Soc. Math. Fr. (to appear)

[St-vD] Steenbrink, J., van Doorn, R.: A supplement to the monodromy theorem. Abh. Math. Sem. Universität Hamburg **59** (1989)

[St-Z] Steenbrink, J., Zucker, S.: Variation of mixed Hodge structure, I. Invent. math. **80** (1985) 489–542

[Su] Sullivan, D.: Infinitesimal computations in topology. Publ. Math. I.H.E.S. **47** (1977) 269–331

[Tan] Tanisaki, T.: Hodge modules, equivariant K-theory and Hecke algebras. Publ. RIMS. Kyoto University **23** (1987) 841–879

[Tat] Tate, J.: Les conjectures de Stark sur les fonctions L d' Artin en $s = 0$, notes d'un cours à Orsay (rédigées par D. Bernardi et N. Schappacher). (Progress in Mathematics, vol. 47). Birkhäuser, 1984

[TTAG] Topics in Transcendental Algebraic Geometry. P. Griffiths, ed. (Ann. of Math. study, vol. 106). Princeton University Press, 1984

[Va 1] Varchenko, A.: The asymptotics of holomorphic forms determine a mixed Hodge structure. Soviet Math. Dokl. **22** (1980) 772–775

[Va 2] Varchenko, A.: Asymptotic Hodge structure in the vanishing cohomology. Math. USSR. Izv. **18** (1982) 469–512

[Va 3] Varchenko, A.: On the monodromy operator in vanishing cohomology and the operator of multiplication by f in the local ring. Soviet Math. Dokl. **24** (1981) 248–252

[Va 4] Varchenko, A.: A lower bound for the codimension of the strata $\mu = \text{const}$ in terms of the mixed Hodge structure. **37-6** (1982)

[Ve 1] Verdier, J.-L.: Stratifications de Whitney et théorème de Bertini-Sard. Invent. math. **36** (1976) 295–312

[Ve 2] Verdier, J.-L.: Catégories dérivées, état 0. Séminaire de Géométrie Algébrique du Bois-Marie (SGA $\frac{1}{2}$). (Lecture Notes in Mathematics, vol. 569). Springer, 1977, pp. 262–311

[Ve 3] Verdier, J.-L.: Extension of a perverse sheaf over a closed subspace. (Systèmes différentiels et Singularités. Astérisque, vol. 130). Soc. Math. Fr., 1985, pp. 210–217

[Z 1] Zucker, S.: Hodge theory with degenerating coefficients: L_2-cohomology in the Poincaré metric. Ann. Math. **109** (1979) 415–476

[Z 2] Zucker, S.: Variation of mixed Hodge structure. II, Invent. math. **80** (1985) 543–565

[Z 3] Zucker, S.: L^2-cohomology of warped products and arithmetic groups. Invent. math. **70** (1982) 169–218

[Z 4] Zucker, S.: Degeneration of mixed Hodge structures. In: Proc. Symp. Pure Math. Amer. Math. Soc., vol. 46, 1987, pp. 283–293

[Z 5] Zucker, S.: The Hodge structures on the intersection homology of varieties with isolated singularities. Duke Math. J. **55** (1987) 603–616

[Z 6] Zucker, S.: L^2-cohomology and intersection homology of locally symmetric varieties III. In: Proc. of the Luminy Conference on Hodge Theory, 1987, Astérisque. Soc. Math. Fr. (to appear)

[Z 7] Zucker, S.: Locally homogeneous variations of Hodge structure. L'enseign. Math. **27** (1981) 243–276

[Z 8] Zucker, S.: The Cheeger-Simons invariant as a Chern class. In: Algebraic analysis, geometry and number theory. (Proc. JAMI Inaugural Conference). JHU Press, 1989, pp. 397–417

Actions of Groups of Holomorphic Transformations

Alan T. Huckleberry

This paper primarily deals with three topics: *Classification results for homogeneous and almost homogeneous spaces*, *complex analytic questions on homogeneous spaces*, and certain types of *actions*, e.g. *of compact Lie groups on complex spaces*. Our goal here is to indicate our own current view of these areas, as opposed to presenting a comprehensive survey.

In Chapt. I we describe in some detail *selected results* which indicate the spirit of the entire paper. We have intended this for non-specialists. The following chapters consist of a summary of results, outlines of their proofs, basic techniques, and open questions.

Contents

Chapter I

Selected Results

In this chapter we discuss results which are typical for several directions of the subject. These should serve as an introduction, but are by no means just preliminary facts.

1. Some Basic Methods for Homogeneous Manifolds

Let X be a connected *compact* complex manifold. It follows that the group Aut(X) of holomorphic automorphisms is a *complex* Lie group acting holomorphically on X ([BM]). The manifold X is said to be *homogeneous* if this group acts transitively. In this case the connected component G of Aut(X) which contains the identity, $G := \mathrm{Aut}(X)^\circ$, also acts transitively and X may be identified with the coset space $G/H := \{gH \mid g \in G\}$, where H is the *isotropy group* of any point $x_0 \in X$, i.e. $H := \{g \in G \mid g(x_0) = x_0\}$.

a) The Tits Fibration

Motivated by the case of a compact complex homogeneous manifold, let G be any connected complex Lie group and H a closed complex subgroup. There is a unique complex structure on the quotient $X = G/H$ so that the left-action of G is holomorphic. Let $N = N_G(H^\circ)$ be the *normalizer of* H° *in* G,

$$N = \left\{g \in G \mid gH^\circ g^{-1} = H^\circ\right\} .$$

Of course N is also a closed complex Lie subgroup of G which contains H and the mapping $G/H \to G/N$, $gH \mapsto gN$, is a *holomorphic fiber bundle*. We refer to this as the *Tits Fibration* of the homogeneous space $X = G/H$. If X is compact, then this fibration does not depend on the choice of the transitive group G (see Theorem 3 below). Otherwise it quite often does depend on this choice.

Let $F := N/H$ be the *fiber* of the Tits Fibration. The group H°, being normal in N, acts trivially on F and we write $F = M/\Gamma$, where $M := N/H^\circ$ and $\Gamma := H/H^\circ$. Thus F is a quotient of a (not necessarily connected) complex Lie group M by a discrete subgroup Γ.

The *base* of the Tits Fibration can be described as a certain G-orbit in an associated Grassmann manifold. For this we first consider the *adjoint representation* $\mathrm{Ad} : G \to \mathrm{GL}(\mathfrak{g})$, where $\mathfrak{g}$ denotes the Lie algebra Lie(G) identified with the tangent space $T_e(G)$:

$$\mathrm{Ad}(g) = \mathrm{int}(g)_* : T_e(G) \to T_e(G) ,$$

where $\mathrm{int}(g)(h) := ghg^{-1}$. Now the Lie algebra $\mathfrak{h} := \mathrm{Lie}(H)$ is a linear subspace of $\mathfrak{g}$ and

$$N = \mathrm{Stab}_G(\mathfrak{h}) = \{g \in G \mid \mathrm{Ad}(g)(\mathfrak{h}) = \mathfrak{h}\} .$$

If we regard $\mathfrak{h}$ as a point in the Grassmann manifold of subspaces having the same dimension as $\mathfrak{h}$, then the orbit $G(\mathfrak{h})$ is the base G/N of the Tits Fibration.

We summarize the above information in

(1.1) Theorem. *Let G be a connected complex Lie group, H a closed complex subgroup, and $N = N_G(H^\circ)$. Then the fibration*

$$G/H \to G/N \hookrightarrow \mathbb{P}_m$$

is a holomorphic G-equivariant fiber bundle with fiber $F = M/\Gamma$, where $\Gamma = H/H^\circ$ is a discrete subgroup of the complex Lie group $M = N/H^\circ$, and base G/N which is realized via a linear representation as a G-orbit in a projective space $\mathbb{P}_m$.

Remarks. (1) Above we only showed that G/N is a G-orbit in a certain Grassmann manifold, but of course the Plücker embedding realizes this in some projective space.

(2) Although the Tits fibration had been implicitely used for some time (see e.g. [Wa 1]), Tits underlined its importance and proved a basic universality property for compact homogeneous manifolds ([T]). At roughly the same time Borel and Remmert emphasized its central role in classification problems ([BR]). We discuss their results in more detail in the next section and the following paragraph.

b) Compact Homogeneous Manifolds

Let X be compact and suppose that a connected complex Lie group G acts transitively on X. The base G/N of the Tits fibration is in this case a *compact* G-orbit in some projective space $\mathbb{P}_m$. Classical results in the theory of Lie groups give us a great deal of information about such orbits. In this regard we begin by recalling the *Levi-Malcev decomposition*.

Let G be a connected (real) Lie group and R *the* maximal, connected, solvable normal subgroup of G. One refers to R as the *radical* of G. It is a closed subgroup and, if G is complex, it is likewise complex. A group G is said to be *semi-simple* if its radical is trivial, i.e. $R = (e)$. It is clear that G/R is always semi-simple.

Levi-Malcev Theorem. *Let G be a simply-connected, connected Lie group with radical R. Then there is a closed semi-simple subgroup S in G so that G is the semi-direct product $G = R \rtimes S$. The group S is unique up to conjugation. If G is complex, then so are R and S.*

Remark. For an arbitrary connected Lie group G we may apply the Levi-Malcev Theorem to its universal covering group. This implies that $G = R \cdot S$, but $R \cap S \neq (e)$ is possible, e.g. $\mathrm{GL}_n(\mathbb{C}) = \mathbb{C}^* \cdot \mathrm{SL}_n(\mathbb{C})$. □

If G is a (not necessarily complex or closed) Lie subgroup of $\mathrm{GL}_n(\mathbb{C})$, then we refer to G as a *linear* group.

Flag Theorem. *Let G be a connected linear group in $\mathrm{GL}_n(\mathbb{C})$. If G is solvable, then it stabilizes a full flag of complex subspaces*

$$\{0\} \subset V_1 \subset V_2 \subset \ldots \subset V_n = \mathbb{C}^n ,$$

where $\dim_{\mathbb{C}} V_k = k$.

As a consequence we have the

Borel Fixed Point Theorem. *Let G be a connected linear solvable group in $\mathrm{GL}_n(\mathbb{C})$ and X be a closed G-stable subvariety in $\mathbb{P}_{n-1}(\mathbb{C})$. Then G has a fixed point in X.* □

Proof. Let $\{p\} \subset V_1 \subset \ldots \subset V_{n-1} = \mathbb{P}_{n-1}(\mathbb{C})$ be a G-stable flag of projective subspaces. Then the connected components of the intersections $W_i := V_i \cap X$ are also G-stable. Since X is compact, some such intersection must consist of finitely many points. □

Remark. This result is true in more general contexts. For example, let G be a connected solvable Lie group acting on a Kähler manifold X. Suppose that G acts trivially on the Albanese torus $\mathrm{Alb}(X)$. Then G has a fixed point in every fiber of $\alpha : X \to \mathrm{Alb}(X)$ (see [So], [Fu]). □

We now return to our considerations of the compact base G/N of the Tits fibration. By the Fixed Point Theorem, the radical R of G has a fixed point in G/N. Hence, since R is normal in G, it fixes every point and we may rewrite G/N as S/P, where S is a semi-simple part of a Levi-Malcev decomposition of G. Now let B be a maximal connected solvable subgroup of S, i.e. a *Borel subgroup* of S. Again applying the Fixed Point Theorem, it follows that B has a fixed point in S/P and therefore P contains a Borel subgroup, i.e. P is a *parabolic subgroup* of S.

The structure theory of complex semi-simple Lie groups and algebras revolves around Borel groups and algebras. In particular it yields a precise description of the isotropy groups P which occur as above. For example, one can prove the following facts:

(i) Any two Borel subgroups of S are conjugate;

(ii) If B is a Borel subgroup of S, then S/B is compact;

(iii) If P is a parabolic subgroup of S, then P is closed, connected, and $N(P) = P$, i.e. S/P is the base of its Tits fibration.

(iv) For any parabolic subgroup P in S there is a complex unipotent subgroup U with $P \cap U = \{e\}$ and $U \cdot P$ Zariski open in S.

Remarks. (1) A complex semi-simple Lie group carries a unique structure of an affine algebraic manifold so that the group multiplication is algebraic. Thus it makes sense to speak of the Zariski topology. Borel subgroups, parabolic subgroups, and groups of the type U are in fact algebraic.

(2) A group U is *unipotent* if it can be realized as a group of upper-triangular matrices with 1's along the diagonal. As affine algebraic manifolds, such groups are algebraically isomorphic to affine space. Thus (*iv*) above shows that S/P contains a Zariski open algebraic copy of $\mathbb{C}^n$, namely the U-orbit of the point corresponding to P in S/P. As a consequence

(v) The homogeneous space S/P is a (simply-connected) projective rational manifold. □

(1.2) Theorem. *Let X be a connected compact complex manifold which is homogeneous with respect to the action of a connected complex Lie group G. Let $X = G/H \to G/N =: Q$ be the Tits fibration. Then*

(a) *The fiber $F = N/H = M/\Gamma$ is a connected complex parallelizable manifold;*

(b) *The base Q is a homogeneous rational manifold.*

Proof. It only remains to show that the fiber is connected. However, if $G/I \to G/J$ is any homogeneous fibration, then there is a unique open subgroup $\tilde{J} < J$ so that $I < \tilde{J}$ and the fibers of $G/I \to G/\tilde{J}$ are connected. Of course, $G/\tilde{J} \to G/J$ is a covering map. So if G/J is simply-connected, as is the case for G/N, then $J = \tilde{J}$. □

Remark. A compact complex manifold F is called *parallelizable* if its complex tangent bundle is holomorphically trivial. Due to the compactness of F, the holomorphic vector fields coming from a trivialization are globally integrable. Since any holomorphic vector field on F is a linear combination (with constant coefficients) of these vector fields, it follows that $\dim_{\mathbb{C}} \mathrm{Aut}(F) = \dim_{\mathbb{C}} F$. Hence $F = M/\Gamma$, where $M := \mathrm{Aut}(X)^\circ$ is a complex Lie group and Γ is a discrete subgroup.

Conversely, the tangent bundle of M/Γ is obviously holomorphically trivial. □

The following result of J. Tits ([T]) gives another indication of the central importance of the Tits fibration.

(1.3) Theorem (Universality of the Tits fibration). *Let X be a homogeneous compact complex manifold and $\phi : X \to Z$ a holomorphic map onto an algebraic variety with* $\mathrm{Alb}(Z) = (*)$. *Suppose that the fibers of ϕ are connected. Then ϕ factors holomorphically through the Tits fibration.*

The main point is to prove

(1.3′) Theorem. *Let X be a parallelizable compact complex manifold, $X = M/\Gamma$. Then there is no complex subgroup J containing Γ such that M/J is a homogeneous rational manifold.*

Tits' proof only uses Lie-theoretic methods, whereas the result, i.e. Theorem 1.3, would seem to be of a complex analytic or algebraic geometric nature. It would be interesting to have a proof along the latter lines.

2. Homogeneous Compact Kähler Manifolds

Let (X, ω) be a Kähler manifold. Recently Dorfmeister and Nakajima ([DN]) proved that if the group $\mathrm{Iso}_{\mathcal{O}}(X)$ acts transitively, then X is biholomorphically equivalent to a product $D \times Q \times T$, where D is a homogeneous bounded domain, Q is a compact homogeneous rational manifold, and $T = \mathbb{C}^k/\Gamma$, where Γ is a discrete additive subgroup. This concluded a long effort during which a number of people made important contributions. For example, in a very interesting paper ([Ma 1]) Matsushima proved the result in the compact case.

The transitivity of the isometry group is of course a very strong condition. However, as we have seen in recent years (see Chapt. III), the Kählerian condition is remarkably strong. The first indication of this, proved at roughly the same time as Matsushima's result, is the

Borel-Remmert Theorem ([BR]). *Let X be a connected compact Kähler manifold which is homogeneous with respect to* $\mathrm{Aut}(X)$. *Then X is a product $T \times Q$, where $T =$* $\mathrm{Alb}(X)$ *and the homogeneous rational manifold Q is the base of its Tits fibration.*

Nowadays this result can be (and has been) proved in a variety of ways. We would like to give a proof along the lines of the original one, but with a small symplectic flair.

The philosophy of the Borel-Remmert Theorem is very simple: Let $G = \mathrm{Aut}(X)^\circ$ and $X = G/H \to G/N$ the Tits fibration. One only needs to show that the fiber is a torus and the bundle is trivial. The result on the fiber was first proved by H.C. Wang ([Wa 2]):

(2.1) Theorem. *Let X be a compact parallelizable Kähler manifold. Then X is a torus, i.e. $X = \mathbb{C}^n/\Gamma$, where Γ is a co-compact, discrete additive subgroup.*

Proof. Let $X = M/\Gamma$ and $\mu_1, \ldots, \mu_n$ be a basis of right-invariant holomorphic 1-forms on M. Since these forms can be regarded as holomorphic 1-forms on the Kähler manifold X, they are closed. Hence for a basis $\mathfrak{X}_1, \ldots, \mathfrak{X}_n$ of invariant vector fields,

$$0 = d\mu_i(\mathfrak{X}_j, \mathfrak{X}_k) = -\mu_i([\mathfrak{X}_j, \mathfrak{X}_k]) ,$$

for all i, j, and k. Consequently $[\mathfrak{X}_j, \mathfrak{X}_k] = 0$ for all j and k, and therefore M is abelian. □

Note that if (X, ω) is a Kähler manifold, $G = \mathrm{Aut}(X)^\circ$, and K is a maximal compact subgroup of G, then, by averaging over K, we may assume that ω is K-invariant:

$$\overline{\omega}(v, w) := \int_{k \in K} \omega\left(k_*(v), k_*(w)\right) d\mu(k) ,$$

where μ is Haar-measure.

(2.2) Lemma. *Let X be a compact complex homogeneous manifold, $G :=$ $\mathrm{Aut}(X)^\circ$, and K_{ss} the semi-simple part of a maximal compact subgroup K of G. Suppose that X possesses a Kähler form ω which is K_{ss}-invariant. Then the Tits fibration is a trivial bundle: $X = F \times Q$.*

Remark. The Levi-Malcev decomposition for a *compact* group is, up to finite intersection, a product $K = Z \cdot K_{ss}$, where Z is the connected component of the identity in the center $Z(K)$ and K_{ss} is semi-simple. □

Before proving Lemma 2.2, we would like to recall some notions from symplectic geometry. In the context of Lemma 2.2, consider the diagram

$$\begin{array}{ccccccc} 0 \longrightarrow \mathbb{R} \longrightarrow & C^\infty_{\mathbb{R}}(X) & \stackrel{\iota}{\longrightarrow} & \mathrm{Ham}_{\mathrm{loc}}(X) \\ & \lambda \nwarrow & \uparrow \kappa & \\ & & \mathfrak{k} & \end{array} ,$$

where $\mathrm{Ham}_{\mathrm{loc}}(X)$ consists of $\mathbb{R}$-vector fields $\mathfrak{X}$ which are "local isometries", i.e. $\mathscr{L}_{\mathfrak{X}}(\omega) = 0$, $\mathfrak{k}$ is the Lie algebra $\mathrm{Lie}(K_{ss})$, $\kappa : \mathfrak{k} \to \mathrm{Ham}_{\mathrm{loc}}(X)$ is the natural inclusion, and $\iota : C^\infty_{\mathbb{R}}(X) \to \mathrm{Ham}_{\mathrm{loc}}(X)$ is defined by *contraction*, $\iota(f) := \mathfrak{X}_f$, where $df = \omega(\mathfrak{X}_f, \cdot)$.

Now $C^\infty_{\mathbb{R}}(X)$ is a Lie algebra with respect to the Poisson structure $\{\ ,\ \}$, where $\{f, g\} := \omega(\mathfrak{X}_f, \mathfrak{X}_g)$. Thus one has the following *lifting problem*: Is there a Lie algebra morphism $\lambda : \mathfrak{k} \to C^\infty_{\mathbb{R}}(X)$ so that the above diagram commutes? This is a general problem which is often interesting. However, for semi-simple algebras, a straight-forward calculation shows that λ does indeed exist (see e.g. [GS]).

Proof of Lemma 2.2. Let $X = G/H \to G/N = S/P = Q$ be the Tits fibration with fiber $F = M/\Gamma$. We will show that P acts trivially on F which of course implies that the bundle is trivial.

After conjugating appropriately, P has the form $L^{\mathbb{C}} \ltimes R_u(P)$, where $R_u(P)$ is the maximal unipotent normal subgroup of P (the *unipotent radical*), L is a maximal compact subgroup of P which is contained in K_{ss}, and $L^{\mathbb{C}}$ is the smallest complex subgroup of P containing L (see e.g. [Hum]).

For every ξ in the Lie algebra Lie(L) we have the *moment function* $f_\xi := \lambda(\xi)$. The definitions show that $\xi(x) = 0$ if and only if $df_\xi(x) = 0$. Applying this statement to the L-action on the fiber F with its induced symplectic structure, it follows that every $\xi \in$ Lie(L) has a zero on F. Since F is parallelizable, it follows that L acts trivially on F and therefore $L^{\mathbb{C}}$ acts trivially as well.

Now let $\rho : P \to M$ be the homomorphism defined by the P-action on F. We have just shown that $L^{\mathbb{C}} < \ker(\rho)$. But the structure theory for parabolic groups shows that P is the only normal subgroup of P which contains $L^{\mathbb{C}}$. Thus $\ker(\rho) = P$ and P acts trivially on F. □

Proof of the Borel-Remmert Theorem. By averaging we may assume that ω is K-invariant. Thus we may apply Lemma 2.2 and $X \cong F \times Q$. By Theorem 2.1, F is a compact complex torus. □

3. Hypersurfaces in Compact Homogeneous Manifolds

A *hypersurface* H in a complex manifold X is a 1-codimensional closed complex analytic subvariety. The set of all such hypersurfaces is denoted by $\mathscr{H}(X)$. The relationships between hypersurfaces (or more generally divisors) and those arising from holomorphic or meromorphic functions underlie many basic problems in complex analysis and algebraic geometry. It is therefore natural to study the possibility of defining equivalence relations via these objects and to analyze the resulting quotients of the manifold X.

Let $\mathcal{O}(X)$ be the algebra of holomorphic functions on X and define an equivalence relation by $x \sim y :\Longleftrightarrow f(x) = f(y)$ for all $f \in \mathcal{O}(X)$. We refer to the quotient $X \to X/_\sim$ as the *holomorphic reduction* of X. It is well-known that unless X has special properties, e.g. holomorphically convex, where $X \to X/_\sim$ is the *Remmert Reduction* (see [CAS]), this quotient makes very little sense. However, if X is homogeneous in the sense that Aut(X) acts transitively, then $X \to X/_\sim$ is an Aut(X)-equivariant holomorphic map onto a holomorphically separable complex manifold. Furthermore, every holomorphic map $X \to Y$ into a holomorphically separable complex space factors through the holomorphic reduction. Similar statements can be made for the bounded holomorphic (resp. meromorphic) reductions (see [Gi 1] resp. [HS 1]).

Grauert and Remmert ([GR 1]) showed that a natural *hypersurface reduction* makes good sense in the case of compact complex homogeneous spaces:

(3.1) Theorem. *Let X be a compact complex homogeneous manifold. For $x \in X$ let $[x] := \bigcap H$, where the intersection is taken over all $H \in \mathscr{H}(X)$ with $x \in H$. Then $x \sim y$, defined by $[x] = [y]$, is an equivalence relation and the quotient $X/_\sim$ carries a unique complex structure with the following properties:*

(i) The map $\pi : X \to X/_\sim$ *is holomorphic and* Aut(X)*-equivariant;*

(ii) $\pi^* \left(\mathscr{H}(X/_\sim)\right) = \mathscr{H}(X)$*;*

(iii) If $\phi : X \to Y$ *is any holomorphic map into a hypersurface separable manifold, i.e.* $[y] = \{y\}$ *for all* $y \in Y$*, then there is a holomorphic map* $\tau_\phi : X/_\sim \to Y$ *with* $\phi = \tau_\phi \circ \pi$.

For simplicity assume that X is connected, let $G = \mathrm{Aut}(X)^\circ$, and write the hypersurface reduction as a homogeneous bundle $X = G/H \to G/J$. Since the base G/N of the Tits fibration is hypersurface separable, it follows that $J < N$, i.e. *the fiber of the hypersurface reduction for a compact homogeneous manifold is complex parallelizable*.

It is also useful to have a hypersurface reduction for non-compact homogeneous manifolds. In [H 1] we considered this for $X = G/H$, where G is a connected complex Lie group and H is a closed complex subgroup. We were not able to show that the equivalence classes used by Grauert and Remmert yield an equivalence relation. However, by weakening the definition of separable to x *is isolated in* $[x]$ (*weakly hypersurface separable*) we were able to provide an equivalence relation with properties exactly analogous to those in Theorem 3.1. This seems to be adequate for most applications. In [OeK] K. Oeljeklaus took another approach and also handled the case where the group is not assumed to be complex, i.e. $X = G/H$, where G is a *real* Lie group of holomorphic transformations. In fact, even for real groups, the hypersurface reduction $G/H \to G/J$ is a *holomorphic* bundle.

In the non-compact case it is in general not clear if meromorphic functions and hypersurfaces are related. For example, we do not know whether or not a hypersurface separable homogeneous manifold is meromorphically separable. Grauert and Remmert's description of the base of the hypersurface reduction resolves this question in the compact case.

(3.2) Theorem. *Let* X *be a connected compact complex homogeneous manifold which is hypersurface separable. Then* X *is biholomorphically equivalent to a product* $T \times Q$*, where* T *is an abelian variety, i.e. a projective algebraic torus, and* Q *is a homogeneous rational manifold.*

In their original proof Grauert and Remmert used the hypersurfaces along with compactness and homogenity to construct an ample bundle on X. Thus X is projective algebraic and, in particular, Kähler. The result then follows from the Borel-Remmert Theorem. We now sketch an alternative proof using methods which are useful in the non-compact case.

Let X be a complex manifold which is homogeneous under the action of a Lie group G of holomorphic transformations. Here we do not assume that G is complex. Suppose further that X is weakly hypersurface separable. By moving hypersurfaces with the automorphisms, given a point $x \in X$ there are hypersurfaces $H_1, \ldots, H_n \in \mathscr{H}(X)$, $n := \dim_{\mathbb{C}} X$, which are coordinate hyperplanes in some local coordinate system at x, i.e. near x there are coordinates $(z_1, \ldots, z_n)$ with $H_i = \{z_i = 0\}$. Let $H := \bigcup H_i$ and define a current T on X by integration over H. By construction, i.e. the nature of the hypersurfaces H_i, T is closed and positive, and is positive definite on an open neighborhood $V = V(x)$.

Let χ be a compactly supported smooth function on G with $\chi \geq 0$, $\int_G \chi = 1$, and let U be the interior of the support of χ. The *smoothed current* T_χ is defined by

$$T_\chi := \int_G \chi(g)\, \ell^*_{g^{-1}}\, T\, dg \,,$$

where $\ell_g : X \to X$ denotes the automorphism of X induced by left-multiplication by $g \in G$ and dg is a left-invariant volume element on G.

The following result was proved for complex groups and discrete isotropy in ([OR 1]) and in general in ([R 1]).

(3.3) Theorem. *The smoothed current* T_χ *is a closed, positive semi-definite* $(1,1)$-*form on* X *which is positive definite on* $U \cdot V$.

Using a suggestion of Berteloot, one can in fact exhaust X with a sequence $\{U_m\}$ and adjust the forms T_{χ_m} which arise from the above theorem so that they converge to a Kähler form on X (see [GOR]). Thus Theorem 3.3 has the following less technical version.

(3.3′) Theorem. *Let* X *be a complex manifold which is homogeneous under the action of a Lie group of holomorphic transformations. If* X *is weakly hypersurface separable, then it is a Kähler manifold.*

Remarks. (1) Richthofer's proof of Theorem 3.3 goes by direct verification. He shows that this can be done in a natural way via fiber integration with respect to $G \to G/H = X$. Due to the technical nature of some of the formulas we do not include it here.

(2) The proof of Theorem 3.2 is now immediate: Since X is hypersurface separable, it follows from Theorem 3.3 that it is Kähler. The result is then a consequence of the Borel-Remmert Theorem. □

The following is also an immediate consequence of Theorem 3.3.

(3.4) Corollary. *Let* $X = G/\Gamma$ *be a compact parallelizable complex manifold. If* X *is hypersurface separable, then* G *is abelian and* X *is an abelian variety.*

Proof. Since X is hypersurface separable, it is Kähler. The result then follows from Theorem 2.1. □

(3.5) Corollary. *Let $X = G/\Gamma$ be a compact parallelizable complex manifold with hypersurface reduction $G/\Gamma \to G/J$. Then G contains the commutator subgroup G', i.e. the base of the hypersurface reduction is an abelian variety.*

Proof. By Theorem 3.2 the base G/J is the product of an abelian variety with a homogeneous rational manifold. The result then follows from the universality of the Tits fibration (Theorem 1.3). □

Example. Let

$$G = \left\{ \begin{pmatrix} 1 & z_1 & z_3 \\ 0 & 1 & z_2 \\ 0 & 0 & 1 \end{pmatrix} \middle| z_i \in \mathbb{C} \right\} \text{ and } \Gamma = \left\{ \begin{pmatrix} 1 & z_1 & z_3 \\ 0 & 1 & z_2 \\ 0 & 0 & 1 \end{pmatrix} \middle| z_i \in \mathbb{Z}\left[\sqrt{-1}\right] \right\}.$$

Then $X = G/\Gamma$ is a compact parallelizable manifold with hypersurface reduction $G/\Gamma \to G/G'\Gamma$, where $G' = \left\{ \begin{pmatrix} 1 & 0 & z \\ 0 & 1 & 0 \\ 0 & 0 & 1 \end{pmatrix} \middle| z \in \mathbb{C} \right\}$. □

Remarks. (1) Using methods developed for the non-compact case (see Chapt. III), it is possible to prove Corollary 3.5 without appealing to the universality of the Tits fibration.

(2) If G is solvable and $X = G/\Gamma$ is compact, then $G'\Gamma$ is closed (see [BO]). Thus $\mathrm{Alb}(X) = G/G'\Gamma$. If G/Γ is not compact, then, even if G is nilpotent and $\mathcal{O}(G/\Gamma) \cong \mathbb{C}$, $G'\Gamma$ may not be closed ([OeK]). **Warning:** In the case of a compact real solv-manifold G/Γ, $G'\Gamma$ may not be closed. However, $N\Gamma$ is closed, where N is the *nil-radical* of G ([Mo 1]). □

4. Linearization of Actions

Suppose $g : X \to X$ is a holomorphic automorphism with a fixed point $x \in X$, i.e. $g(x) = x$. In order to understand the local behavior of g near x, one begins by studying the differential

$$L_x(g) : T_xX \to T_xX .$$

Example. Let $X = \mathbb{P}_1$ and $g(z) = z + a$ be an arbitrary translation. Then $\infty \in \mathbb{P}_1$ is a fixed point, but $L_\infty(g) = \mathrm{Id}$. □

Of course one can gain more information by considering the analogous maps of higher order jets. Here, however, we restrict our attention to first order phenomena.

a) Local Linearization of Compact Group Actions

One of the most beautiful ideas at the (historical) beginning of the study of holomorphic transformations is that of H. Cartan for proving his "Identity Theorem":

Let D be a bounded domain in $\mathbb{C}^n$ and $g : D \to D$ a holomorphic map. Suppose that there is a point $z_0 \in D$ with $g(z_0) = z_0$ and $L_{z_0}(g) = \mathrm{Id}$. Then g is the identity, i.e. $g(z) \equiv z$.

Proof. Except for superficial complications with indices, the proof in several variables is the same as in one variable. So we assume that $D \Subset \mathbb{C}$ and $z_0 = 0$. Suppose that g is not the identity and consider its power series

$$g(z) = z + a_k z^k + O(k+1)$$

at the origin, where a_k is the first higher order non-vanishing coefficient.

Now let g_n be the n-th iterate, i.e. $g^n(z) = g(g(\ldots g(g(z))\ldots))$, and note that

$$g^n(z) = z + na_k z^k + O(k+1)\ .$$

Thus

$$\lim_{n\to\infty} \frac{d^k}{dz^k}\bigg|_{z=0} g^n(z) = \infty\ .$$

Since g^n is uniformly bounded, this violates the Cauchy-Inequality for the k-th derivative. □

Let D be a bounded domain, $z_0 \in D$, and $I_{z_0} := \{g \in \mathrm{Aut}(D) \mid g(z_0) = z_0\}$. As a consequence of the Identity Theorem, it follows that

$$L_{z_0} : I_{z_0} \to \mathrm{GL}(T_{z_0}D)$$

is an injective homomorphism. The following *compactness* result of Cartan helps to put this in a more general framework:

Let $\{g_n\}$ be a sequence in $\mathrm{Aut}(D)$. Then there is a convergent subsequence $g_{n_k} \to g$, where g maps D into the closure $\overline{D}$. Furthermore, either g is an automorphism or $g(D) \subset \partial D$.

This shows that I_{z_0} is compact and hence $L_{z_0}(I_{z_0})$ lies in a unitary group. In fact the faithful *linearization of compact isotropy groups* holds in complete generality.

Let K be a compact Lie group of holomorphic automorphisms of a complex space X. Let $x \in X$ and suppose that $K(x) = x$. Let $U = U(x)$ be a K-invariant neighborhood of x which is small enough so that there is a biholomorphic map $\phi : U \to T_x$ of U onto a subvariety of an open neighborhood of the origin in the Zariski tangent space T_x.

Let μ be the normalized Haar measure on K and define $\tau : U \to T_x$ by

$$\tau(z) = \int_{k \in K} L(k^{-1}) \circ \phi \circ k(z)\, d\mu(k)\ .$$

It is clear that τ is K-equivariant, i.e. $\tau(k(z)) = L(k)(\tau(z))$, and after shrinking U appropriately it is biholomorphic onto its image. □

The linearization theorem of Cartan ("Identity Theorem"), in the smooth case, and W. Kaup [Ka 1], as proved above, has a slight improvement which is useful for actions of complex groups (see [HO 1]):

Let G be a connected complex Lie group with $G(x) = x$. Let K be a maximal compact subgroup of G and $\hat{K}$ the smallest complex Lie subgroup which contains K. Then there is an open neighborhood W of K in $\hat{K}$ and a neighborhood $U' = U'(x) \subset U$ so that for all $g \in W$

$$\begin{array}{ccc} U' & \xrightarrow{g} & U \\ \tau \downarrow & & \downarrow \tau \\ T_x & \xrightarrow[L(g)]{} & T_x \end{array}$$

is commutative. □

b) Slice Theorems

A second type of *local* linearization phenomenom is suggested by the *Slice Theorem for differentiable actions*:

Let M be a differentiable manifold equipped with an action of a compact group K of diffeomorphisms. Equip M with a K-invariant Riemannian metric. Then, for every K-orbit $K(m) = K/L$, $L = \mathrm{Iso}_K\{m\} := \{g \in K \mid g(m) = m\}$, there is an L-splitting of the tangent space

$$T_m M = T_m K(m) \oplus V\ ,$$

where $V := T_m K(m)^{\perp}$. Accordingly the tangent bundle splits,

$$TM|K(m) = TK(m) \oplus N\ ,$$

where $N = K \times_L V$ is the normal bundle of $K(m)$, and there is a K-invariant tubular neighborhood T of the 0-section in N which is K-equivariantly diffeomorphic via the exponential map to a neighborhood of $K(m)$ in M. More precisely, there is an L-invariant open neighborhood U of $0 \in V$ so that $T = K \times_L U$. The piece of manifold $\exp(U)$ is called a *slice* at the point m (see e.g. [Bre]). □

We refer to the slice theorem as a local theorem, because it is really only a local linearization of the isotropy group L transversal to the orbit. One of the most useful tools for complex group actions is *Luna's Slice Theorem* for reductive groups ([L]). It is a local result analogous to that above, but, since it is algebraic geometric in nature, it is almost global. We now give a brief description of Luna's Theorem.

1. Reductive Groups. Let K be a compact (real) Lie group with Lie algebra $\mathfrak{k}$. It is only a formal matter to complexify the Lie algebra: $\mathfrak{k}^{\mathbb{C}} = \mathfrak{k} \oplus i\mathfrak{k}$. If G is the simply-connected complex Lie group corresponding to $\mathfrak{k}^{\mathbb{C}}$, then K may not lie in G. This is the case, for example, if K is not connected or not simply-connected. Nevertheless, for every compact Lie group K there is a *universal complexification* $K^{\mathbb{C}}$ with the following properties (see [Ho]):

(i) $K^{\mathbb{C}}$ is a complex Lie group with maximal compact subgroup K;

(ii) K is a totally real submanifold of $K^{\mathbb{C}}$ with

$$\dim_{\mathbb{R}} K = \dim_{\mathbb{C}} K^{\mathbb{C}}$$

(iii) Every homomorphism $\phi : K \to G$ into a complex Lie group extends to a holomorphic homomorphism of $K^{\mathbb{C}}$ into G. □

Complex Lie groups of the type $K^{\mathbb{C}}$ are referred to as (linear) *reductive groups*.

Remark. Connected reductive groups are of the form $(\mathbb{C}^*)^k \cdot S$, where the $(\mathbb{C}^*)^k$ factor is central and S is semi-simple. A reductive group possesses a unique structure as an affine algebraic group. □

2. The Categorical Quotient. In order to understand Luna's Slice Theorem it is useful to first take a look at the *categorical quotient*. This is of course one of *the* basic results in invariant theory (see [K, M]).

Let X be an affine algebraic variety equipped with an algebraic action of a reductive group G. Denote by $\mathcal{O}_{\text{alg}}(X)$ the ring of regular function on X, i.e. restrictions of polynomials on some closed embedding, and $R := \mathcal{O}_{\text{alg}}(X)^G$ the ring of invariant functions.

Consider the equivalence relation $x \sim y :\Longleftrightarrow f(x) = f(y)$ for all $f \in R$. The quotient of X by this relation is denoted by $X/\!/G$ and called the *categorical quotient*. In this case, i.e. reductive actions, the *ring R of invariant functions is finitely generated*. Thus $X/\!/G = \text{Spec}(R)$ is an affine algebraic variety having the desired universality properties.

In order to motivate Luna's Theorem we consider one particular property of the categorical quotient. Let $\pi : X \to X/\!/G$ be the quotient map and consider a fiber $\pi^{-1}(y) =: F$. Now F is G-stable and the action is algebraic. Therefore every G-orbit in F is Zariski open in its closure. In particular, some orbit is closed. Suppose there are two closed orbits V_1 and V_2 in F. They are likewise closed in X and, since X is affine, there is a regular function $f \in \mathcal{O}_{\text{alg}}(X)$ with $f|V_i \equiv i$, $i = 1, 2$. Define a new function by averaging over K:

$$\overline{f}(x) = \int_{k \in K} f\,(k(x))\; d\mu(k)\ ,$$

where μ is normalized Haar measure.

Now $\bar{f}$ is a K-invariant regular function. By the identity principle it is therefore $K^{\mathbb{C}}$-invariant. But $\bar{f}|V_i \equiv i$, $i = 1, 2$, contrary to V_1 and V_2 being in the same fiber of the categorical quotient. Thus

each fiber of the categorical quotient contains exactly one closed G-orbit.

3. Closed Orbits. Before going further it is useful to understand the group-theoretic nature of the closed orbits. They are of course affine algebraic.

(4.1) Theorem (Matsushima [Ma 4]). *Let G be a reductive complex Lie group, H a closed complex subgroup, and $Z := G/H$. Then the following are equivalent:*

(i) Z is Stein;

(ii) Z is affine algebraic;

(iii) H is reductive.

Proof. (ii) ⇒ (i). This is clear.

(i) ⇒ (iii). Suppose that Z is Stein. Then in particular Z is holomorphically separable and therefore H is an algebraic subgroup of G (see e.g. Theorem III.1.1). Let $H = L^{\mathbb{C}} \ltimes U$ be a Levi-decomposition of H, where U is the *unipotent radical* of H and $L^{\mathbb{C}}$ is a maximal reductive subgroup (see e.g. [Hum]).

Now consider the U-principle bundle $G/L^{\mathbb{C}} \to G/H$. Since $U \cong \mathbb{C}^k$, the base and total space have the same homology. But the total space is of the form $K^{\mathbb{C}}/L^{\mathbb{C}}$ which is retractable to the orbit K/L. Since none of the statements is changed by going to a finite cover, we may assume that K/L is orientable. Thus the homology $H_n(G/L^{\mathbb{C}}, \mathbb{C})$, $n := \dim_{\mathbb{C}} G/L^{\mathbb{C}}$, is non-zero. If U is positive-dimensional, then this implies that $H_n(G/H, \mathbb{C}) \neq (0)$, where $n > \dim_{\mathbb{C}} G/H$. However this is contrary to $Z = G/H$ being Stein. Hence $U = (e)$ and $H = L^{\mathbb{C}}$ is reductive.

(iii) ⇒ (ii). The right action of H on G is a reductive action on an affine algebraic manifold. In this case, since *all* H-orbits are closed, the geometric quotient G/H is the same as the categorical quotient. Thus $Z = G/H$ is affine algebraic. □

Luna's Slice Theorem. Although Luna's construction is completely functorial, we describe it here in an ad hoc fashion (see [L] for complete details). The basic example is that of a linear representation, where $G = K^{\mathbb{C}}$ is a group of complex linear transformations of a complex vector space V. We may equip V with a K-invariant hermitian inner-product so that K is contained in the unitary group.

Let B be a closed G-orbit in V, $\phi : B \to \mathbb{R}^+$ the restriction of the norm-function to B, and $b \in B$ be a point where ϕ attains its minimum. In fact the set of *all* points where ϕ attains a local minimum is just the orbit $K(b) = K/L$, $L = \mathrm{Iso}_K\{b\}$, and $B = G(b) = K^{\mathbb{C}}(b) = K^{\mathbb{C}}/L^{\mathbb{C}}$.

Now let W be the orthogonal complement to the (affine) tangent space of B at b in V. It follows that W is an $L^{\mathbb{C}}$-invariant linear subspace of V and we have the *normal bundle* $N := K^{\mathbb{C}} \times_{L^{\mathbb{C}}} W$. The map $\alpha : W \to W$, $w \mapsto w + b$, is $L^{\mathbb{C}}$-equivariant and induces

$$\mathscr{L} : N \to V , \qquad [g, w] \mapsto g\alpha(w) .$$

It is easy to check that $\mathscr{L}$ is a well-defined G-equivariant algebraic morphism which maps the 0-section biregularly onto the closed orbit B. Furthermore, $\mathscr{L}$ is generically locally biholomorphic, in particular at points of the 0-section, and is an etale map on a Zariski open G-invariant subset.

For the deeper properties of $\mathscr{L}$, consider the following diagram which results from the functoriallity of the categorical quotient:

$$\begin{array}{ccc} N & \xrightarrow{\mathscr{L}} & V \\ \pi_N \downarrow & & \downarrow \pi_V \\ N/\!/G & \xrightarrow[\mathscr{L}/\!/]{} & V/\!/G \end{array} .$$

Let p be the neutral point in N with $\mathscr{L}(p) = b$. Then *there are Zariski (resp. Stein) open neighborhoods* Q_N *(resp.* Q_N^ϵ*) of* $\pi_N(p)$ *and* Q_V *(resp.* Q_V^ϵ*) of* $\pi_V(b)$ *with* $\Omega_N := \pi_N^{-1}(Q_V)$ *and* $\Omega_V := \pi_V^{-1}(Q_V)$ *so that*

(1) $\mathscr{L} : \Omega_N \to \Omega_V$ *is etale and* $\mathscr{L} : \Omega_N^\epsilon \to \Omega_V^\epsilon$ *is biholomorphic.*

(2) $\mathscr{L}_{/\!/} : Q_N \to Q_V$ *is etale and* $\mathscr{L}_{/\!/} : Q_N^\epsilon \to Q_V^\epsilon$ *is biholomorphic.*

In order to obtain a "slice theorem" in the classical sense, let $S := \Omega_V^\epsilon \cap W$. It then follows that

> *the closed orbit* $B = G(b)$ *possesses a G-stable open neighborhood which (via the map* $\mathscr{L}$*) is biholomorphically equivalent to* $K^{\mathbb{C}} \times_{L^{\mathbb{C}}} S$.

Now suppose that X is an affine algebraic variety equipped with an algebraic G-action. Since the linear span of every G-orbit in $\mathscr{O}_{\text{alg}}(X)$ is finite-dimensional, one easily finds a closed G-equivariant embedding of X in a G-vector space V. This of course yields the commutative diagram

$$\begin{array}{ccc} X & \hookrightarrow & V \\ \downarrow & & \downarrow \\ X/\!/G & \hookrightarrow & V/\!/G \end{array} .$$

Thus in order to obtain the above type results for the G-action on X, one restricts the vector space results to the image of X in V. □

Luna's Slice Theorem is an extremely important tool in the theory of algebraic reductive group actions and of course it has numerous direct applications. For example, *suppose that* X *is a smooth connected affine manifold, that* $\text{Fix}(G) \neq \emptyset$, *and that* $\mathscr{O}_{\text{alg}}(X)^{\mathbb{C}} \cong \mathbb{C}$. *Then* $\text{Fix}(G)$ *consists of exactly one point* x_0 *and there is a biregular equivariant map* $\phi : X \to V$, $\phi(x_0) = 0$, *onto a G-vector space* V.

As far as we know, the first application in the complex analytic theory was D. Snow's construction of the categorical quotient for Stein spaces ([Sn 1]).

Let X be a Stein space equipped with a holomorphic action of a reductive group G. If one could realize X as a closed subspace of some vector space V where the action is the restriction of a linear action on V, then the desired Stein structure on $X /\!\!/ G$ would be that induced from $V /\!\!/ G$. However, even for X smooth and connected, such a realization is not always possible ([Hei 4]). Nevertheless, by constructing an injective holomorphic G-equivariant immersion of a G-stable neighborhood of a given closed G-orbit in X, Snow was able to extract enough information from the Luna slice in order to define the Stein structure on $X /\!\!/ G$. Using the existence and functorial properties on $X \to X /\!\!/ G$, one can then prove an embedding theorem ([Hei 4]):

There exists a G-equivariant closed holomorphic embedding of a Stein manifold X in a G-vector space if and only if, up to conjugation in K, the underlying K-action has finitely many isotropy subgroups. □

c) Global Linearization

The following classical observation is a typical global linearization statement:

Let D be a bounded domain in the complex plane, and G a group of holomorphic transformations of D. Suppose there is a point $z_0 \in D$ *so that* $\mathrm{Iso}_G\{z_0\}$ *is infinite. Then there is a biholomorphic map* $\phi : D \to \Delta$ *onto the unit disk, where* $\phi(z_0) = 0$ *and the stabilizer of* 0 *acts as a (linear) group of rotations.*

There are primarily two reasons for such strong results in the 1-dimensional setting:

(1) Positive dimensional orbits of a (real) Lie group of holomorphic transformations are either open or 1-codimensional;

(2) The Uniformization Theorem.

The several variable situation is substantially different.

1. Domains in $\mathbb{C}^n$. In ([Car]) H. Cartan began a study of domains in $\mathbb{C}^2$ and in certain situations proved results similar to that above (see [Gal] for a modern explanation of this paper). For example, modulo small improvements which automatically follow from the modern theory, he proved the following fact:

Let X be a connected 2-dimensional Stein manifold equipped with an S^1*-action of holomorphic transformations. Suppose that* S^1 *has a fixed point* $x_0 \in X$ *so that the characters of the local linearization in* $T_{x_0}X$ *have the same sign, i.e.* $t(z, w) = (t^n \cdot z, t^m \cdot w)$*, where* $n, m > 0$ *or* $n, m < 0$*. Then there is a biholomorphic map* $\phi : X \twoheadrightarrow D \subset \mathbb{C}^2$ *with* $\phi(x_0) = 0$ *such that the induced* S^1*-action on D is linear.* □

Cartan also considered domains with 1-dimensional fixed point sets and proved under certain additional assumptions that they are "semi-circular" domains. Instead of stating his result, we state a more general version due to Heinzner.

(4.2) Theorem ([Hei 1]). *Let X be a connected Stein manifold equipped with a K-action of holomorphic transformations. Assume that the fixed point set $F := \{x \in X \mid K(x) = x\}$ is non-empty. For $x \in F$ let $T_xX = T_xF \oplus N_x$ be the K-splitting of the tangent space and let N be the normal bundle of F equipped with the induced K-action. Suppose that $\mathcal{O}(N_{x_0})^K \cong \mathbb{C}$ for some $x_0 \in F$. Then there is a biholomorphic K-equivariant map $\phi : X \to N$ onto an open neighborhood of the 0-section of N. In particular, F itself is connected.*

One might hope that it would be possible to prove a stronger linearization result, particularly for bounded domains. However, without correspondingly strong assumptions, this is not possible.

Example. Let C be a smooth elliptic curve in $\mathbb{P}_2$, $H \cong \mathbb{P}_1$ a hyperplane in $\mathbb{P}_2$ which intersects C in a triple point, and $V := C \setminus H \subset \mathbb{P}_2 \setminus H \cong \mathbb{C}^2$. Let B be a large Euclidean ball in $\mathbb{C}^2$ so that $F := V \cap B$ is topologically the same as $C \setminus H = C \setminus \{p\}$. Let T be a tubular neighborhood of the 0-section in the normal bundle of F which can be biholomorphically embedded onto a domain D in B. The S^1-action along the fibers of T yields a holomorphic S^1-action on D. If this action could be linearized in the strong sense, i.e. there is an embedding of D so that the induced S^1-action is linear, then F would be realized as the intersection of a complex line with D. However, F is not even topologically planar. □

Even if the fixed point set is discrete, there are still many open problems. For example, the following still seems to be unresolved.

Question. *Let D be a connected Stein domain in $\mathbb{C}^2$ which is equipped with an S^1-action of holomorphic transformations. Suppose that there is a fixed point p so that the signs of the characters in the local linearization at p are different, e.g. $t(z,w) = (t \cdot z, t^{-1}w)$. Can the action be globally linearized?*

In fact very little is known about linearization of automorphisms of domains in $\mathbb{C}^n$ or $\mathbb{C}^n$ itself. For example, the following has been the subject of much attention in certain areas of invariant theory in recent years.

Question. *Let G be a reductive (or compact) group of holomorphic automorphisms of $\mathbb{C}^n$. Is there a holomorphic coordinate system for $\mathbb{C}^n$ so that G is acting linearly?*

Of course invariant theorists are usually more interested in algebraic actions. Recently G. Schwarz provided an example of an algebraic action of

$\mathbb{C}^* \rtimes \mathbb{Z}_2$ on $\mathbb{C}^4$ which *cannot* be algebraically linearized, but which *can* be holomorphically linearized.

It probably shouldn't be expected that a reductive group acting on $\mathbb{C}^n$ is linearizable. Thus it would seem reasonable to consider (seemingly) simpler problems.

Question. *Let A be a finite abelian group of holomorphic automorphisms of a contractible Stein domain in $\mathbb{C}^n$. Does A have a fixed point?*

Consider again the domain D in the above example. It fibers over the S^1-fixed point set which is by no means a candidate for a fixed point set coming from a linear action. Note that this projection is exactly the quotient defined by the equivalence relation $x \sim y :\Longleftrightarrow f(x) = f(y)$ for every S^1-invariant holomorphic function f. In fact if there exist invariant functions, then, without further information, there is very little hope of proving global linearization theorems. □

2. The Categorical Quotient for Compact Group Actions. As mentioned in Section b above, the existence of a categorical quotient for reductive algebraic group actions on affine varieties is a now classical result of invariant theory. The analogous theorem for reductive actions on Stein spaces was proved more recently ([Sn 1]). Now, when one considers actions on manifolds with many bounded holomorphic functions, e.g. bounded domains, one *never* has a non-trivial action of a complex Lie group (Liouville's Theorem applied to the orbits of 1-parameter complex subgroups). Hence in this case one studies actions of totally real groups and in particular compact groups. Recently P. Heinzner proved the existence of a categorical quotient for compact group actions.

Let X be a Stein space equipped with a compact group K of holomorphic transformations. Define $x \sim y :\Longleftrightarrow f(x) = f(y)$ for all $f \in \mathcal{O}(X)^K$ and let $\pi : X \to X/\!\!/K$ be the quotient defined by this equivalence relation.

(4.3) Theorem ([Hei 2]). *The quotient $X/\!\!/K$ carries a unique complex structure as a Stein space so that $\pi : X \to X/\!\!/K$ is holomorphic and for any K-invariant holomorphic map $F : X \to Y$ into a complex space Y there is a holomorphic map $\tau_F : X/\!\!/K \to Y$ with $F = \tau_F \circ \pi$.*

Since the base of the categorical quotient can in principle be *any* Stein space, the only realistic linearization questions are on the linearization of the actions on the fibers. One way of approaching this type of problem is to enlarge X so that the universal complexification $K^{\mathbb{C}}$ acts.

3. The Universal Complexification. We begin with a simple but informative

Example. Let X be an annulus in the complex plane and let $K = S^1$ act on X in the usual way as a group of rotations. Then one can embed X as an open Runge subset of the punctured plane $X^{\mathbb{C}}$ and extend the action of K to a holomorphic action of $K^{\mathbb{C}} = \mathbb{C}^*$ on $X^{\mathbb{C}}$. □

In general, let X be a complex space equipped with a K-action of holomorphic transformations. A complex space $X^{\mathbb{C}}$ with a holomorphic action of $K^{\mathbb{C}}$ and a K-equivariant holomorphic map $\iota : X \to X^{\mathbb{C}}$ is called a *universal complexification of the K-action on X* if the following property holds:

Let Y be a complex space equipped with a holomorphic action of $K^{\mathbb{C}}$ and $\phi : X \to Y$ a holomorphic K-equivariant map. Then there is a $K^{\mathbb{C}}$-equivariant holomorphic map $\tau_\phi : X^{\mathbb{C}} \to Y$ with $\phi = \tau_\phi \circ \iota$. □

(4.4) Theorem ([Hei 3]). *Let X be a Stein space equipped with a K-action of holomorphic transformations. Then there exists a unique universal complexification $X^{\mathbb{C}}$, it is Stein, $\iota : X \to X^{\mathbb{C}}$ is an open embedding, and the image $\iota(X)$ is Runge in $X^{\mathbb{C}}$. Furthermore, $X/\!\!/K$ and $X^{\mathbb{C}}/\!\!/K^{\mathbb{C}}$ are naturally isomorphic and*

$$\begin{array}{ccc} X & \overset{\iota}{\hookrightarrow} & X^{\mathbb{C}} \\ \downarrow & & \downarrow \\ X/\!\!/K & \cong & X^{\mathbb{C}}/\!\!/K^{\mathbb{C}} \end{array}$$

commutes.

4. Linearization Under the Condition $\mathcal{O}(X)^K \cong \mathbb{C}$. The fibers of the categorical quotient of the universal complexification of a K-action are affine algebraic varieties where $K^{\mathbb{C}}$ is acting algebraically ([Sn 1]). In particular such a fiber can be realized as a closed subvariety of some $\mathbb{C}^n$ where the $K^{\mathbb{C}}$-action is the restriction of a *linear* action. Without further assumptions, this is the best way to study the K-action on X.

Example. Recall that the 3-sphere S^3 is $\mathbb{R}$-analytically equivalent to the unitary group SU_2. Let $K := \mathrm{SU}_2 \times \mathrm{SU}_2$ act on S^3 via this identification, i.e. one factor on the left and the other on the right. Now S^3 can be $\mathbb{R}$-analytically embedded as a totally real submanifold of $\mathbb{C}^3$ (see [AR]). On the other hand $K^{\mathbb{C}}$ acts transitively on the 3-dimensional complex affine quadric Q_3 and the minimal K-orbit in Q_3 is a totally real submanifold M which is likewise $\mathbb{R}$-analytically equivalent to S^3. Therefore the map $M \cong S^3 \hookrightarrow \mathbb{C}^3$ can be extended to a biholomorphic K-equivariant map $\phi : U \twoheadrightarrow D \Subset \mathbb{C}^3$ of a K-invariant neighborhood U of M in Q_3 onto a bounded domain in D. In this case $D \hookrightarrow D^{\mathbb{C}} = Q_3$ is the natural realization of D and of course there is no embedding of D in $\mathbb{C}^3$ so that the induced K-action is linear. □

Remark. In the above example, *every* relatively compact K-invariant neighborhood of M in Q_3 can be realized as a bounded domain in $\mathbb{C}^3$. Other than Euclidean balls in $\mathbb{C}^n$, these are the only bounded domains which admit a K-action with generic 1-codimensional orbits. This is a consequence of a result of [MN] which states that, except for the possibility of tubular neighborhoods of the 0-sections of the tangent bundles of S^3 and S^7, the only examples are Euclidean balls. It was shown in ([SZ]) that S^7 can *not* be realized as a totally real submanifold of $\mathbb{C}^7$. □

Manifolds of the type D in the above example are of central importance for the study of K-actions when $\mathcal{O}(X)^K \cong \mathbb{C}$.

Let X be a Stein manifold equipped with a K-action of holomorphic transformations and suppose that $\mathcal{O}(X)^K \cong \mathbb{C}$. Let B be a minimal (in the sense of dimension) K-invariant analytic subset of X (B takes the place of the closed $K^{\mathbb{C}}$-orbit in the case of reductive actions). In follows that B is unique and smooth. In the *local sense* $K^{\mathbb{C}}$ acts transitively on B, i.e. for every $x \in B$ the complex vector space generated by $T_{\mathbb{R}}K(x)$ is the entire tangent space T_xB.

(4.5) Theorem ([Hei 3]). *There exists a point $x \in B$ with* $\text{Iso}_K\{x\} =: L$ *such that*

(i) The universal complexification $B^{\mathbb{C}}$ is the affine algebraic homogeneous space $K^{\mathbb{C}}/L^{\mathbb{C}}$;

(ii) The induced representation of $L^{\mathbb{C}}$ on T_xX contains an $L^{\mathbb{C}}$-stable complex vector space V which is transversal to B at x such that $X^{\mathbb{C}}$ is $K^{\mathbb{C}}$-equivariantly biholomorphic to the normal bundle $K^{\mathbb{C}} \times_{L^{\mathbb{C}}} V$. □

(4.6) Corollary. *Let X be a Stein manifold equipped with a K-action of holomorphic transformations. Suppose that $\mathcal{O}(X)^K \cong \mathbb{C}$ and that* $\text{Fix}(K) \neq \emptyset$. *Then* $\text{Fix}(K) = \{x_0\}$ *and there is a biholomorphic map $\phi : X \to D$ onto a domain in $\mathbb{C}^n$ such that $\phi(x_0) = 0$ and the induced K-action is linear.*

(4.7) Corollary. *Let X be a Stein manifold equipped with a $K = (S^1)^k$-action and suppose that $\mathcal{O}(X)^K \cong \mathbb{C}$. Then there is a biholomorphic map $\phi : X \to D$ onto a domain D in $\mathbb{C}^n$ so that the induced action is linear.*

Proof. In this case $B^{\mathbb{C}} \cong (\mathbb{C}^*)^l$, $L^{\mathbb{C}} \cong (C^*)^{k-l}$, and the normal bundle $(\mathbb{C}^*)^k \times_{L^{\mathbb{C}}} \mathbb{C}^{n-k}V$ is equivariantly trivial. □

Remarks. (1) Note that for manifolds which possess a Stein envelope of holomorphy, e.g. domains in $\mathbb{C}^n$, one has a version of most results above.

(2) Although Heinzner developed a number of technical methods for proving the above results, his basic principles are of an almost classical geometric nature: *orbit convexity and geometric slice theorems*.

(3) Corollary 4.6 above can be proved directly with elementary methods ([Hei 2]). The special case of an $(S^1)^n$-action on an n-dimensional manifold X was first proved by Barrett, Bedford and Dadok ([BBD]). □

Chapter II

Fibration Methods and Applications to Classification Problems

Let X be a connected normal complex space, G a connected (real) Lie group, and $G \times X \to X$ a G-action by holomorphic transformations. If this action is transitive, then we refer to X as a *homogeneous G-space* and write $X = G/H$. If G has an open orbit in X, then we refer to X as an *almost homogeneous* G-space. If G is complex, the action is holomorphic, and X is almost homogeneous, then there is a *unique* open G-orbit Ω. Its complement $E := X \setminus \Omega$ is a closed thin analytic subset.

Our approach for studying a given almost homogeneous space X is to consider equivariant holomorphic or meromorphic maps of X onto lower-dimensional spaces. Such maps, which arise from a mixture of group-theoretic and complex-analytic considerations, are called *G-fibrations*. At the level of homogeneous spaces, these are holomorphic maps of the form $G/H \to G/J$.

Warning. If G is only a real Lie group of holomorphic transformations, the C^∞-fibration $G/H \to G/J$ is of course a G-equivariant holomorphic map but not necessarily a holomorphic bundle. □

Our goal here is to present some of the basic fibration techniques and give typical applications in the classification theory in cases where the technical details are not too cumbersome. Although many of the methods are classical, our viewpoint of the subject was developed in a relatively recent work with E. Oeljeklaus ([HO 1]).

1. Compact Almost Homogeneous Spaces

Let X be a connected compact complex manifold with *Albanese map* $\alpha : X \to \mathrm{Alb}(X)$. Recall that $\mathrm{Alb}(X)$ is a compact complex torus and α is a holomorphic mapping which satisfies the following *universality* property (see [Bl] for the construction): *For every holomorphic map $\phi : X \to T$ into a compact complex torus T there is an affine map A_ϕ : $\mathrm{Alb}(X) \to T$ with $\phi = A_\phi \circ \alpha$.* It follows that the map α is $\mathrm{Aut}(X)$-equivariant.

Suppose that G is a connected complex Lie group and that X is G-almost homogeneous. Consider the induced action on $\mathrm{Alb}(X)$ given by $\alpha_* : \mathrm{Aut}(X)^\circ \to \mathrm{Aut}(\mathrm{Alb}(X))^\circ \cong \mathrm{Alb}(X)$. It follows that $G/\ker(\alpha_*) \cap G$ acts freely on $\mathrm{Alb}(X)$, and, since some G-orbit in $\alpha(X)$ is open, G acts transitively on the image $\alpha(X)$. In particular $\alpha(X)$ is itself a torus and the universality property implies that $\alpha(X) = \mathrm{Alb}(X) = G/H$, where H is the stabilizer of some fiber F.

Albanese Theorem. *Let X be a connected compact complex manifold which is almost homogeneous with respect to a complex Lie group G. Then $\alpha : X \to$ Alb(X) is surjective and G acts transitively on the base so that the Albanese map is a homogeneous bundle*

$$X = G \times_H F \to G/H = \mathrm{Alb}(X) ,$$

where $F = \alpha^{-1}(\alpha(x))$ is connected and H is its stabilizer in G.

Proof. It only remains to show that F is connected. This follows from the fact that Stein-factorization of a G-equivariant map is G-equivariant, that a finite cover of a torus is a torus, and from the universality of the Albanese. □

a) Kähler Manifolds

Suppose that X is Kähler with Albanese $\alpha: X \to \mathrm{Alb}(X)$, then $\dim_{\mathbb{C}} \mathrm{Alb}(\mathbb{C}) = h^{1,0}(X) = \frac{1}{2} b_1(X)$. More precisely, every holomorphic 1-form on X is a lift of a linear form on Alb(X). If X is in addition almost homogeneous with Albanese fiber F, then $\pi_1(F) = 1$ and F is *projective algebraic* ([Oe 1]). Furthermore, unless $q = 0$, $h^{0,q}(F) = 0$ ([Oe 1]). In particular, $H^1(F, \mathcal{O}^*) \cong H^2(F, \mathbb{Z})$. Now let $G = \mathrm{Aut}(F)^\circ$ and note that, since $H^2(F, \mathbb{Z})$ is discrete, G stabilizes every line bundle over F. As a consequence, there is a G-equivariant embedding of F in some projective space and in particular *G is an algebraic group acting algebraically*. This can be generalized in a useful way: $\ker(\alpha_* : \mathrm{Aut}(X)^\circ \to \mathrm{Alb}(X))$ *acts on F as an algebraic group* (see [BOe] for the almost homogeneous case and [Fu] for the general setting). If follows that if *X is an almost homogeneous compact Kähler manifold then X is projective algebraic if and only if* Alb(X) *is projective algebraic*.

If X is an almost homogeneous compact Kähler manifold, then *a finite unramified cover of the Albanese fibration is topologically trivial* ([BOe]). This is not true *holomorphically*. For example, let L be a topologically trivial line bundle over an elliptic curve T so that L^n is *not* trivial for all $n \in \mathbb{N}^+$. Let X be the compactification of L as a $\mathbb{P}_1$-bundle. Then X is almost homogeneous and, since the Albanese is just the $\mathbb{P}_1$-fibration $X \to T$, no finite cover is holomorphically trivial.

If X is an almost homogeneous *projective algebraic* manifold, then we have the following

Question. *Is the Albanese $\alpha : X \to$* Alb(X) *meromorphically trivial?*

b) Some Non-Kählerian Examples

From a certain algebraic-geometric point of view, the natural context for the combinatorial study of compact almost homogeneous varieties, e.g. *torus embeddings*, $\mathrm{SL}_2(\mathbb{C})$-*embeddings*, ..., is that of an *algebraic space*. In this way it is not unusual to encounter Moišezon spaces which are not Kählerian. These spaces are of such a combinatorial nature that one can often calculate very precise information. For example, let X be a smooth SL_2-embedding, i.e. X is a smooth compact Moišezon space with an algebraic $SL_2(\mathbb{C})$-action with open orbit $SL_2/\{e\} \cong SL_2$. Then *X is Kähler if and only if there are no complex curves in X which are homologous to zero* ([MJ]).

It is also not unusual to encounter compact almost homogeneous spaces which are very far away from being Moišezon. Of course the generic compact torus, or generic manifold of the form $X = G/\Gamma$, has no meromorphic functions other than the constants. However, this phenomenon even occurs when the open orbit and its complement are algebraic.

Example. Let $\tilde{X} = \mathbb{C}^2 \setminus \{(0,0)\}$ and consider the *non-linear contraction* $\gamma \in \mathrm{Aut}(\tilde{X})$ defined by $\gamma(z_1, z_2) = (\alpha^m z_1 + \lambda z_2^m, \alpha z_2)$, where $|\alpha| < 1$, $\lambda \in \mathbb{C}^*$, and $m \in \mathbb{N}^+$. For $(\xi, \eta) \in \mathbb{C}^* \times \mathbb{C}$ define $g(z_1, z_2) = (\xi^m z_1 + \eta z_2^m, \xi z_2)$. This yields a $\mathbb{C}^* \times \mathbb{C}$-action on $\tilde{X}$ which centralizes $\Gamma := \langle\gamma\rangle_{\mathbb{Z}}$ and induces a $\mathbb{C}^* \times \mathbb{C}^*$-action on the quotient $X = \tilde{X}/\Gamma$. The open orbit is $\mathbb{C}^* \times \mathbb{C}^*$ itself and the complement E is the elliptic curve corresponding to $\{z_2{=}0\}$. In fact E is the *only* complex curve in X and, other than the constants, X posesses *no* meromorphic functions (see [Kod]). □

Remarks. (1) F. Lescure ([Les 1]) has recently analyzed equivariant compactifications of abelian groups. In particular he discusses the *Moišezon question* in great detail.

(2) If X is a compact almost homogeneous manifold with respect to a complex Lie group G and the open orbit is contractible, then *X is Moišezon* ([Gell]). □

2. The $\mathfrak{g}$-anticanonical Fibration

If G is a complex Lie group and $X = G/H$ is homogeneous, then one may consider the *normalizer fibration* $G/H \to G/N$, where $N = N_G(H^\circ)$. The base can be realized as a G-orbit in some projective space and the fiber is of the form M/Γ, where M is a complex Lie group and Γ is a discrete subgroup (see I.1). It is useful to look at this fibration from another viewpoint.

Let X be a complex manifold, K^{-1} its *anti-canonical* bundle, and $\Gamma(X, K^{-1})$ the associated space of sections. Suppose that G is a (real) Lie group of holomorphic transformations of X and V is a G-stable finite dimensional subspace of $\Gamma(X, K^{-1})$. Then the meromorphic map $\phi_V : X \to \mathbb{P}(V^*)$, $x \mapsto [f_x]$, where $f_x(\sigma) = \sigma(x)$, is G-equivariant.

Let $\Gamma_{\mathcal{O}}(X, TX)$ be the space of *real* vector fields on X which are holomorphic, i.e. $Z \in \Gamma_{\mathcal{O}}(X, TX) :\iff Z^{\mathbb{C}} = \frac{1}{2}(Z - iJZ)$ is a holomorphic $(1,0)$-field. We may assume that G is acting almost effectively on X and thus we have the natural injection $\mathfrak{g} \hookrightarrow \Gamma_{\mathcal{O}}(X, TX)$, $v \mapsto \mathfrak{X}_v$. In this way we have the associated complex Lie subalgebra $\hat{\mathfrak{g}} := \mathfrak{g} + J\mathfrak{g} \subset \Gamma_{\mathcal{O}}(X, TX)$. By complexification, i.e.

$Z \mapsto Z^{\mathbb{C}}$, we realize $\hat{\mathfrak{g}}$ as finite-dimensional Lie algebra of holomorphic (1,0)-fields on X. Now, let $n := \dim_{\mathbb{C}} X$, and let $V_{\mathfrak{g}}$ be the subspace of $\Gamma(X, K^{-1})$ which is generated by sections of the form $\sigma = Z_1 \wedge \ldots \wedge Z_n$, $Z_i \in \hat{\mathfrak{g}}$. Note that if the tangent space $T_x(G(x))$ of a G-orbit generates the full tangent space $T_x X$ as a *complex* vector space, then there exists $\sigma \in V_{\mathfrak{g}}$ with $\sigma(x) \neq 0$.

Of course $V_{\mathfrak{g}}$ is G-stable. We refer to the associated G-equivariant meromorphic map $\phi_{\mathfrak{g}} : X \to \mathbb{P}(V_{\mathfrak{g}}^*)$ as the $\mathfrak{g}$-*anticanonical map*. The following summarizes the basic properties of $\phi_{\mathfrak{g}}$ in the almost homogeneous case.

(2.1) Theorem ([HO 1]). *Let X be a connected complex manifold which is homogeneous with respect to a real Lie group G of holomorphic transformations. Let $\phi_{\mathfrak{g}} : X = G/H \to G/N$ be the $\mathfrak{g}$-anticanonical fibration. Equip G/H° with the G-invariant structure as a covering space of G/H. Then*

1) $N = \{n \in N_G(H^\circ) \mid \text{Right-action of } n \text{ on } G/H^\circ \text{ is holomorphic}\}$
2) *The induced fibration $G/H^\circ \to G/N$ is a holomorphic principal bundle with complex Lie group N/H° as fiber*
3) *The fibration $\phi_{\mathfrak{g}} : X = G/H \to G/N$ is a holomorphic fiber bundle with fiber M/Γ, where $M = N/H^\circ$ is a complex Lie group and $\Gamma = H/H^\circ$ is a discrete subgroup.*

(2.2) Corollary. *Let X be almost homogeneous with respect to the holomorphic action of a complex Lie group G. Then the normalizer fibration $\Omega = G/H \to G/N$ of the open orbit is G-equivariantly meromorphically extendible to a map of X into some projective space.*

Proof. Since G is complex, the right-action of every element $n \in N_G(H^\circ)$ is holomorphic. Thus the normalizer fibration agrees with the $\mathfrak{g}$-anticanonical fibration of the open orbit. □

Remarks. (1) If $\phi_{\mathfrak{g}} : X \to Z \hookrightarrow \mathbb{P}_m$ is the anti-canonical map, then the smallest *complex* subgroup $\hat{G}$ which contains the image of G in the linear transformations of $\mathbb{P}_m$ locally stabilizes Z. In particular, if X is homogeneous, then the base G/N is open in the $\hat{G}$-orbit $\hat{G}/\hat{N}$. Thus we may use the fibration theory for complex groups to obtain fibrations of G/N.

(2) The main difficulty in dealing with homogeneous complex manifolds of a *real* Lie group of holomorphic transformations is that most interesting holomorphic equivariant fibrations $G/H \to G/J$ are *not* holomorphic fiber bundles. In this sense the map $\phi_{\mathfrak{g}} : G/H \to G/N$ is an important exception. □

3. Fibrations of Orbits of Linear Groups

The $\mathfrak{g}$-anticanonical map brings us from the abstract setting to the situation of a linear action. Even in the case of actions of real Lie groups of holomorphic transformations we have the complex linear group $\hat{G}$. Hence, fibrations of $\hat{G}$-orbits induce holomorphic maps of the G-orbits.

In this section we only consider complex groups. So, in order to simplify the notation, let G be a connected complex linear group and $G(x) = G/H$ an orbit in some $\mathbb{P}_m$.

a) The Commutator Fibration

An orbit of a linear group G may not be Zariski open in its closure. However, if $G(x) = G/H$ is an orbit and N is a normal complex subgroup of G which is embedded as an algebraic linear group, then NH is closed in G and we have the homogeneous fibration $G/H \to G/NH$. Thus it is clear that the following result is extremely useful.

(3.1) Theorem (Chevalley ([Ch])). *Let G be a connected linear group. Then the commutator group G' is an algebraic subgroup of the full linear group.*

Consequently, for every G-orbit $G(x) = G/H$ in $\mathbb{P}_m$, we have *the commutator fibration* $G/H \to G/G'H$. Of course the base $G/G'H$ is a complex abelian Lie group, but it is not an arbitrary one.

Note that the orbit $G(x) = G/H$ is contained in the orbit $\overline{G}(x) = \overline{G}/\overline{H}$ of the complex algebraic closure of G (By abusing notation, $\overline{H}$ denotes the $\overline{G}$-isotropy at the point x). Now $\overline{G}' = G'$ (see [Bo 1]). Thus the homogeneous space $G/G'H$ is a G-orbit in the *algebraic abelian group* $\overline{G}/G'\overline{H}$. Thus G/HG' is holomorphically separable and therefore $G/HG' = \mathbb{C}^k \times (\mathbb{C}^*)^l$.

(3.2) Theorem. *Let $X = G(x) = G/H$ be an orbit of a connected complex linear group in some $\mathbb{P}_m$. Then the commutator fibration*

$$X = G/H \to G/HG' = \mathbb{C}^k \times (\mathbb{C}^*)^l$$

realizes X as an algebraic fiber bundle over an abelian group $\mathbb{C}^k \times (\mathbb{C}^*)^l$.

Proof. It remains to clarify the notion of *algebraic fiber bundle*. For this note that the fiber $G'H/H =: F$ is a G'-orbit and is therefore Zariski open in its closure $\overline{F}$. Thus the algebraic (Zariski) closure $\overline{HG'}$ acts on $\overline{F}$ with open orbit F. □

If U is a connected complex linear *unipotent* group, then U is algebraic and every U-orbit $U(x)$ is algebraically isomorphic to some affine space. Now, if G is a solvable group, then G' is unipotent and the commutator fibration realizes $X = G(x) = G/H$ as an algebraic $\mathbb{C}^n$-bundle over $\mathbb{C}^k \times (\mathbb{C}^*)^l$. In fact this bundle is trivial (see [Sn 2]) and we have the following

(3.3) Corollary. *Let $X = G/H$ be an orbit of a connected complex linear solvable group G. Then $X \cong \mathbb{C}^m \times (\mathbb{C}^*)^n$ as a complex manifold.*

Remark. The above Corollary 3.3 does not reflect the real situation for solv-manifolds X. The interesting ones are of the form G/Γ, where Γ is discrete and *highly* non-abelian. □

b) Non-Fiberable Linear Orbits

We say that an orbit $G(x) = G/H$ can *not* be fibered if there is no closed proper complex subgroup J of G with $H < J$ and $\dim_{\mathbb{C}} G/H > \dim_{\mathbb{C}} G/J$.

(3.4) Theorem. *Suppose $X = G(x) = G/H$ is an orbit of a connected complex linear group in some $\mathbb{P}_m$. Assume that $\dim_{\mathbb{C}} X > 1$ and that X cannot be fibered. Then either*

(i) G is semi-simple, H is maximal and therefore G/H is affine or compact rational

or

(ii) $X \cong \mathbb{C}^n$ and the algebraic group G' contains an algebraic subgroup of the form $A = T \rtimes M$, where $T = (\mathbb{C}^n, +)$ acts freely by translations and M acts irreducibly on T.

Proof. If $G' < H$, then X is an abelian group. Thus, as a manifold $X \cong \mathbb{C}^m \times (\mathbb{C}^*)^n$, and, unless $\dim_{\mathbb{C}} X = 1$, it can be fibered. Therefore we may assume that G' acts non-trivially on X. Since it can't be fibered, it follows that G' acts transitively. Consequently $X = \overline{G}(x)$, where $\overline{G}$ is the algebraic closure of G.

Let $\overline{G} = U \rtimes L$ be a Levi-decomposition of $\overline{G}$ and first consider the case $U \neq (e)$. Since X cannot be fibered, it follows that every non-trivial characteristic complex subgroup of U acts transitively on X. In particular, $X \cong \mathbb{C}^n$ and, since it acts transitively on X, G' is not reductive, because for topological reasons a maximal compact subgroup has a fixed point. Arguing as we did for $\overline{G}$, it follows that the center $T = (\mathbb{C}^n, +)$ of the unipotent radical of G' acts freely and transitively on X and the maximal reductive subgroup M of G' must act irreducibly on T.

If $\overline{G} = L$ is reductive, then G is reductive and in fact $G = \overline{G}$. Let $Z = (\mathbb{C}^*)^k$ be the connected component of the center of G. Since X cannot be fibered, if $k > 0$ then $k = 1$ and $\dim_{\mathbb{C}} X = 1$. Thus we may assume that G is semi-simple.

If G is semi-simple and H is not reductive, we consider the Levi-decomposition $H^\circ = U \rtimes L$. Let $\tilde{U}$ be the unipotent radical of the normalizer $N_G(U)$. If $\tilde{U} = U$, then $N_G(U)$ is parabolic (see e.g. [Hum]). Otherwise $G/H \to G/\tilde{U}H$ is a fibration. Since X cannot be fibered, it therefore follows that $H = N_G(U)$ is a (maximal) parabolic subgroup of G. □

Remark. We would like to emphasize the procedure used in the last step of the above proof. Let G be semi-simple and $H = U \rtimes L$ non-reductive. Define U_1 to be the unipotent radical of $N_G(U)$ and $H_1 := HU_1$. Now repeat this, obtaining sequences $U < U_1 < \ldots$ and $H < H_1 < \ldots$ until $U_i = U_{i+1}$. At this point $N_G(U_i) = P$ is parabolic. Hence, if H is not reductive, then we have a concrete procedure for fibering X, i.e. $G/H \to G/H_1 \to \ldots \to G/P$. □

c) Eliminating the Radical

Let $G = R \cdot S$ be a connected complex linear group and $G(x) = G/H$ an orbit in some projective space.

(3.5) Lemma. *Suppose that the radical orbits are (analytically) Zariski dense in $G(x)$, i.e. RH is contained in no proper closed complex subgroup of G. Then R acts transitively.*

Proof. Note first that if A is a characteristic normal complex subgroup of G which is acting as a linear algebraic subgroup on the ambient projective space, $G/AH \hookrightarrow \overline{G}/\overline{AH}$ is likewise a linear orbit. If the commutator R' acts non-trivially, then we have just such a fibration $G/H \to G/R'H$ and the result follows by induction. Hence we may assume that R is abelian, $R = \mathbb{C}^n \times (\mathbb{C}^*)^n$. The reductive part $(\mathbb{C}^*)^m$ of R is acting as an algebraic group. Therefore, if $m > 0$, then we may apply the same argument as above to complete the proof. Thus we may assume that $R = \mathbb{C}^n$.

Note that the unipotent radical $R_u(G')$ is algebraic and characteristic. Thus if $R_u(G') \neq (e)$, then again the result follows by induction. Hence, we may assume that G' is reductive which in particular implies that S is a normal subgroup of G, i.e. $G = \mathbb{C}^n \times S$.

Let $A = R \cap HS$ be the projection of the isotropy H into the radical $\mathbb{C}^n$, i.e. A consists of the elements of R which stabilize the fiber F of $G/H \to G/HS$. Since the R-orbits are dense in G/H, it follows that the A-orbits are dense in F. Now F is an S-orbit. Hence it is Zariski open in its closure and consequently the algebraic closure $\overline{A}$ acts transitively on F, since $\overline{A}$ normalizes S. But $\overline{A}$ is solvable. So $F \cong \mathbb{C}^k \times (\mathbb{C}^*)^l$, which is contrary to F being an S-orbit. Therefore this last case does not occur. □

Applying the above Lemma in the obvious way, we can *eliminate the radical*.

(3.6) Theorem. *Let G be a connected complex linear group and $G(x) = G/H$ a G-orbit in some projective space. Let $J := \overline{RH}^\circ$ be the connected component of the complex closure of RH in G. Then the radical R_J acts transitively on the orbit $\overline{RH}/H$. In particular $\overline{RH}/H \cong \mathbb{C}^k \times (\mathbb{C}^*)^l$.*

4. Fibrations in the Case of Discrete Isotropy

If we begin with an abstract homogeneous or almost homogeneous space and the g-anticanonical fibration maps it to a point, then the above methods for linear orbits are of very little use. In this case, however, the open orbit is of the form G/Γ, where G is a complex Lie group and Γ is discrete. We now discuss the possibilities for fibering such spaces.

a) Reductive Groups

If G is a complex reductive Lie group, then there is a unique affine algebraic structure on G so that multiplication is a morphism. Thus, if H is any subgroup, then we have a well-defined Zariski closure $\overline{H}$. The algebraic homogeneous space $G/\overline{H}$ can be realized as an orbit in some projective space.

Let H be a complex subgroup of G and suppose that $\overline{H} = G$. Note that $N_G(H^\circ)$ is an algebraic group. Hence $H^\circ \triangleleft G$. Since we generally can reduce to the case where G is acting almost effectively on G/H, this indicates that many problems can be reduced to considerations of Zariski dense discrete groups. □

b) Abelian Groups

A connected complex abelian Lie group is a product $\mathbb{C}^l \times (\mathbb{C}^*)^m \times \mathbb{C}^k/\Gamma$, where Γ is a discrete additive group in $\mathbb{C}^k$ with $\mathcal{O}(\mathbb{C}^k)^\Gamma \cong \mathbb{C}$. The only interesting case is $G = \mathbb{C}^k/\Gamma$, i.e. just the last factor. Unless G is itself a *compact simple* torus, one can always construct a fibration (*homomorphism*) onto a lower-dimensional compact torus. This can be done by explicit calculations with period matrices (see e.g. [AG]).

For the moment, let us look at the complex geometry of such groups. In general, if G is a complex Lie group with $\mathcal{O}(G) \cong \mathbb{C}$, then, since $\mathrm{Ad}(G)$ is holomorphically separable, it follows that G is abelian. In particular $G = V/\Gamma$, where V is a complex vector space and Γ is a discrete additive group with $\operatorname{rank}\Gamma > n := \dim_{\mathbb{C}} V$. Let W be the *real* vector subspace of V which is generated by Γ. The condition $\mathcal{O}(G) \cong \mathbb{C}$ immediately implies that $W + \sqrt{-1}W = V$.

Consider $M := W \cap \sqrt{-1}W$, i.e. the maximal complex subspace of W, and note that the M-orbits in G are contained in translates of the maximal compact torus $W/\Gamma =: T$. In fact the condition $\mathcal{O}(G) \cong \mathbb{C}$ is exactly the condition that the M-orbits are topologically dense in T (see e.g. [HMar]). □

c) Nilpotent Groups

Let G be a connected, simply-connected, complex nilpotent Lie group. Using the fact that $\exp : \mathfrak{g} \to G$ is biholomorphic along with induction type proofs involving the center $Z(G)$, one easily derives a number of useful first results. For example, if H is a connected complex subgroup, then $G/H \cong \mathbb{C}^n$ as a complex manifold. The normalizer of any connected subgroup is again connected. Hence, the base of the normalizer fibration $G/H \to G/N$ is $\mathbb{C}^n$ and therefore the bundle is holomorphically trivial. So, from the point of view of complex analysis, we only need to study the fiber, i.e. nil-manifolds of the form G/Γ, where Γ is discrete.

It follows from the result of Malcev-Matsushima ([Ma 2]) that there is a *unique connected* real subgroup $G(\Gamma)$ in G so that $G(\Gamma) > \Gamma$ and $G(\Gamma)/\Gamma$ is compact. Let $\mathfrak{g}(\Gamma)$ be its Lie algebra and define $\mathfrak{g}_{\mathbb{C}}(\Gamma)$ to be the smallest complex Lie algebra in $\mathfrak{g}$ which contains $\mathfrak{g}(\Gamma)$. Finally, let $G_{\mathbb{C}}(\Gamma)$ be the connected complex group corresponding to $\mathfrak{g}_{\mathbb{C}}(\Gamma)$. Again, $G/G_{\mathbb{C}}(\Gamma) \cong \mathbb{C}^n$ and therefore it is enough to consider the case where $G_{\mathbb{C}}(\Gamma) = G$.

Primarily due to the existence of the *compact model* $G(\Gamma)/\Gamma$, complex nil-manifolds are relatively easy to work with. On the one hand, their complex-geometry is quite analogous to that of abelian groups, e.g. the behaviour of the complex group corresponding to $\mathfrak{m} := \mathfrak{g}(\Gamma) \cap i\mathfrak{g}(\Gamma)$ determines the holomorphic function theory. On the other hand, the various characteristic subgroups of G and $G(\Gamma)$ yield interesting fibrations of G/Γ.

The following is a moreorless general principle. Suppose we have a normal subgroup Λ in Γ so that $G(\Lambda)$ is a normal subgroup in G. Thus $G_{\mathbb{C}}(\Lambda) \lhd G$ and conjugation yields a map

$$k : G \to \mathrm{Aut}\,(G_{\mathbb{C}}(\Lambda)) \ .$$

Now $k(\Gamma)$ is contained in $\mathrm{Aut}_\Lambda\,(G_{\mathbb{C}}(\Lambda))$, i.e. the automorphisms which stabilize Λ. But $\mathrm{Aut}_\Lambda\,(G_{\mathbb{C}}(\Lambda))$ is a discrete subset of $\mathrm{Aut}\,(G_{\mathbb{C}}(\Lambda))$. Thus $k^{-1}\,(k(\Gamma))$ is closed in G, i.e. *$\Gamma \cdot Z_G(\Lambda)$ is a closed complex subgroup.*

A simple fibration is given by $G/\Gamma \to G/N(\Gamma)$. Since $Z(G)$ is positive dimensional, the normalizer $N(\Gamma)$ is likewise positive-dimensional. If $N(\Gamma) = G$, then of course Γ is central. This however implies that $G_{\mathbb{C}}(\Gamma)$ is central. □

d) Solvable Groups

From our point of view, homogeneous manifolds of solvable complex Lie groups are extremely complicated; in any case, much more difficult to handle than nil-manifolds. Nevertheless, they are always fiberable.

(4.1) Theorem. *Let G be a connected, non-abelian, solvable complex Lie group and let Γ be a discrete subgroup. Then there is a fibration $G/\Gamma \to G/J$ with* $\dim_{\mathbb{C}} G > \dim_{\mathbb{C}} J \geq 1$.

Proof. If $\Gamma \ntriangleleft G$ and $N_G(\Gamma)$ is positive-dimensional, then let $J := N_G(\Gamma)$. We discuss the case $\Gamma \lhd G$ at the end. Otherwise, $N_G(\Gamma)$ is discrete and in particular we may assume that the center Z_G is likewise discrete.

We first consider the case where Γ is non-abelian. In particular $\Gamma \cap G' \neq (e)$ and we have the real span $G'(\Gamma \cap G')$ in the nilpotent group G' (Note that we may assume that G is simply-connected, so that $\pi_1(G') = 1$). Let S denote the complex group $G'_{\mathbb{C}}(\Gamma \cap G')$. Since Γ normalizes S, if $S \ntriangleleft G$, then we may take $J := N_G(S)$.

If $S \lhd G$, then, as in the above discussion on nilpotent groups, it follows that $\Gamma \cdot Z_G(\Gamma \cap G')$ is closed. Thus we need only consider the case where $\Gamma \cap G'$ is central. But this would imply that S is central, contrary to $Z_G^\circ = (e)$. Therefore the non-abelian case is complete.

If Γ is abelian, then for each $\gamma \in \Gamma$ we consider the map $\phi_\gamma : G \to G'$, $g \mapsto g\gamma g^{-1}\gamma^{-1}$. Since $\dim_{\mathbb{C}} G > \dim_{\mathbb{C}} G'$, it follows that for all $\gamma \in \Gamma$ the centralizer $Z_G(\gamma)$ is positive-dimensional. Unless Γ is itself central, we can choose $J := Z_G(\gamma)$ for an appropriate γ. If Γ is central, then $\Gamma < \ker(\mathrm{Ad})$ and $G/\ker(\mathrm{Ad})$ is a linear group which can always be fibered. □

e) The Mixed Case

Here we consider a simply-connected complex Lie group $G = R \rtimes S$, where R and S are both positive-dimensional, and a discrete group Γ in G. *Let H be the connected component of the topological closure of $R\Gamma$. Then H is solvable* ([Aus]). As was pointed out by B. Gilligan ([Gi 2]), if $\hat{H}$ is the connected complex subgroup corresponding to $\hat{\mathfrak{h}} := \mathfrak{h} + i\mathfrak{h}$, then $\Gamma < N_G(\hat{H})$ and we have a fibration $G/\Gamma \to G/N_G(\hat{H})$. Of course, if $N_G(\hat{H}) = G$, then $\hat{H} = R$, i.e. $R\Gamma$ was closed in the first place and we have the fibration $G/\Gamma \to G/R\Gamma$. □

Collecting the various results in the case of discrete isotropy, we now have the following statement.

(4.2) Theorem. *Let G be a connected complex Lie group with* $\dim_{\mathbb{C}} G > 1$, $\Gamma < G$ *a discrete subgroup, and $X = G/\Gamma$. Then there is a fibration $X = G/\Gamma \to G/J$,* $\dim_{\mathbb{C}} G > \dim_{\mathbb{C}} J > 1$, *except in the following cases:*

(a) *G is abelian and X is a simple torus*

(b) *G is semi-simple and Γ is a Zariski dense subgroup with $G_i\Gamma$ non-closed for all factors G_i.* □

5. Some Typical Applications

One of the first *fine classification* results for almost homogeneous spaces is the following **characterization of projective space** ([Oe 2]): *Let X be a connected compact complex manifold which is almost homogeneous with respect to a complex Lie group G. Suppose E contains an isolated point. Then $X \cong \mathbb{P}_n$.*

By applying the fibration methods one can prove a version of this result for *non-compact* and/or *singular spaces.*

Cone Theorem. *Let X be an irreducible almost homogeneous complex space. Assume that E contains an isolated point x_0. Then X is locally irreducible and its normalization $\hat{X}$ is either a projective or affine homogeneous cone over a homogeneous rational manifold Q. Moreover*

1) *if X is non-compact then $E = \{x_0\}$*

2) *if X is compact, then either $E = \{x_0\}$ or $E = \{x_0\} \cup Q$.*

If X is non-singular, then it is either $\mathbb{C}^n$ or $\mathbb{P}_n$.

Remarks. (1) This result was proved in ([HO 2]). There is also a version for real groups (see [HO 1]). In the context of algebraic groups this was first proved by D. Ahiezer ([A 1]).

(2) If X is non-compact or if E contains more that the isolated point, then the open orbit $X \setminus E = \Omega = G/H$ has at least two ends ($e(\Omega) \geq 2$) in the sense of Freudenthal (see e.g. [Gi 2] for basic definitions). In such a situation the fibration methods are extremely strong. Thus, to give an elementary example of how to use these methods, we will now prove the Cone Theorem using a result on homogeneous spaces with more than one end. □

a) Homogeneous Manifolds with More Than One End

Let X be the homogeneous space of a complex Lie group, $X = G/H$, and assume that $e(X) \geq 2$. Suppose further that X possesses a non-constant holomorphic function, i.e. $\mathcal{O}(X) \ncong \mathbb{C}$. Then we have the *holomorphic reduction* $X = G/H \to G/J$.

(5.1) Theorem ([Gi 2]). *The fiber J/H is connected and compact, and the base G/J is an affine cone over a homogeneous rational manifold with its vertex removed.*

Remark. This type of result can be proved for real groups, meromorphic functions, analytic hypersurfaces, ... (see [GOR]).

Proof. For any fibration $\pi : X \to B$ with connected fiber F of a manifold X with $e(X) \geq 2$ we have the following topological information ([GH 1]): *Either F or B is compact. If F is compact, then $e(B) = e(X)$. If B is compact, then $e(F) \geq e(X)$.*

Suppose that the holomorphic reduction $G/H \to G/J$ has positive-dimensional fiber and consider the associated bundle $G/H \to G/\tilde{J}$ with connected fiber. Then, since the base has non-constant holomorphic functions, it is non-compact with $e(G/\tilde{J}) \geq 2$. The result follows by induction on dimension. Thus it is enough to consider the case where $G/H \to G/J$ has discrete fibers, i.e. $\mathcal{O}(X)$ *has maximal rank*.

Let $G/H \to G/N$ be the normalizer fibration. If the fiber is positive-dimensional and $N \neq G$, then, since $\mathcal{O}(X)$ has maximal rank, G/N is compact, i.e. N is parabolic. Let P be a minimal parabolic group containing H and consider the fibration $G/H \to G/P$. Apply the induction assumption to the fiber, it follows that P/H is an affine cone with vertex removed, e.g. there is a fibration $P/H \xrightarrow{\mathbb{C}^*} P/\tilde{P}$ onto a homogeneous rational manifold. But P was chosen to be minimal. Hence we have the $\mathbb{C}^*$-bundle $G/H \to G/P$. Since $\mathcal{O}(G/H)$ has maximal rank, it follows that the associated line bundle is positive. Therefore, as desired, X is an affine cone (with vertex removed) over the homogeneous rational manifold G/P.

Note that the above argument only uses the existence of a parabolic group containing H. Hence, even if N/H is discrete, it is enough to fiber the base G/N which is an orbit of a linear group. Now, since $e(X) \geq 2$, G/H is not Stein. Thus, if N/H is discrete, the base G/N is also not Stein. It is also not compact and therefore, if N/H is discrete, we have a fibration $G/N \to G/P$ with a positive-dimensional fiber (Theorem 3.4). Thus P is parabolic and we apply the same argument as above. □

The only remaining case is that in which $N = G$. Therefore the result is a consequence of the following

(5.2) Lemma. *Let $X = G/\Gamma$, where G is a connected complex Lie group and Γ is a discrete subgroup. Suppose $e(X) \geq 2$ and that $\mathcal{O}(X)$ has maximal rank. Then $X \cong \mathbb{C}^*$.*

Proof (By induction on dimension). If G is semi-simple, then the assumption that $\mathcal{O}(X)$ has maximal rank implies that Γ is algebraic (see e.g. Theorem III.1.1), i.e. Γ is finite and therefore X is Stein. But $e(X) \geq 2$. Hence X is 1-dimensional. However, since G is semi-simple, this is in fact a contradiction.

Since G is not semi-simple and X is not compact, if $\dim_{\mathbb{C}} X > 1$, we have a fibration $G/\Gamma \to G/J$ with connected fiber J/Γ (Theorem 4.2). We choose $J \neq G$ to be maximal in the sense of dimension. Now $\mathcal{O}(X)$ has maximal rank. Therefore the fiber is non-compact and has more than one end. Applying induction, $J/\Gamma \cong \mathbb{C}^*$. The base can *not* be fibered. Thus we are in one of the following two situations (Theorem 4.2):

(i) *G/J is a simple torus;*

(ii) *the radical R of G is contained in J and $G/J = S/\Lambda$, where Λ is a cocompact discrete subgroup of the semi-simple group S.*

In *(i)*, since it is homogeneous, the $\mathbb{C}^*$-bundle $G/H \to G/J$ over the torus G/J is *topologically trivial* ([Ma 3]). But this is contrary to $\mathcal{O}(X)$ having maximal rank. In *(ii)* we obtain a similar contradiction, because if $f \in \mathcal{O}(X)$, then it has a Laurent series development

$$f = \sum_{n \in \mathbb{C}} s_n z^n$$

in the $\mathbb{C}^*$-bundle $G/\Gamma \to G/J = S/\Lambda$. Here z is a fiber coordinate and $s_n \in \Gamma(X, L^{-n})$, where L is the associated line bundle. Recall that the homogeneous space S/Λ has *no* analytic hypersurfaces, e.g. $\{s_n{=}0\} = \emptyset$, $n \in \mathbb{Z}$. This is contrary to $\mathcal{O}(X)$ having maximal rank. □

b) Proof of the Cone Theorem

Using the fibration methods along with the above result in the case $e(X) \geq 2$, we now prove the Cone Theorem. It is convienient to organize this in a sequence of lemmas.

(5.3) Lemma. *Under the assumption of the Cone Theorem, let $\Omega = G/H$ be the open orbit. Suppose there is a fibration $\phi : G/H \to G/J$, $J \neq G$, with connected fiber F. Then $Y := F \cup \{x_0\}$ is a connected analytic subset of $\Omega \cup \{x_0\}$ and G/J is compact.*

Proof. First, note that we have three immediate consequences of the *Theorem of Remmert and Stein*:

(i) The topological closure $\overline{F}$ of F in $\Omega \cup \{x_0\}$ is an analytic subset of $\Omega \cup \{x_0\}$;
(ii) The topological closure Z of the graph of $G/H \to G/J$ in $(\Omega \cup \{x_0\}) \times G/J$ is an analytic subset of the product which contains the graph as a Zariski open subset;
(iii) Let B be a connected Stein neighborhood of x_0 in $\Omega \cup \{x_0\}$. Then every fiber of ϕ which intersects B must also intersect ∂B.

As a consequence of *(iii)* x_0 is in the domain of definition of the meromorphic map ϕ defined by the extended graph Z in (ii). Furthermore, $\phi(\{x_0\}) \subset \phi(\partial B)$. Hence $\phi(\{x_0\})$ is a compact G-stable subset of G/J, i.e. $G/J = \phi(\{x_0\})$ is compact and $\{x_0\}$ is contained in the closure of every fiber of the holomorphic fibration $\phi : G/H \to G/J$. □

(5.4) Lemma. *Under the assumption of the Cone Theorem, let $\Omega = G/H$ be the open orbit and assume further that G is solvable. Then X is 1-dimensional.*

Proof. Since Ω is non-compact and G is solvable, unless $\dim_{\mathbb{C}} X = 1$, we always have a fibration $G/H \to G/J$. Assuming $\dim_{\mathbb{C}} X > 1$, we choose $J \neq G$ to be maximal in the sense of dimension. By Lemma 5.3 we may apply induction to the closure of the fiber J/H in $\Omega \cup \{x_0\}$. Thus $J/H = \mathbb{C}$ or $\mathbb{C}^*$. By the maximality of J, the base is either 1-dimensional or G/J is a simple torus. In the first case Ω would be a Stein manifold, which is of course not true. Hence we may assume that $G/H \to G/J$ is a $\mathbb{C}$- or $\mathbb{C}^*$-bundle over a simple torus. Its extension Y as a $\mathbb{P}_1$-bundle has a section Σ which can be blown down to $\{x_0\}$ as in the proof of Lemma 5.3.

Now Y is an almost homogeneous $\mathbb{P}_1$-bundle over a torus. In particular such a bundle must be topologically trivial (see [HO 1]). Hence no section can be blown down. □

(5.5) Lemma. *Under the assumptions of the Cone Theorem, let $\Omega = G/\Gamma$ and assume that Γ is discrete. Then* $\dim_{\mathbb{C}} X = 1$.

Proof. Suppose that G contains a *semi-simple* subgroup S. Then, in the sense of I.4, we can linearize S at the fixed point x_0. Let B be a Borel subgroup in S. Now B stabilizes a full flag in the Zariski tangent space $T_{x_0}X$. Since the linearization is *locally faithfull*, there is a germ of a complex curve through x_0 which is B-stable. This is a contradiction, because B is at least 2-dimensional and is acting with discrete isotropy on Ω. Thus G is solvable and the result follows from Lemma 5.4. □

We now complete the *proof of the Cone Theorem*. Consider the normalizer fibration $G/H \to G/N$ of the open orbit. By Lemma 5.5, we may assume that $N \neq G$. The base is an orbit of a linear group. So we choose a fibration $G/N \to G/I$, $I \neq G$, with $\dim_{\mathbb{C}} I/N$ maximal. We claim that G/I is not Stein. If $\Omega \cup \{x_0\}$ is compact, i.e. $X = \Omega \cup \{x_0\}$ is compact, then this is obvious, because in this case $\mathcal{O}(\Omega) \cong \mathbb{C}$. If $\Omega \cup \{x_0\}$ is non-compact, then $e(\Omega) \geq 2$. If G/I were Stein, then $\mathcal{O}(\Omega) \not\cong \mathbb{C}$ and Theorem 5.1 is applicable. Using this, it follows that $G/I = \mathbb{C}^*$ and the fiber I/H is compact. But this is contrary to Lemma 5.3. So G/I is in fact not Stein and therefore I is parabolic (Theorem 3.4).

Now let P be a minimal parabolic subgroup containing H and consider the fibration $\Omega = G/H \to G/P$ with fiber $F = P/H$. Since $\overline{F} = F \cup \{x_0\} \subset \Omega \cup \{x_0\}$ satisfies the assumptions of the Cone Theorem, induction plus the minimality of P implies that F is 1-dimensional.

Consider the associated $\mathbb{P}_1$-bundle $\Omega \subset B \xrightarrow{\mathbb{P}_1} Q = G/P$. It contains a section Σ which can be blown down to $\{x_0\}$. Let L be the normal bundle of Σ. It follows that $L^* > 0$. But $\Sigma \cong Q$ is homogeneous rational. So in particular $H^1(\Sigma, L^*) = (0)$, Hence $B \xrightarrow{\mathbb{P}_1} Q$ has the structure of a compactified positive line bundle. In this special case, i.e. a positive line bundle over a homogeneous rational manifold, this bundle is very ample. So, by blowing down the appropriate section of $B \to Q$, we obtain the desired cone. □

c) Meromorphic Functions on Almost Homogeneous Manifolds

Let X be a connected compact almost homogeneous manifold and $\mathcal{M}(X)$ its field of meromorphic functions. Of course if X is homogeneous or $\operatorname{codim}_{\mathbb{C}} E = 1$, then it is possible that $\mathcal{M}(X) \cong \mathbb{C}$ (see 1.b above). For $\operatorname{codim}_{\mathbb{C}} E > 1$ this is not the case.

(5.6) Theorem ([HO 1]). *Let X be a compact almost homogeneous manifold with open orbit $\Omega = G/H$. Assume that $E := X \backslash \Omega \neq \emptyset$. Then there exists a parabolic subgroup P of G which contains H so that*

$$\dim_{\mathbb{C}} P/H \leq \dim E + 1 \ .$$

The map $\Omega = G/H \to G/P =: Q$ extends to a meromorphic map $\phi : X \rightharpoonup Q$. In particular,

$$\operatorname{tr} \deg_{\mathbb{C}} \mathcal{M}(X) \geq \operatorname{codim}_{\mathbb{C}} E - 1 \ .$$

Proof. It is enough to consider the case where $\operatorname{codim}_{\mathbb{C}} E \geq 2$. Consider the anti-canonical map $\phi_g : X \to Z \hookrightarrow \mathbb{P}_m$, i.e. the meromorphic extension of the normalizer fibration $\Omega = G/H \to G/N$. If $N = G$, then $E = \{\sigma{=}0\}$, where

$\sigma = \mathfrak{X}_1 \wedge \ldots \wedge \mathfrak{X}_n$ is an anti-canonical section coming from the G-action. This is of course contrary to $\operatorname{codim}_{\mathbb{C}} E \geq 2$. Hence $N \neq G$ and we may apply fibrations to the base G/N which is an orbit of a linear group.

Observe that $\mathcal{O}(\Omega) \cong \mathbb{C}$. Hence, if we choose $G/N \to G/J$, $J \neq G$, to be maximal in the sense of dimension, then J is parabolic (Theorem 3.4). Let P be a minimal parabolic subgroup which contains H. We have the following situation:

$$X \supset \Omega = G/H \to G/P =: Q\,, \qquad \operatorname{codim}_{\mathbb{C}} X \setminus \Omega \geq 2\,.$$

The map $\phi : \Omega \to Q$ is defined by sections in some very ample line bundle on Q. As divisors, these may be continued across the analytic set E which is at least 2-codimensional. Thus ϕ extends to a meromorphic map $\phi : X \rightharpoonup Q$.

It remains to show that $\dim_{\mathbb{C}} P/H \leq \dim_{\mathbb{C}} E + 1$. Suppose to the contrary that the fiber P/H is bigger. Let Y be its topological closure in X which is of course a complex space which is almost homogeneous with respect to P and $\operatorname{codim}_{\mathbb{C}} E_Y \geq 2$. However, since Y may be singular, we cannot apply induction.

Note that if $H^\circ \ntriangleleft P$, then, since $\mathcal{O}(P/H) \cong \mathbb{C}$, the base of the normalizer fibration of the fiber, i.e. $P/H \to P/\tilde{N}$, $\tilde{N} = N_P(H^\circ)$, may be fibered onto a compact rational manifold (Theorem 3.4). This is contrary to the minimality of P. So in fact $H^\circ \triangleleft P$.

Let $k := \dim_{\mathbb{C}} P/H$. Choose vector fields $\mathfrak{X}_1, \ldots, \mathfrak{X}_k \in \mathfrak{p}$ so that $\mathfrak{X}_1 \wedge \ldots \wedge \mathfrak{X}_k$ vanishes nowhere on P/H. Then choose complementary fields $\mathfrak{X}_{k+1}, \ldots, \mathfrak{X}_n \in \mathfrak{g}$, $n := \dim_{\mathbb{C}} X$, so that $\sigma := \mathfrak{X}_1 \wedge \ldots \wedge \mathfrak{X}_n$ does not vanish in a neighborhood of the fiber P/H. Thus $\{\sigma{=}0\} \cap Y = E_Y$, i.e. $\operatorname{codim}_{\mathbb{C}} E_Y = 1$, contrary to assumption. □

Remarks. (1) We do not know how to prove the above theorem in the singular case. The essential question is the following: *Let X be an almost homogeneous complex space with open orbit $\Omega = G/\Gamma$, where Γ is discrete. Is it then true that* $\operatorname{codim}_{\mathbb{C}} E = 1$? This is of course a special case of the Lipmann-Zariski conjecture.

(2) Using fibration methods such as those above, D. Snow and the author ([HS 1, HS 2]) proved a fine classification of strongly pseudoconcave homogeneous manifolds. It would be interesting to prove the above theorem under the concavity conditions which are analogous to $\operatorname{codim}_{\mathbb{C}} E = k > 1$. □

d) The Curve Theorem

As a special case of Theorem 5.6, if $\dim_{\mathbb{C}} E = 1$, then there is a G-equivariant meromorphic map $X \rightharpoonup Q = G/P$ so that the fibration $\Omega = G/H \to G/P$ of the open orbit satisfies $\dim_{\mathbb{C}} P/H \leq 2$. By carefully studying this map it is in fact possible to prove a complete classification.

Curve Theorem ([HO 1]). *Let X be a compact manifold which is almost homogeneous with respect to a complex Lie group G. Assume that $n := \dim_{\mathbb{C}} X \geq 3$ and* *that E is purely 1-dimensional. Then the following is the list of all possibilities which occur.*

1) $X = \mathbb{P}_n$, *E is a complex line in X;*
2) $X = C \times \mathbb{P}_{n-1}$, *where C is an elliptic or rational curve, $E = C \times \{p\}$, and $p \in \mathbb{P}_{n-1}$ is arbitrary;*
3) $X = \mathbb{P}_3$, *E is the union of two disjoint complex lines in X;*
4) $X = Q_3$ *(the 3-dimensional quadric), E is a complex line in X;*
5) *X is a desingularized Segre cone and E is the rational curve over the vertex;*
6) *X is a homogeneous Hopf manifold over $\mathbb{P}_{n-1}$ and E is any fiber.*

Remarks. (1) The technicalities which arise in the proof are substantially more complicated than in the Cone Theorem.

(2) Lescure has proved the Curve Theorem for singular spaces ([Les 2]). In particular, he discovered an interesting infinite series $\Theta_{k,l}$ of rational almost homogeneous spaces with $E \cong \mathbb{P}_1$.

(3) One of the major steps in Lescure's proof is the following result on $SL_2(\mathbb{C})$.

Lescure's Lemma. *Let X be an irreducible 3-dimensional complex space. Suppose that $SL_2(\mathbb{C})$ acts holomorphically on X with an open orbit Ω such that $X \setminus \Omega =: E \neq \emptyset$. Then E is purely 2-codimensional.*

e) Low-Dimensional Homogeneous Spaces

The fibration methods are quite useful in the classification of low-dimensional homogeneous spaces. This is particularly true if one additionally assumes that the group in question is complex and is acting holomorphically (see e.g. [HL]). However, when one only has a transitive *real* Lie group of holomorphic transformations, even in the two-dimensional case there are interesting complex-geometric problems.

Classification of Homogeneous Surfaces ([OR 2]). *The following is a list of all 2-dimensional connected complex manifolds which are homogeneous under the action of a real Lie group of holomorphic transformations.*

I. *Products of homogeneous Riemann surfaces.*

II. *Surfaces which are homogeneous under the holomorphic action of a complex Lie group*

   1) *G/Γ, where G is either the additive group $(\mathbb{C}^2, +)$ or the unique simply-connected non-abelian two-dimensional complex Lie group and Γ is discrete* (see [ES] for discrete groups in dimension 2 and 3)
   2) *Any principal $\mathbb{C}^*$-bundle over $\mathbb{P}_1$;*
   3) *The k-th power H^k of the hyperplane bundle over $\mathbb{P}_1$, $k \geq 0$;*
   4) *The affine quadric Q_2;*
   5) $\mathbb{P}_2 \setminus Q$, *where Q is a quadric curve;*

6) *Rational homogeneous surfaces:* $\mathbb{P}_2$, $\mathbb{P}_1 \times \mathbb{P}_1$;
7) *Any principal bundle over* $\mathbb{P}_1$ *with fiber an elliptic curve.*

III. *Surfaces which are homogeneous under a real Lie group of holomorphic transformations, but not under a complex group.*
1) *Bounded domains:* $\Delta_2 := \{(z,w) \mid |z| < 1,\ |w| < 1\}$, $\mathbb{B}_2 := \{(z,w) \mid |z|^2 + |w|^2 < 1\}$;
2) $\mathbb{P}_2 \setminus \overline{\mathbb{B}}_2$;
3) $\mathbb{P}_2 \setminus \overline{\mathbb{B}}_2 \cup L$, *where* L *is a projective line in* $\mathbb{P}_2$ *tangent to* $\mathbb{B}_2$;
4) $\mathbb{P}_2 \setminus \mathbb{P}_2(\mathbb{R})$, *where* $\mathbb{P}_2(\mathbb{R})$ *denotes the set of real points in* $\mathbb{P}_2$;
5) $\mathbb{P}_1 \times \mathbb{P}_2 \setminus \mathbb{P}_1^{\mathbb{R}}$, *where* $\mathbb{P}_1^{\mathbb{R}} := \{(p,\overline{p}) \in \mathbb{P}_1 \times \mathbb{P}_1\}$;
6) $\mathbb{C}^2 \setminus \mathbb{R}^2$ *or any covering space of* $\mathbb{C}^2 \setminus \mathbb{R}^2$.

The strategy for proving such a result is the following: Let $X = G/H$ and consider the g-anticanonical map $G/H \to G/N$. If $N = G$, then $X = G/\Gamma$, where G is complex and Γ is discrete. This case must be handled by special methods (see [ES]). If $\dim_{\mathbb{C}} G/N = 1$, then we have a holomorphic fiber bundle over a 1-dimensional base. This is easy to classify. Thus it remains to consider the case where $\dim_{\mathbb{C}} G/N = 2$. In fact it is enough to consider the case where $X = G/H$ is a linear orbit, where we have the complex linear group $\hat{G}$ and $X = G/H \underset{\text{open}}{\hookrightarrow} \hat{G}/\hat{H}$.

If $\mathfrak{g} \cap i\mathfrak{g} =: \mathfrak{m} \neq (0)$, then one can construct a *holomorphic* fibration via $M := \exp(\mathfrak{m})$. Hence one immediately reduces to the case where G is a real form of $\hat{G}$. Since $\dim_{\mathbb{C}} X = 2$, there are very few possibilities for the action of the semi-simple part $\hat{S}$. Thus, in fact one reduces rather quickly to the case where G is solvable.

Up to this point the *strategy* requires only group-theoretical techniques. However, in order to complete the job one must invest some complex geometry. The following is typical of the type of argument which can be used.

At a certain point there is a fibration of the complex orbit,

$$\mathbb{C}^2 = \hat{G}/\hat{H} \xrightarrow{\mathbb{C}} \hat{G}/\hat{J} = \mathbb{C}\,,$$

where $\hat{J} = \hat{H}Z_{\hat{G}'}$, which induces a fibration $G/H \to G/J = \mathbb{C}$ with fiber J/H being the upper half-plane H^+. Of course one cannot expect this to be a holomorphic bundle. However it does extend to a fibration of the boundary of X in $\mathbb{C}^2$, $\partial X = G/L \xrightarrow{\mathbb{R}} G/J = \mathbb{C}$, and explicit calculations via this fibration show that the Levi-form of ∂X is non-degenerate and the commutator group G' acts freely and transitively on ∂X. In other words, the structure on ∂X equips the Heisenberg group G' with its unique invariant CR-structure, i.e. the punctured structure of the sphere $S^3 \setminus \{p\} \subset \mathbb{C}^2$.

Now regard S^3 in the usual way as a hypersurfaces in $\mathbb{P}_2$. By Tanaka's Theorem, the G-action on $S^3 \setminus \{p\}$ extends uniquely to a group of linear transformations on $\mathbb{P}_2$. Thus our original manifold is either the inside or the outside: $X = \mathbb{B}_2$ or $\mathbb{P}_2 \setminus \overline{\mathbb{B}}_2 \cup L$. □

The classification of 3-dimensional homogeneous complex manifolds has been carried out by J. Winkelmann ([Wi 1, Wi 2]). Here there are 34 classes, each described with the same amount of detail as in the case of surfaces. There are many new phenomena and many interesting examples, even "new" domains in $\mathbb{C}^3$. The basic strategy of Winkelmann's proof is the same as in the 2-dimensional case, i.e. fibration methods. However, it was necessary to introduce many new techniques, both group-theoretical and complex-geometric.

In may ways Winkelmann's work in the 3-dimensional case is quite general. For example, the following is a useful observation.

Fibrations of Bounded Domains ([Wi 3]). *Let X be a connected complex manifold which is homogeneous under the action of a Lie group G of holomorphic transformations. Let $\phi : X \to Y$ be a holomorphic map which is a G-equivariant fibration, i.e. $X = G/H \to G/J = Y$. Suppose that the base and fiber are bounded domains. Then X is a bounded domain.*

With the exception of one missing link, the fibration methods, particularly with Winkelmann's refinements, should be strong enough to handle virtually any classification problem. As suggested by B. Gilligan, we have attempted to formulate this missing link in terms of the Kobayashi pseudometric (see [Wi 3, Gi 1] for partial results). This may indeed be the main problem, but nevertheless we have the following

Question. *Let X be a complex homogeneous manifold. Consider the fibration $X = G/H \to G/J = Y$ which fibers out the degeneracy directions of the Kobayashi-pseudometric. What can be said about this fibration? For example, is there a complex structure on the base so that the map $X \to Y$ is holomorphic? Is the base hyperbolic? ...*

f) Homogeneous CR-hypersurfaces

In closing this section we would like to mention that the fibration methods are also useful for studying real submanifolds. For example, let $M = G/H$ be a homogeneous CR-hypersurface. Then M can be embedded as a real hypersurface in a complex manifold $M^{\mathbb{C}}$ so that the CR-structure is the induced structure and $\mathfrak{g} \subset \Gamma_{\mathcal{O}}(M^{\mathbb{C}}, TM^{\mathbb{C}})$. Therefore, after perhaps shrinking $M^{\mathbb{C}}$, the anticanonical map $\phi_{\mathfrak{g}} : M^{\mathbb{C}} \to \mathbb{P}(V_{\mathfrak{g}}^*)$ is holomorphic and, restricted to M, is the CR-homogeneous fibration $M = G/H \to G/N_{\mathrm{CR}}$. Here N_{CR} consists of the elements of the normalizer $N_G(H^\circ)$ which act as CR-automorphisms of G/H° on the right (see [R 2, AHR, HR] for details).

Analogous to the case of complex homogeneous manifolds, we consider the complex group $\hat{G}$ acting on $\mathbb{P}(V_{\mathfrak{g}}^*)$ so that the orbit G/N_{CR} is a real hypersurface in the complex $\hat{G}$-orbit or $G/N_{\mathrm{CR}} = \hat{G}/\hat{N}$. Using the fibration methods for linear orbits, such hypersurfaces can be *completely* classified ([AHR]). Furthermore, the fiber of $G/H \to G/N$ is Levi-flat. So, if for example the Levi-form of M is non-degenerate, then either $N/H = S^1$ or N/H is finite. This leads to a complete classification of such CR-hypersurfaces by using only elementary group-theoretic methods (see [MN], [R], [BS] for related earlier results). In fact it is possible to completely classify the CR-homogeneous hypersurfaces M under the assumption that there is a *Kählerian complexification* $M^{\mathbb{C}}$ ([R 2]).

Of course Richthofer makes use of the fibration methods. But there are numerous complications, e.g. with discrete groups. □

6. Spherical Spaces

Applications of the fibration methods sometimes only serve as a reduction to the case of algebraic groups. Quite often we obtain an algebraic almost homogeneous space of a reductive group. Although there are some general principles, a fine description of all such varieties is of course not possible. However certain classes of these varieties are of particular interest. One such class is that of "spherical embeddings" (see e.g. [A 2, B, BLV, BP, KV]).

Definition. *Let X be an irreducible algebraic space which is almost homogeneous with respect to an algebraic action of a reductive group G. Let B be a Borel subgroup of G. The open orbit $\Omega = G/H$ is said to be a spherical homogeneous space whenever B has an open orbit on Ω. In this case X is called a spherical embedding of Ω.* □

The following conditions are equivalent:

(i) G/H is a spherical homogeneous space;

(ii) Every G-almost homogeneous space with open orbit $\Omega = G/H$ has only finitely many G-orbits;

(iii) Every G-almost homogeneous space with open orbit $\Omega = G/H$ has only finitely many B-orbits;

(iv) For every G-line bundle L on G/H the G-representation on $\Gamma(G/H, L)$ is multiplicity-free.

In brief, spherical homogeneous spaces and embeddings have so many beautiful properties that it is possible to have detailed understanding of them. On the other hand, this seems to be a very large class of varieties which occurs in numerous applications (see e.g. [HW] for applications in complex analysis).

We would now like to indicate two directions where spherical varieties play an underlying role.

a) Two-Orbit Varieties

Let X be an irreducible compact algebraic space and G a reductive group acting algebraically on X. The space X is called a *two-orbit variety* if G has exactly two orbits, i.e. one orbit is open and *the* other one, E, is closed. Of course such varieties occur as closures of orbits in non-transitive actions. Surprisingly, however, there are not many such varieties. The complete classification for the cases $\operatorname{codim}_{\mathbb{C}} E = 1$ ([A 3, HS 3]) and $\operatorname{codim}_{\mathbb{C}} E = 2$ ([Feld]) is known. In particular, all examples are *spherical embeddings*. It would be very interesting to have a direct general proof of such a result. We optimistically formulate this as a

Conjecture. *Every two-orbit variety is spherical.*

b) Symmetric Spaces in the Sense of A. Borel

Recall that classically the symmetries of a "symmetric space" are required to be isometries of an underlying Riemannian structure. In ([Bo 2]) Borel observed that in the complex analytic setting it is possible to drop the isometry condition and still have an interesting class of spaces.

Definition. *Let X be an irreducible complex space and G a (not necessarily connected) Lie group of holomorphic transformations of X. The manifold X is said to be complex symmetric with respect to G if for every $x \in X$ there exists $g \in G$ with $x \in \operatorname{Fix}(g)$ so that $(dg)_x = -I$.* □

Reminiscent of spherical varieties, if X is *complex symmetric then G has at most countable many orbits in X* ([Bo 2]). In particular, X is almost homogeneous with respect to G, e.g. if X is compact, then G has only finitely many orbits. There are numerous (non-artificial) examples which are indeed not homogeneous (see e.g. [Le 1]). However, if G° is reductive, this is not the case.

(6.1) Theorem. *Let X be a connected compact complex manifold which is complex symmetric with respect to the Lie group G. Suppose that X is minimal in the sense that the only proper modifications $X \to Y$ onto manifolds are biholomorphic maps. Then, if G° is reductive, it follows that X is hermitian symmetric.*

Remarks. (1) D. Ahiezer ([A 4]) first proved this in the case where G° is a simple group. The above result, due to R. Lehmann ([Le 2]), is not at all an immediate consequence of the result in the simple case. Methods from the theory of spherical embeddings were extremely useful in simplifying his proof. It should be remarked that Lehmann has also classified (singular) symmetric torus embeddings ([Le 3]).

(2) It would be extremely interesting to understand complex symmetric spaces without the "reductive" assumption. Even if G° is acting algebraically, this seems to be quite difficult. However, even a partial understanding of these spaces might lead to a notion of "spherical" in the non-reductive case. □

Chapter III

Complex Analysis on Homogeneous Manifolds

Here we consider homogeneous manifolds of the type $X = G/H$, where G is a connected complex Lie group and H is a closed complex subgroup. We are primarily interested in non-compact manifolds and questions of existence or non-existence of basic objects, e.g. holomorphic/meromorphic functions, analytic hypersurfaces.

1. Holomorphic Functions

One always has the representation of G on the Fréchet space $\mathcal{O}(X)$ of holomorphic functions. A function $f \in \mathcal{O}(X)$ is called *G-finite* if the vector space $\langle G(f) \rangle$ generated by the orbit $G(f)$ in $\mathcal{O}(X)$ is finite-dimensional. In some philosophical sense, the G-finite functions should be regarded as the "algebraic functions" with respect to the G-action.

If $G = K^{\mathbb{C}}$ is reductive and G acts holomorphically on a complex space X, *then the G-finite functions are dense in* $\mathcal{O}(X)$ ([Har]). Thus one expects the function theory of homogeneous spaces of reductive groups to be of an algebraic nature.

a) Holomorphic Functions on Homogeneous Spaces of Reductive Groups

The following result confirms the above remarks.

(1.1) Theorem. *Let $X = G/H$ be the homogeneous space of a reductive group. Suppose $\mathcal{O}(X)$ has maximal rank. Then H is an algebraic subgroup of G and X is G-equivariantly quasi-affine.*

Proof. Let $f_1, \dots, f_n \in \mathcal{O}(X)$, $n := \dim_{\mathbb{C}} X$, be such that $df_1 \wedge \dots \wedge df_n(x) \neq 0$ for some $x \in X$. We may approximate these functions with G-finite functions $g_1, \dots, g_n \in \mathcal{O}(X)$ so that $dg_1 \wedge \dots \wedge dg_n(x) \neq 0$. Let V be the finite-dimensional G-stable subspace of $\mathcal{O}(X)$ which is generated by $g_1, \dots, g_n$. Since V locally separates points, the associated equivariant map $\phi : X \to V^*$ has discrete fibers. As a homogeneous fibration, ϕ is of the form $G/H \to G/J$. Since G is reductive and the base is a linear orbit in V^*, it follows that J is algebraic and G/J is quasi-affine. Since J/J° is finite, the map ϕ is a finite map and consequently X is holomorphically separable ([GR 2]). In particular, we may enlarge V by including a G-finite function g which separates a fixed ϕ-fiber. The resulting map $X \to V^*$ realizes X as a quasi-affine G-orbit. □

Remark. It is possible to give a group-theoretic characterization of the subgroups H of a reductive group G so that $X = G/H$ is quasi-affine ([HM 1]).

b) Stein Homogeneous Spaces (Connected Isotropy)

Let $X = G/H$, where G is a connected complex Lie group. Recall that if G is reductive, then X is a Stein manifold if and only if H is reductive (see Chapter I). In general, i.e. G arbitrary, there is no known satisfactory characterization of Stein homogeneous spaces. There are a number of partial results, which we will discuss in the next section. These generally involve discrete groups. Our goal in the present section is to outline the *problem* in the case of connected isotropy. Note that if H/H° is finite, then G/H is Stein if and only if G/H° is Stein.

For simplicity let $X = G/H$, where G is a algebraic group and H is a connected algebraic subgroup (see [Sn 2] for the complex analytic case). Let $H = U \rtimes L$ be a Levi-decompostion of H and consider the L-principal bundle $\pi : Z = G/U \to G/H = X$. Since $X = Z/\!/L$, if Z is Stein then X is Stein. Conversely, since $\pi : Z \to X$ is a principal bundle and L is Stein, if X is Stein, then so is Z. Thus it is enough to consider the special case of unipotent isotropy.

Question. *Let $X = G/U$, where G is an algebraic group and U is a unipotent subgroup. Under what conditions is X Stein (affine)?*

We would of course hope that the "conditions" would be of a group-theoretic nature.

Suppose $X = G/U$ is Stein and consider the action of a maximal reductive subgroup M of G. Every M-orbit is Zariski open in its closure. In particular, the minimal orbits are closed and therefore Stein. Let $M(x) = M/H$ be such an orbit. On the one hand, since U is unipotent, so is H. On the other hand, since M is reductive and M/H is Stein, H is reductive. Hence $H = \{e\}$ is the trivial group. Thus there is a necessary condition:

If X is Stein, then every maximal reductive subgroup M of G acts freely and properly on X.

Restricting to the class of manifolds satisfying this condition, it follows that $X = G/U$ is Stein if and only if the quotient $X/\!/M = M \setminus G/U$ is Stein.

Note that $M \setminus G$ is algebraically equivalent to affine space $\mathbb{A}^n$, where U is acting freely, properly, and algebraically. In this setting, if $\dim_{\mathbb{C}} U = 1$, i.e. if we begin with a 1-parameter unipotent group U in G and consider its action on $M \setminus G = \mathbb{A}^n$, then $\mathbb{A}^n/U = \mathbb{A}^{n-1}$ ([Sn 2]). This leads to the following special, but very closely related,

Question. *Let U be a unipotent group of affine transformations which is acting freely and properly on $\mathbb{A}^n$. Under what conditions is the quotient $\mathbb{A}^n/U$ affine?*

Winkelmann ([Wi 4]) has observed that there are affine actions where $\mathbb{A}^n/U$ is not affine. For example, there is an action of $U = (\mathbb{A}^2, +)$ by affine transformations on $\mathbb{A}^6$ so that the quotient is $Q_4 \setminus E$, where Q_4 is the 4-dimensional affine quadric and E is a 2-dimensional subspace. □

c) Stein Homogeneous Spaces (Discrete Isotropy)

Here we consider homogeneous spaces of solvable complex Lie groups. The essential problems occur when $X = G/\Gamma$ and Γ is discrete.

1. Nilpotent Groups. Let G be a simply-connected complex nilpotent Lie group. If Γ is a discrete subgroup of G, let $G(\Gamma)$ denote the uniquely determined connected real subgroup of G with $G(\Gamma) > \Gamma$ and $G(\Gamma)/\Gamma$ compact. Let $\mathfrak{g}(\Gamma)$ be its Lie algebra, $\mathfrak{m} := \mathfrak{g}(\Gamma) \cap i\mathfrak{g}(\Gamma)$ and $M = \exp(\mathfrak{m})$. Let $X = G/\Gamma$ and note that the M-orbit of the neutral point is contained in the compact manifold $G(\Gamma)/\Gamma$. Hence, if X is holomorphically separable or $\mathcal{O}(X)$ has maximal rank, then $M = (e)$.

(1.2) Theorem ([GH 2]). *Let G be a simply-connected complex nilpotent Lie group, Γ a discrete subgroup, and $X = G/\Gamma$. Then the following conditions are equivalent:*

(i) X is a Stein manifold;

(ii) $\mathcal{O}(X)$ has maximal rank;

(iii) $M = (e)$.

Proof. It is enough to prove *(iii)⇒(i)*. Note as usual we may assume that $G_{\mathbb{C}}(\Gamma) = G$. Let Z be a real central subgroup of $G(\Gamma)$ so that $\Gamma \cdot Z$ is closed (see e.g. [Mo 2] for existence). Let $Z^{\mathbb{C}}$ be the connected group corresponding to $\mathfrak{z}^{\mathbb{C}} = \mathfrak{z} \oplus i\mathfrak{z}$. It follows that $\Gamma Z^{\mathbb{C}}$ is closed, and we consider the fibration

$$G/\Gamma \to G/\Gamma Z^{\mathbb{C}} = \hat{G}/\hat{\Gamma} \ ,$$

where $\hat{G} = G/Z^{\mathbb{C}}$ and $\hat{\Gamma} = \Gamma/\Gamma \cap Z$.

Since $G(\Gamma)$ is totally real in G and $Z^{\mathbb{C}} \cap G(\Gamma) = Z$, i.e. the *real* points of $Z^{\mathbb{C}}$, it follows that $\hat{G}(\hat{\Gamma})$ is totally real in $\hat{G}$. By induction it follows that the base $\hat{G}/\hat{\Gamma}$ is Stein. Of course the fiber is also Stein, because it is simply $(\mathbb{C}^*)^k$ for some k. Since $Z^{\mathbb{C}}$ is central, the bundle is in fact a principal bundle. Thus the total space is Stein. □

Remark. In the early 1950's Serre formulated the following problem: *Let $F \to E \to B$ be a holomorphic fiber bundle with fiber F and base B being Stein manifolds. Is the total space E a Stein manifold?* If the structure group of the bundle is a finite-dimensional complex Lie group with finitely many components, then E is indeed a Stein manifold ([MM]). There are examples where E is not Stein ([Sko, deM]), some of which are very important for analysis on homogeneous spaces ([Loe]). We discuss this in the next section.

Returning to nil-manifolds, we note that in this special case it is possible to prove the optimal result on holomorphic functions.

(1.3) Theorem ([GH 2]). *Suppose $X = G/H$ is the homogeneous space of a complex nilpotent Lie group G and let $X = G/H \to G/J = Y$ be its holomorphic reduction. Then Y is Stein and the fiber $F = J/H$ is connected with $\mathcal{O}(F) = \mathbb{C}$.*

Proof. We may assume that G is simply-connected. Recall that, since $N := N_G(H^\circ)$ is connected, it follows that the normalizer fibration

$$G/H \to G/N = \mathbb{C}^k$$

is trivial. So it is enough to consider the fiber, i.e. we may assume that $H = \Gamma$ is discrete. Finally, since $G/G_{\mathbb{C}}(\Gamma) = \mathbb{C}^l$, we may also assume that $G_{\mathbb{C}}(\Gamma) = G$.

Now consider the holomorphic reduction $G/\Gamma \to G/J$. Since $N_G(J^\circ)$ is connected and contains Γ, it follows that $J^\circ \triangleleft G$. So we write the base $G/J = \hat{G}/\hat{\Gamma}$. By Theorem 1.2, since it is holomorphically separable, $\hat{G}/\hat{\Gamma}$ is Stein. We always have an open subgroup $\tilde{J} < J$ with $\Gamma < \tilde{J}$ and $\tilde{J}/\Gamma$ connected. Since G/J is Stein, it follows that $G/\tilde{J}$ is also Stein. So $G/\Gamma \to G/\tilde{J}$ is holomorphic reduction, i.e. $\tilde{J} = J$ and the fiber F is connected.

Consider the holomorphic reduction $J/\Gamma \to J/I$ of the fiber. This yields a fibration $G/\Gamma \to G/I$. By the same argument as above $I^\circ \triangleleft G$, so we write $G/I = \tilde{G}/\tilde{\Gamma}$. Now consider the complex group $\tilde{M} \subset \tilde{G}(\tilde{\Gamma})$. Since the base is Stein, the $\tilde{M}$-orbit of the neutral point must be contained in the fiber of $G/I \to G/J$. But this fiber is likewise holomorphically separable. Hence $\tilde{M} = (e)$ and G/I is Stein. So $G/H \to G/I$ is the holomorphic reduction, i.e. $I = J$ and $\mathcal{O}(F) \cong \mathbb{C}$. □

d) Solvable Groups

Homogeneous spaces of solvable complex Lie groups are much more difficult to deal with than nil-manifolds. We begin with an example which is even less complicated than usual, because, as in the case of nil-manifolds, we have the "real form $G(\Gamma)$".

Example. Let $A = (\mathbb{C}^2, +)$ and $B = (\mathbb{C}, +)$. Define the semi-direct product $G = A \rtimes_\rho B$ by a 1-parameter group $\rho : B \to \mathrm{Aut}(A) = \mathrm{GL}_2(\mathbb{C})$. We choose $\rho(z) = e^{Mz}$, where M is a matrix so that $e^M \in \mathrm{SL}_2(\mathbb{Z})$. In this way we have $G_{\mathbb{R}} := \mathbb{R}^2 \rtimes_\rho \mathbb{R}$ and $\Gamma := G_{\mathbb{Z}} = \mathbb{Z}^2 \rtimes_\rho \mathbb{Z}$ so that $G_{\mathbb{R}}/\Gamma$ is compact. Let $X = G/\Gamma$.

Of course we don't choose M to be trivial, e.g. we choose it general enough so that $G' = B$. Note that we have a fibration

$$\pi : X = G/\Gamma \to G/\Gamma G' = \mathbb{C}^* \equiv Z ,$$

with fiber $G'/G' \cap \Gamma = \mathbb{C}^* \times \mathbb{C}^*$.

Such a setting is not unknown in representation theory, particularly in the context of "entire vectors" for unitary representation, e.g. the $G_{\mathbb{R}}$-representation on $\mathscr{L}^2(G_{\mathbb{R}}/\Gamma)$. For example, if the eigenvalues of M are not purely imaginary (including 0), then it has been known for quite some time that $\pi^*\mathcal{O}(Z) = \mathcal{O}(X)$ (see R. Goodman [G] for particular cases and R. Penny [P] for a general theory). This phenomenon, e.g. a negative example for the question of Serre, was first pointed out to complex analysts by J.J. Loeb ([Loe]). Furthermore, he observed important relationships with $G_{\mathbb{R}}$-invariant pluri-subharmonic functions on G. As a consequence, he showed that X is not even a Kähler manifold. □

We now give a precise statement of Loeb's Theorem. Suppose $G_{\mathbb{C}}$ is a simply-connected solvable Lie group, $G_{\mathbb{R}}$ is a real form of $G_{\mathbb{C}}$ and $\Gamma < G_{\mathbb{R}}$ is cocompact, i.e. $G_{\mathbb{R}}/\Gamma$ compact.

(1.4) Theorem ([Loe]). *Let $X = G_{\mathbb{C}}/\Gamma$. Then the following conditions are equivalent:*

(i) X is Stein;
(ii) X is Kähler;
(iii) Spec($\mathfrak{g}_{\mathbb{R}}$) *is purely imaginary.*

Remark. Condition *(iii)* means the following: For every $\mathfrak{X} \in \mathfrak{g}_{\mathbb{R}}$,

$$\operatorname{Spec}(\operatorname{ad}(\mathfrak{X})) \subset i\mathbb{R}\,. \qquad \square$$

At first glance Loeb's result seems negative and furthermore assumes the existence of the real form $G_{\mathbb{R}}$ with $G_{\mathbb{R}}/\Gamma$ compact. This is of course not realistic. However, using at least the philosophy of Loeb's result, E. Oeljeklaus and the author ([HO 3]) were able to prove a technical lemma which has a number of applications.

(1.5) Lemma. *Let G be a connected complex solvable Lie group, Γ a discrete subgroup, and $X = G/\Gamma$. Suppose that X is Kähler and that Γ contains a regular element of G. Then Γ contains a nilpotent subgroup $\tilde{\Gamma}$ of finite index.*

Remark. We should recall that an element $g \in G$ is called *regular* if the algebraic closure of the cyclic group generated by Ad(g) contains a maximal torus in the algebraic closure of Ad(G). □

As an application we have the following

(1.6) Theorem ([HO 3]). *Let X be the homogeneous space of a complex solvable Lie group G. Suppose that $\mathcal{O}(X)$ has maximal rank. Then X is Stein and the fundamental group $\pi_1(X)$ contains a nilpotent subgroup of finite index.*

Proof. As usual one reduces to the case of discrete groups by using the normalizer fibration. Here we only carry out the proof for $X = G/\Gamma$.

We may assume that G is simply-connected and consider its abstract *algebraic hull* $G_a = (\mathbb{C}^*)^k \ltimes G$ ([HM 2]). Let $G_a(\Gamma)$ be the algebraic closure of Γ in G_a.

As a complex manifold $X_a := G_a/\Gamma = (\mathbb{C}^*)^k \times G/\Gamma$. In particular $\mathcal{O}(X_a)$ has maximal rank. Using this fact along with its homogeneity, one can construct a holomorphic immersion $\phi : X_a \to \mathbb{C}^N$ to produce a Kähler-form on X_a.

Now apply the Lemma: $\Gamma \cap G_a(\Gamma)^\circ$ contains a regular element of $G_a(\Gamma)^\circ$, because it is Zariski-dense. Thus $G_a(\Gamma)^\circ$ is nilpotent. Since $\mathcal{O}(X_a)$ has maximal rank, it follows that $G_a(\Gamma)/\Gamma$ is Stein (Theorem 1.2). Now we have the fibration

$$X_a = G_a/\Gamma \to G_a/G_a(\Gamma)$$

with Stein fiber, Stein base, and *algebraic* structure group. Thus X_a is Stein ([MM]). Finally, as a closed submanifold of X_a, G/Γ is likewise Stein. The nilpotent group of finite index is of course the group $\Gamma \cap G_a(\Gamma)^\circ$. □

2. Analytic Hypersurfaces, Kählerian Structures

As we have indicated above, it is possible to derive information about the holomorphic functions on a homogeneous space by using induced Kähler structures. This is also the case for analytic hypersurfaces: *Let $X = G/H$ be a homogeneous space which is locally hypersurface separable. Then X is Kähler (I.3.3).*

a) Hypersurfaces in Homogeneous Manifolds of Semi-Simple Lie Groups

Several years ago G. Margulis and the author ([HMar]) proved the following negative result.

(2.1) Theorem. *Let S be a complex semi-simple Lie group, H a closed complex Zariski dense subgroup, and $X = S/H$. Then X posesses no analytic hypersurfaces, i.e. $\mathcal{H}(X) = \emptyset$.*

Remarks. (1) Formulated in terms of invariant hypersurfaces, this is equivalent to the following: *Let H be any Zariski dense subgroup of S. Then the space $\mathcal{H}(S)^H$ of H-invariant hypersurfaces is empty.*

(2) D. Ahiezer ([A 5]) proved the first result in this direction (at least for X non-compact). His result implies for example that $X := \mathrm{SL}_2(\mathbb{C})/\mathrm{SL}_2(\mathbb{Z})$ has no analytic hypersurfaces.

(3) Here we give an elementary proof using *Kählerian properties*.

Proof. Let $S/H \to S/J$ be the hypersurface reduction of X. Since H is Zariski-dense, so is J and therefore $J^\circ \lhd S$. Hence we may assume that $X = S/\Gamma$ is hypersurface separable and Γ is discrete Zariski dense. This will yield a contradiction. Hence $J = S$ and $\mathscr{H}(X) = \emptyset$.

Since Γ is Zariski-dense, there is a semi-simple element $\gamma \in \Gamma$ which generates a free abelian group $\Lambda < S$. Let A be the Zariski closure of Λ in S. We may assume that $A \cong (\mathbb{C}^*)^k$ is connected and, since $\overline{\Lambda} = A$, it follows that A/Λ is an abelian group having no non-constant holomorphic functions. Let T be the maximal (real) torus in A/Λ and $M \cong \mathbb{C}$ the maximal complex subgroup of T.

Now X is hypersurface separable and therefore Kähler. Consequently the covering space $\tilde{X} = S/\Lambda$ is also Kähler. Note that $b_2(\tilde{X}) = (0)$ and therefore the Kähler form ω on $\tilde{X}$ is exact: $\omega = d\mu$. Note also that $\tilde{X}$ comes equipped with a holomorphic T-action (from the right). Therefore, by averaging ω and μ, we may assume that they are T-invariant. In particular, the pull-back of μ to M is an invariant form on $M \cong \mathbb{C}$. Consequently, $\omega = d\mu$ vanishes on $\mathbb{C}$, contrary to ω being positive-definite. □

The following *final* result of F. Berteloot and K. Oeljeklaus ([BeO]) cannot be proved with such elementary tools. In particular one needs the *analytic* SL_2*-Lemma of Berteloot* ([Be]) and the *minimum principle of Kiselmann* ([Ki]).

(2.2) Theorem. *Let S be a complex semi-simple Lie group, H a closed complex subgroup, and $X = S/H$. Then the following are equivalent:*

(i) X is Kähler;

(ii) X is locally hypersurface separable;

(iii) H is algebraic, i.e. X is a quasi-projective orbit of S.

b) Hypersurfaces in Homogeneous Spaces of Solvable Lie Groups

As we have seen above, Kählerian structures and analytic objects, e.g. holomorphic functions, analytic hypersurfaces, ..., are very much intertwined. The weakest assumption is that the homogeneous space $X = G/H$ is a Kähler-manifold. Thus it is appropriate to first attempt to provide a structure theory in the Kähler case. Just as the above result is the final one for semi-simple groups, it would seem that the following one is best possible in the solvable case.

(2.3) Theorem ([OR 1]). *Let X be a homogeneous manifold of a complex solvable Lie group. Assume that X is a Kähler-manifold. Then, if $X = G/H \to G/J =: Y$ is the holomorphic reduction of X, it follows that*

(i) Y is Stein

and

(ii) the fiber $F = J/H$ is an abelian group with $\mathcal{O}(F) \cong \mathbb{C}$.

Furthermore, $\pi_1(X)$ contains a nilpotent subgroup of finite index.

This result is of course reminiscent of Theorem 1.6 above. The authors do indeed make use of the "technical lemma" as well as similar arguments with algebraic hulls. In addition, however, they make use of delicate arguments relating the group- and the Kähler-structure. *Reminder:* The Kählerian structure is not assumed to be G-invariant. □

c) Some Final Remarks

In summary if X is the homogeneous space of a solvable or of a semi-simple complex Lie group, then, in a rough sense, we understand the analytic objects on X. For example, if the group is semi-simple, then the analytic objects are very closely related to algebraic ones. In the solvable case one reduces to the study of Stein manifolds, which are *very* close to nil-manifolds, and abelian groups with only constant holomorphic functions. Here Θ-functions play a dominant role.

One would hope to make some sense out of the general case, i.e. no assumption on the nature of the group. We consider the following to be the basic *working questions* in this regard:

Let G be a complex Lie group, Γ a discrete subgroup, and $X = G/\Gamma$. Assume X is holomorphically separable. Is X Stein? If X is Stein, what does this impose on Γ and the semi-direct product structure on G?[1]

In another direction, D. Ahiezer has posed a very interesting question:

Let $X = G/H$ and suppose that $\mathcal{O}(X) \cong \mathbb{C}$ and that H is contained in no parabolic subgroup of G. Let $X = G/H \to G/J = Y$ be the hypersurface reduction of X. Is Y an abelian group, i.e. $G' < J$?

In other words, if a hypersurface doesn't come from a holomorphic function, an algebraic variety, or something closely related to a Θ-function, then where does it come from?

Ahiezer answered this positively for nilpotent groups ([A 6]). For semi-simple (resp. solvable) groups, this is a consequence of Theorem 2.1 (resp. Theorem 2.3). If X is compact, then this follows from the Theorem of Grauert and Remmert (I.3.2). □

References

[A 1] Ahiezer, D.N.: Dense orbits with two ends. Math. USSR, Izvestija II, 293–307 (1977)

[A 2] Ahiezer, D.N.: Actions with a finite number of orbits. Funct. Anal. Appl. **19** (1) (1985) 1–4

[1] *Added in proof.* K. Oeljeklaus has just provided an example of $X = G/\Gamma$ which is holomorphically separable but *not* Stein.

[A 3] Ahiezer, D.N.: Algebraic groups acting transitively in the complement of a homogeneous hypersurface. Sov. Math. Dokl. **20** (1975) 278–281

[A 4] Ahiezer, D.N.: On algebraic varieties that are symmetric in the sense of Borel. Sov. Math. Dokl. **30** (1984) 579–582

[A 5] Ahiezer, D.N.: Invariant meromorphic functions on complex semi-simple Lie groups. Invent. math. **65** (1982) 325–329

[A 6] Ahiezer, D.N.: Invariant analytic hypersurfaces in complex nilpotent Lie groups. Ann. Glob. Analysis and Geometry **2** (1984) 129–140

[AG] Andreotti, A., Ghiardelli, F.: Estensioni commutative di Varieta' Abeliane. Quaderno manoscritto del Centro di Analisi Globale del CNR, Firenze (1972) 1–48

[AHR] Azad, H., Huckleberry, A.T., Richthofer, W.: Homogeneous CR-manifolds. J. reine angew. Math. **358** (1985) 125–154

[AR] Ahern, P., Rudin, W.: Totally real embeddings of S^3 in $\mathbb{C}^3$. Proc. of AMS **94** (3) (1985) 460–462

[Aus] Auslander, L.: On radicals of discrete subgroups of Lie groups. Am. J. Math. **85** (1963) 145–150

[B] Brion, M.: Quelques propriétés des espaces homogènes sphériques. Manuscr. math. **55** (1986) 191–198

[BBD] Barrett, D.E., Bedford, E., Dadok, J.: T^n-actions on holomorphically. Math. Z. (1989)

[Be] Berteloot, F.: Fonctions plurisubharmonique sur SL(2, $\mathbb{C}$) invariant par un sous-groupe monogène. J. d'analyse Math. **48** (1987) 267–276

[BeO] Berteloot, F., Oeljeklaus, K.: Invariant plurisubharmonic functions and hypersurfaces on semi-simple complex Lie groups. Math. Ann. **281** (1988) 513–530

[Bl] Blanchard, A.: Sur les variétés analytiques complexes. Ann. Sci. Ec. Norm. Sup. **73** (1956) 157–202

[BLV] Brion, M., Luna, D., Vust, Th.: Espaces homogènes sphériques. Invent. math. **84** (1986) 617–632

[BM] Bochner, S., Montgomery, D.: Groups on analytic manifolds. Ann. Math. **48** (1947) 659–669

[BO] Barth, W., Otte, M.: Über fast-uniforme Untergruppen komplexer Liegruppen and auflösbare komplexe Mannigfaltigkeiten. Comm. Math. Helv. **44** (1969) 269–281

[Bo 1] Borel, A.: Linear algebraic groups. Benjamin, New York 1969

[Bo 2] Borel, A.: Symmetric compact complex spaces. Arch. Math. **33** (1979) 49–56

[BOe] Barth, W., Oeljeklaus, E.: Über die Albanese-Abbildung einer fast homogenen Kähler-Mannigfaltigkeit. Math. Ann. **211** (1974) 47–62

[BP] Brion, M., Pauer, F.: Valuations des espaces homogènes sphériques. Comment. Math. Helv. **62** (1987) 265–285

[BR] Borel, A., Remmert, R.: Über kompakte homogene Kählersche Mannigfaltigkeiten. Math. Ann. **145** (1962) 429–439

[Bre] Bredon, G.E.: Introduction to compact transformation groups. Academic Press, New York London 1972

[BS] Burns, D., Shnider, S.: Spherical hypersurfaces in complex manifolds. Invent. math. **33** (1976) 223–246

[Car] Cartan, H.: Sur les transformations analytiques des domaines cerclés et semi-cardés bornés. Math. Ann. **106** (1932) 540–573

[CAS] Grauert, H., Remmert, R.: Coherent analytic sheaves (Grundlehren der mathematischen Wissenschaften, vol. 265). Springer, Berlin Heidelberg 1984

[Ch] Chevalley, C.: Théorie des groupes de Lie II: Groupes algébriques. Hermann, Paris 1951

[deM] de Mailly, J.-P.: Un example de fibré holomorphe non de Stein à fibre $\mathbb{C}$ ayant pour base le disque ou le plan. Invent. math. **48** (1978) 293–302

[DN] Dorfmeister, J., Nakajima, K.: The fundamental conjecture for homogeneous Kähler manifolds. Acta math. **161** (1988) 23–70

[ES] Erdmann-Snow, J.: On the classification of solv-manifolds in dimension 2 and 3. 10. Revue de l'Institut Elie Cartan, Nancy, 57–103 (1986)

[Feld] Feldmüller, D.: Two-orbit varieties with smaller orbit of codimension two. (To appear in Arch. Math.)

[Fu] Fujiki, A.: On automorphism groups of compact Kähler manifolds. Invent. math. **44** (1978) 225–258

[G] Goodman, R.: Analytic and entire vectors for representations of Lie groups. Trans. A.M.S **143** (1969) 55–76

[Gal] Galemann, B.: Beschränkte Gebiete über $\mathbb{C}^2$ mit holomorphen S^1-Wirkungen. Diplomarbeit, Ruhr-Universität Bochum 1989

[Gell] Gellhaus, C.: Äquivariante Kompaktifizierungen des $\mathbb{C}^n$. (To appear in Math. Z. 1990)

[GH 1] Gilligan, B., Huckleberry, A.T.: Complex homogeneous manifolds with two ends. Mich. J. Math. **28** (1981) 183–196

[GH 2] Gilligan, B., Huckleberry, A.T.: On non-compact complex nilmanifolds. Math. Ann. **238** (1978) 39–49

[Gi 1] Gilligan, B.: On bounded holomorphic reductions of homogeneous spaces. C.R. Math. Rep. Acad. Sci. Canada, vol. VI (1984) 175–178

[Gi 2] Gilligan, B.: Ends of complex homogeneous manifolds having non-constant holomorphic functions. Arch. Math. **37** (1981) 544–555

[GOR] Gilligan, B., Oeljeklaus, K., Richthofer, W.: Homogeneous complex manifolds with more than one end. Can. J. Math. vol. XLI, No. 1 (1989) 163–177

[GR 1] Grauert, H., Remmert, R.: Über kompakte homogene komplexe Mannigfaltigkeiten. Arch. Math. **13** (1962) 498–507

[GR 2] Grauert, H., Remmert, R.: Komplexe Räume. Math. Ann. **136** (1958) 245–318

[GS] Guillemin, V., Sternberg, S.: Symplectic techniques in physics. Cambridge Univ. Press 1984

[H 1] Huckleberry, A.T.: Hypersurfaces in homogeneous spaces. Journées Complexes, Nancy, Institut Elie Cartan (1982)

[Har] Harish-Chandra: Discrete series for semi-simple Lie groups II. Acta Math. **116** (1966) 1–111

[Hei 1] Heinzner, P.: Fixpunktmengen kompakter Gruppen in Teilgebieten Steinscher Mannigfaltigkeiten. J. reine angew. Math. **402** (1989) 128–137

[Hei 2] Heinzner, P.: Kompakte Transformationsgruppen Steinscher Räume. Math. Ann. **285** (1989) 13–28

[Hei 3] Heinzner, P.: Invariantentheorie in der komplexen Analysis. Habilitationsarbeit, Ruhr-Universität Bochum 1990

[Hei 4] Heinzner, P.: Linear äquivariante Einbettungen Steinscher Räume. Math. Ann. **280** (1988) 147–160

[HL] Huckleberry, A.T., Livorni, L.: A classification of complex homogeneous surfaces. Can. J. Math. **33** (1981) 1096–1109

[HM 1] Hochschild, G., Mostow, G.D.: Affine embeddings of complex analytic homogeneous spaces. Am. J. Math. **87** (1965) 807–839

[HM 2] Hochschild, G., Mostow, G.D.: On the algebra of representative functions of an analytic group, II. Am. J. Math. **86** (1964) 869–887

[HMar] Huckleberry, A.T., Margulis, G.A.: Invariant analytic hypersurfaces. Invent. math. **71** (1983) 235–240

[Ho] Hochschild, G.: The structure of Lie groups. Holden-Day 1965

[HO 1] Huckleberry, A.T., Oeljeklaus, E.: Classification theorems for almost homogeneous spaces. Revue De l'Institut Elie Cartan, Numéro 9, Janvier 1984

[HO 2] Huckleberry, A.T., Oeljeklaus, E.: A characterization of complex homogeneous cones. Math. Z. **170** (1980) 181–194

[HO 3] Huckleberry, A.T., Oeljeklaus, E.: On holomorphically separable complex solvmanifolds. Annales de l'Institut Fourier, Tome XXXVI - Fascicule 3, 57–65 (1986)

[HR] Huckleberry, A.T., Richthofer, R.: Recent developments in homogeneous CR-hypersurfaces, Contributions to several complex variables. Aspects of Math. E9, Vieweg, Braunschweig, 149–177 (1986)

[HS 1] Huckleberry, A.T., Snow, D.: Pseudoconcave homogeneous manifolds. Ann. Scuola Norm. Sup. Pisa, Série IV, vol. VII 29–54 (1980)

[HS 2] Huckleberry, A.T., Snow, D.: A classification of strictly pseudoconcave homogeneous manifolds. Ann. Scuola Norm. Sup. Pisa, Série IV, vol. VIII, 231–255 (1981)

[HS 3] Huckleberry, A.T., Snow, D.: Almost-homogeneous Kähler-manifolds with hypersurface orbits. Osaka J. Math. **19** (1982) 763–786

[Hum] Humphreys, J.E.: Linear algebraic groups. (Graduate Texts in Mathematics, vol. 21). Springer, New York 1975

[HW] Huckleberry, A.T., Wurzbacher, T.: Multiplicity-free complex manifolds. Math. Ann. (to appear)

[K] Kraft, H.: Geometrische Methoden in der Invariantentheorie. Vieweg-Verlag, Braunschweig-Wiesbaden 1985

[Ka 1] Kaup, W.: Reelle Transformationsgruppen und invariante Metriken auf komplexen Räumen. Invent. math. **3** (1967) 43–70

[Ki] Kiselman, C.O.: The partial Legendre transformation for plurisubharmonic functions. Invent. math. **49** (1978) 137–148

[Kod] Kodaira, K.: On compact analytic surfaces II. Ann. Math. **77** (1963) 563–626

[KV] Kimelfeld, B.N., Vinberg, E.B.: Homogeneous domains on flag manifolds and spherical subgroups of semi-simple Lie groups. Funct. Analysis and its Appl. **12.3** (1978) 12–19

[L] Luna, D.: Slices etales. Bull. Soc. Math. Fr. Mem. **33** (1973) 81–105

[Le 1] Lehmann, R.: Complex-symmetric spaces. Dissertation, Ruhr-Universität Bochum 1988

[Le 2] Lehmann, R.: Complex-symmetric spaces. Ann. Inst. Fourier **39.2** (1989) 373–416

[Le 3] Lehmann, R.: Singular complex-symmetric torus embeddings. (To appear)

[Les 1] Lescure, F.: Sur les compactifications équivariantes des groupes commutatifs. Annales de l'Institut Fourier **38.4** (1988) 93–120

[Les 2] Lescure, F.: Compactifications équivariantes par des courbes. Bull. Soc. Math. France **115** (1987) Mém. 26

[Loe] Loeb, J.: Actions d'une forme de Lie réelle d'un groupe de Lie complexe sur les fonctions plurisousharmoniques. Annales de l'Institut Fourier **35-4** (1985) 59–97

[M] Mumford, D., Fogarty, J.: Geometric Invariant Theory. Ergebnisse der Mathematik, vol. 34 (Second Enlarged Edition). Springer, Berlin Heidelberg 1982

[Ma 1] Matsushima, Y.: Sur les espaces homogènes kählériens d'un groupe de Lie réductif. Nagoya Math. J. **11** (1957) 53–60

[Ma 2] Matsushima, Y.: On discrete subgroups and homogeneous spaces of nilpotent Lie groups. Nagoya Math. J. **2** (1951) 95–110

[Ma 3] Matsushima, Y.: Fibrés holomorphes sur un tore complexe. Nagoya Math. J. **14** (1959) 1–24

[Ma 4] Matsushima, Y.: Espaces homogènes de Stein des groupes de Lie complexes I. Nagoya Math. J. **16** (1960) 205–218

[MJ] Moser-Jauslin, L.: The Chow rings of smooth complete SL(2)-embeddings. (Preprint)

[MM] Matsushima, Y., Morimoto, A.: Sur certains espaces fibrés holomorphes sur une variété de Stein. Bull. Soc. Math. France **88** (1960) 137–155

[MN] Morimoto, Y.; Nagano, T.: On pseudo-conformal transformations of hypersurfaces. J. Math. Soc. Japan **14** (1963) 289–300

[Mo 1] Mostow, G.D.: Some applications of representative functions to solv-manifolds. Am. J. Math. **93** (1971) 11–32

[Mo 2] Mostow, G.D.: Factor spaces of solvable groups. Ann. Math. **60** (1954) 1–27

[Oe 1] Oeljeklaus, E.: Fast homogene Kählermannigfaltigkeiten mit verschwindender erster Bettizahl. Manuskr. Math. **7** (1972) 175–183

[Oe 2] Oeljeklaus, E.: Ein Hebbarkeitssatz für Automorphismengruppen kompakter Mannigfaltigkeiten. Math. Ann. **190** (1970) 154-166

[OeK] Oeljeklaus, K.: Hyperflächen und Geradenbündel auf homogenen komplexen Mannigfaltigkeiten. Schriftenreihe des Mathematischen Instituts der Universität Münster, Ser. 2, Heft 36, Münster 1985

[OR 1] Oeljeklaus, K., Richthofer, W.: On the structure of complex solvmanifolds. J. Diff. Geom. **27** (1988) 399–421

[OR 2] Oeljeklaus, K., Richthofer, W.: Homogeneous complex surfaces. Math. Ann. **268** (1984) 273–292

[P] Penny, R.: Entire vectors and holomorphic extension of representations. Trans. AMS. **198** (1974) 107–121

[R] Rossi, H.: Homogeneous strongly pseudoconvex hypersurfaces. Rice Studies **59** (3) (1973) 131–145

[R 1] Richthofer, W.: Currents in homogeneous manifolds. (To appear)

[R 2] Richthofer, W.: Homogene CR-Mannigfaltigkeiten. Dissertation, Ruhr-Universität Bochum 1985

[Sko] Skoda, H.: Fibrés holomorphes à base at à fibre de Stein. Invent. math. **43** (1977) 97-107

[Sn 1] Snow, D.M.: Reductive group action on Stein spaces. Math. Ann. **259** (1982) 79-97

[Sn 2] Snow, D.M.: Stein quotients of connected complex lie groups. Manuscripta math. **49/50** (1985) 185–214

[So] Sommese, A.J.: Extension theorems for reductive group actions on compact Kähler manifolds. Math. Ann. **218** (1975) 107–116

[SZ] Stout, E.L., Zame, W.R.: A Stein manifold topologically but not holomorphically equivalent to a domain in $\mathbb{C}^n$. Adv. in Math. **60** (1986) 154–160

[T] Tits, J.: Espaces homogènes complexes compacts. Comm. Math. Helv. **37** (1962) 111–120

[Wa 1] Wang, H.C.: Closed manifolds with homogeneous complex structure. Am. J. Math. **76** (1954) 1–32

[Wa 2] Wang, H.C.: Complex parallelisable manifolds. Proc. Am. Math. Soc **5** (1954) 771–776

[Wi 1] Winkelmann, J.: Classification des espaces complexes homogènes de dimension 3 I. C.R.A.S. Paris **306** (1988) Série I, 231–234

[Wi 2] Winkelmann, J.: Classification des espaces complexes homogènes de dimension 3 II. C.R.A.S. Paris **306** (1988) Série I, 405–408

[Wi 3] Winkelmann, J.: The Kobayashi pseudometric on homogeneous complex manifolds. (To appear)

[Wi 4] Winkelmann, J.: On Stein homogeneous manifolds and free holomorphic $\mathbb{C}$-actions on $\mathbb{C}^n$. Math. Ann. (To appear in Math. Ann.)

Stable Bundles, Instantons and C^∞-Structures on Algebraic Surfaces

Ch. Okonek and A. Van de Ven

Contents

Introduction

The ties between topology and algebraic geometry go back to the nineteenth century, but specific questions about surfaces being homeomorphic or not, have only come up in the fifties and sixties. Severi asked in 1954 if every algebraic surface homeomorphic to the projective plane, is biregularly equivalent to it. The (positive) answer came three decades later, as a result of Yau's work on the Calabi conjecture. In 1965 Kodaira constructed examples of elliptic surfaces with the homotopy type of a $K3$-surface, and he posed the question, whether they are homeomorphic to such a surface. At the time, the only theorem known was a result of J.H.C. Whitehead, stating that two simply-connected surfaces are of the same homotopy type if and only if they have isomorphic (integer-valued) intersections forms. Given the known structure theorems for indefinite forms, it was a simple matter to verify this in Kodaira's case. The affirmative answer to Kodaira's question and many similar ones is an immediate consequence of Freedman's fundamental theorem from 1982, saying in particular that the homeomorphism type of a compact, oriented, simply-connected *differentiable* 4-fold is completely determined by its intersection form.

Prior to Freedman's work, only limited attention was paid to the differentiable equivalence of algebraic surfaces. This changed as soon as there was a simple criterion for surfaces to be homeomorphic. But in the beginning little could be done, for the new results in differential topology obtained about fifteen years earlier, concerned only manifolds of dimension at least 5.

So it was a great day when Donaldson in 1985 came up with an example of two algebraic surfaces which are homeomorphic but not diffeomorphic. The surfaces in question are the projective plane blown up in nine points, and a certain elliptic surface, namely the Dolgachev surface of type (2,3). The result was quickly extended, by Friedman and Morgan on the one hand and by the present authors on the other, to the case of all Dolgachev surfaces of type $(2,p)$, p odd; they were shown to be all different as differentiable manifolds (and also different from the blown-up plane).

That day in 1985 was of course a great day for differential topology too, since for the first time a break-through was obtained in dimension 4. It brought things never heard of in higher dimensions, like that infinite number of distinct differentiable structures on a compact topological manifold.

Also for the people working on algebraic vector bundles there was reason for joy. Since the Leiden réveil of the early seventies much of their efforts had gone to stable bundles; in particular to the construction of moduli spaces for such bundles. And now, some of these moduli spaces became of vital importance for questions about the differential topology of surfaces. In all fairness, it should be said that relations with gauge theory, the base for Donaldson's construction, had already existed for several years.

The new ideas introduced by Donaldson have many more applications. For example, a question which had been studied for some time (mainly by

Mandelbaum and Moishezon), both from the topological and the differential point of view, is the problem of the decomposability of 4-manifolds and, in particular, of algebraic surfaces. Is a 1-connected surface homeomorphic or even diffeomorphic to a connected sum of simple building blocks, like the projective plane, the quadric or $K3$-surfaces, possibly with orientation reversed?

Freedman's work implies that, from the topological point of view, the answer is yes. Donaldson has shown that, from the point of view of differential topology, the situation is completely different.

In this survey we would like to discuss not only the preceding applications, together with some other ones, but also the construction of the invariants used to obtain them. This means that we have to start with stable bundles, gauge theory and Hermite-Einstein bundles. There are very few proofs, but we have tried to give precise references everywhere.

As to language and notations, a manifold will always be connected, and an algebraic surface will always be smooth. If X is an oriented manifold, then $\overline{X}$ will be X with reversed orientation. If X and Y are oriented manifolds of the same dimension, then $X\#Y$ will denote a (differentiable) connected sum of X and Y. (They are all diffeomorphic.) Furthermore, kX means $X\#X\#\ldots\#X$ (k times). *We only consider connected sums* $X_1\#X_2$ *with* $b_2(X_1) \neq 0$, $b_2(X_2) \neq 0$.

If X is a 4-manifold, then S_X denotes its cup (or intersection) form, whereas $b_+(b_-)$ is the number of positive (negative) eigenvalues of $S_X \otimes \mathbb{R}$.

We denote the canonical divisor class of a surface by K_X, whereas $\mathcal{K}_X = \mathcal{O}_X(K)$ is the canonical line bundle.

In Chap. V we shall write k_X for $-c_1(X) = c_1(\mathcal{K}_X)$ (X an algebraic surface). If $b_1(X) = 0$, in particular if X is 1-connected, then linear and homological equivalence are the same on X, so it will not lead to confusion if we call k_X the canonical class in this case.

If $|L|$ is a linear system of divisors, then $Bs|L|$ will be the *set* of base points.

If X is an algebraic surface, then $\widehat{X}(p_1, \ldots, p_k)$ denotes X blown up in the points $p_1, \ldots, p_k$.

It is a pleasure to thank M. Lübke for useful discussions.

I. Stable Bundles

I.1 Background

Let X be a complex surface. The C^∞-classification of complex 2-bundles on X is the same as the topological classification. Namely, there is a 1-1 correspondence between equivalence classes of 2-bundles and ordered pairs (c_1, c_2) with $c_1 \in H^2(X, \mathbb{Z})$ and $c_2 \in H^4(X, \mathbb{Z})$. This correspondence, which does not hold in higher dimensions, is obtained by attaching to any 2-bundle E its Chern classes $c_1(E)$ and $c_2(E)$. So, classifying holomorphic 2-bundles on X boils down to classifying the holomorphic 2-bundles with a given pair of Chern classes. Of course, we would like these holomorphic structures to form a complex space, in other words, it would be wonderful if the set of holomorphic structures on a given 2-bundle E would have a coarse or even a fine moduli space. Unfortunately, "jumping structures" prohibit the existence of such moduli spaces. For example, it is not difficult to construct on $\mathbb{P}_2$ a family of holomorphic 2-bundles $(\mathscr{E}_s)_{s\in\Delta}$, parametrised by the unit disk Δ, with $\mathscr{E}_s \cong \mathcal{O}_{\mathbb{P}_2}^{\oplus 2}$ for $s \neq 0$, but $\mathscr{E}_s$ not trivial for $s = 0$ ([B], p. 108, [ST 1], p. 393). However, there is a large class of bundles which do have nice moduli spaces. As it turns out, these moduli spaces of *stable* bundles play a vital role in applying Donaldson's theory to the differentiable structure of algebraic surfaces. Before we describe these special classes of bundles, we first want to recall a general construction, which later will be needed.

I.2 The Serre Construction

Let $\mathscr{E}$ be a holomorphic 2-bundle on X and s a section, which vanishes on a zero-dimensional subscheme Z of X, with ideal sheaf $\mathscr{I}_Z$. Then we have an exact sequence

$$0 \to \mathcal{O}_X \to \mathscr{E} \to \mathscr{I}_Z \otimes \det(\mathscr{E}) \to 0\,,$$

in other words we obtain $\mathscr{E}$ as an extension of $\mathscr{I}_Z \otimes \det(\mathscr{E})$ by $\mathcal{O}_X$. Given any zero-dimensional subscheme $Z \subset X$, and any line bundle $\mathscr{L}$ on X, it is natural to ask when it is possible to find a 2-bundle $\mathscr{E}$ with a section, vanishing on Z, such that $\mathscr{L} \cong \det(\mathscr{E})$ and the above sequence holds. It turns out that for the existence, a certain obstruction element in $H^2(X, \mathscr{L}^\vee)$ has to vanish. This element admits a very geometric interpretation if $Z = \{z_1, \ldots, z_m\}$ consists of m simple points ([G/H], p. 470).

Definition. *A subset $Z = \{z_1, \ldots, z_m\}$ of m simple points on X has the Cayley-Bacharach property with respect to a linear system $|L|$ if no curve in $|L|$ contains exactly $m-1$ points of Z.*

Theorem I.2.1. *Let be given on a complex surface X a subset Z, consisting of m simple points, and a line bundle $\mathscr{L}$. Then there exists a holomorphic 2-bundle $\mathscr{E}$ with a section s, with $\mathscr{L} \cong \det(\mathscr{E})$, $Z = Z(s)$, if and only if Z has the Cayley-Bacharach property with respect to $|K_X + L|$, where $\mathscr{L} = \mathcal{O}_X(L)$. In this case $\mathscr{E}$ can be given by an extension*

$$0 \to \mathcal{O} \to \mathscr{E} \to \mathscr{I}_{Z/X} \otimes \mathscr{L} \to 0$$

and $m = c_2(\mathscr{E})$.

As an illustration we solve the existence problem for 2-bundles on algebraic surfaces, a result which goes back to Schwarzenberger ([S 1], p. 618). We recall that the Néron-Severi group $\mathrm{NS}(X) \subset H^2(X, \mathbb{Z})$ is the dual of the subgroup of $H_2(X, \mathbb{Z})$, generated by algebraic cycles.

Example. *Let X be an algebraic surface, $c_1 \in H^2(X,\mathbb{Z})$, $c_2 \in H^4(X,\mathbb{Z})$. Then there exists a holomorphic 2-bundle $\mathscr{E}$ on X with $c_1(\mathscr{E}) = c_1$, $c_2(\mathscr{E}) = c_2$ if and only if $c_1 \in \mathrm{NS}(X)$.*

Proof. The condition is obviously necessary. To prove sufficiency, let D be a divisor with homology class dual to c_1, and let H be an ample divisor, with dual class $h \in H^2(X,\mathbb{Z})$. Then, for $\alpha \in \mathbb{N}$ large, we have
a) $\alpha^2 h^2 - \alpha c_1 h + c_2 > 0$;
b) $|K_X + D - 2\alpha H| = \emptyset$.

If $Z \subset X$ consists of $m = \alpha^2 h^2 - \alpha c_1 h + c_2$ simple points, then Z has the Cayley-Bacharach property with respect to $D - 2\alpha H$. By Theorem I.2.1, there exists a holomorphic 2-bundle $\mathscr{E}$ on X with $c_1(\mathscr{E}) = c_1 - 2\alpha h$ and $c_2(\mathscr{E}) = m$. The bundle $\mathscr{E} \otimes \mathscr{O}_X(\alpha H)$ has Chern classes c_1 and c_2.

Remark 1. The proof uses in an essential way the projectivity of X, and, in fact, the statement is false for non-algebraic surfaces ([B/L], p. 2).

Remark 2. To the best of our knowledge there still is no example of a differentiable vector bundle on any algebraic manifold with algebraic Chern classes but no algebraic structure. This is not possible on $\mathbb{P}_3$, but on $\mathbb{P}_4$ there should be plenty of examples in dimension 2 (see [V 1]).

I.3 Simple Bundles and Stable Bundles

Definition. A holomorphic bundle $\mathscr{E}$ over X is simple if its only endomorphisms are constant multiples of the identity.

Using Banach-analytic methods ([N, K/O]) or techniques from differential geometry ([K 2, L/O 1, F/S]) one can construct a locally Hausdorff complex space $\mathscr{M}^s_X(c_1,c_2)$(1) parametrising isomorphism classes of simple holomorphic structures with given c_1, c_2, such that the germs $(\mathscr{M}^s_X(c_1,c_2), [\mathscr{E}])$ of this space are the bases of the universal deformations at all points $[\mathscr{E}] \in \mathscr{M}^s_X(c_1,c_2)$. These moduli spaces of simple bundles may be very complicated globally ([L/O 1] p. 673).

Let X be a projective-algebraic surface polarised by a very ample divisor H.

Definition (Mumford, Takemoto). *A holomorphic r-bundle $\mathscr{E}$ on X is μ-stable relative to H if every coherent subsheaf $\mathscr{F} \subset \mathscr{E}$ with $0 < \mathrm{rk}(\mathscr{F}) < r$ satisfies*

$$\frac{c_1(\mathscr{F}) \cdot H}{\mathrm{rk}(\mathscr{F})} < \frac{c_1(\mathscr{E}) \cdot H}{r} \, .$$

(1) Since a differentiable 2-bundle on a surface is completely determined by its Chern classes, we can write $\mathscr{M}^H_X(E)$ (or $\mathscr{M}^s_X(E)$) instead of $\mathscr{M}^H_X(c_1,c_2)$ (or $\mathscr{M}^s_X(c_1,c_2)$), where E denotes a *fixed* differentiable bundle. Both notations will be used.

Stability is preserved if the bundle is tensored with a line bundle or replaced by its dual.

Furthermore, μ-stable bundles are simple ([O/S/S] p. 172) and form open subfamilies in any family of bundles ([M 1] p. 42).

Example. Let $X = \mathbb{P}_2$ polarised by a line $H \subset \mathbb{P}_2$. For a fixed integer $l > 0$ choose a set $Z = \{z_1, \ldots, z_m\}$ of m simple points in $\mathbb{P}_2$ such that no curve of degree $l-3$ contains exactly $m-1$ of them. This is possible for $m > \binom{l-1}{2}$ and it means that Z has the Cayley-Bacharach property with respect to the linear system $|K + lH|$. Consider a 2-bundle $\mathscr{E}$ defined by the extension

$$0 \to \mathcal{O}_{\mathbb{P}_2} \to \mathscr{E} \to \mathscr{J}_{Z/\mathbb{P}_2}(l) \to 0 \,.$$

A necessary and sufficient condition for $\mathscr{E}$ to be μ-stable relative to H is that Z should not be contained in a curve $C \subset \mathbb{P}_2$ of degree $\deg(C) \leq \frac{l}{2}$. This is clearly an open condition and can be met if and only if $m > \binom{[\frac{l}{2}]+2}{2} - 1$.

Consider now the functor $\mathscr{M}_X^H(c_1, c_2)(-)$ which associates to a complex space S the set of equivalence classes of families $\mathfrak{E}/S \times X$, such that

(i) $\mathfrak{E}_s$ has Chern classes c_1, c_2 for every $s \in S$ (so the underlying differentiable bundle E is fixed);

(ii) $\mathfrak{E}_s$ is μ-stable relative to H for every point $s \in S$.

Theorem I.3.1 (Maruyama, Gieseker). *Let (X, H) be an algebraic surface together with a very ample divisor H. Then there exists a quasi-projective variety whose underlying complex space $\mathscr{M}_X^H(c_1, c_2)$ is a coarse moduli space for the functor $\mathscr{M}_X^H(c_1, c_2)(-)$.*

For bundles of rank 2 the proof can be found in ([M 1] p. 55); the general case follows from results of Gieseker ([G] p. 46) and Maruyama ([M 2, M 3]).

These authors construct a projective variety, parametrising certain equivalence classes of semi-stable sheaves, whose underlying complex space contains $\mathscr{M}_X^H(c_1, c_2)$ as an open subset.

Let us now briefly discuss the most important properties of the moduli spaces $\mathscr{M}_X^H(c_1, c_2)$.

1) Universal Deformation

The germ of $\mathscr{M}_X^H(c_1, c_2)$ at any point $[\mathscr{E}]$ is the base of the universal deformation of $\mathscr{E}$ ([WE], p. 69).

2) Universal Bundles

Locally, there always exist on $\mathscr{M}_X^H(c_1, c_2)$ universal bundles. In general these bundles do not fit together to a global universal bundle over $\mathscr{M}_X^H(c_1, c_2) \times X$

([LP] p. 218). There is however a useful sufficient criterion due to Maruyama ([M 3] p. 598). Write the Hilbert polynomial $\chi(\mathscr{E}(m))$ of $\mathscr{E}$ as

$$\chi(\mathscr{E}(m)) = \sum_{k=0}^{2} a_k \binom{m+k}{k}$$

with integer coefficients $a_k \in \mathbb{Z}$.

Criterion. A universal bundle $\mathfrak{E}/\mathscr{M}_X^H(c_1, c_2) \times X$ exists if g.c.d. $(a_i) = 1$.

Example. For a 2-bundle $\mathscr{E}$ with Chern classes $c_i = c_i(\mathscr{E})$ this criterion reads

$$\text{g.c.d.}\ \left(2H^2, \chi(\mathscr{E}), H^2 + c_1 \cdot H - H \cdot K\right) = 1\ .$$

If a universal bundle exists, $\mathscr{M}_X^H(c_1, c_2)$ is said to be a fine moduli space; in this case $\mathscr{M}_X^H(c_1, c_2)$ represents the functor $\mathscr{M}_X^H(c_1, c_2)(-)$.

3) Local Structure

By general deformation theory, the Zariski tangent space of $\mathscr{M}_X^H(c_1, c_2)$ at a point $[\mathscr{E}]$ is $H^1(X; \operatorname{End}\mathscr{E})$.

From the decomposition $\operatorname{End}\mathscr{E} \cong \operatorname{End}_0\mathscr{E} \oplus \mathscr{O}_X$ we obtain

$$H^1(X; \operatorname{End}\mathscr{E}) = H^1(X; \mathscr{O}_X) \oplus H^1(X; \operatorname{End}_0\mathscr{E})\ .$$

The space $H^1(X; \operatorname{End}_0\mathscr{E})$ is the Zariski tangent space of the (base of) the versal deformation of the projective bundle $\mathbb{P}(\mathscr{E})$ associated to $\mathscr{E}$.

In fact, much more is true: $\mathscr{M}_X^H(c_1, c_2)$ is locally a product of the base of this versal deformation with $\operatorname{Pic}^0(X)$ ([E/F] p. 38). The base of the versal deformation of $\mathbb{P}(\mathscr{E})$ can be described as the inverse image of $0 \in H^2(X; \operatorname{End}_0\mathscr{E})$ under the Kuranishi map

$$K_{[\mathscr{E}]} : H^1(X; \operatorname{End}_0\mathscr{E}) \to H^2(X; \operatorname{End}_0\mathscr{E})\ .$$

This is a (germ of a) holomorphic map which is at least quadratic ([K 2] p. 254). Putting everything together we get an isomorphism of germs

$$\left(\mathscr{M}_X^H(c_1, c_2), [\mathscr{E}]\right) \cong \left(K_{[\mathscr{E}]}^{-1}(0) \times \operatorname{Pic}^0(X)\ ,\ \ (0,0)\right)\ .$$

4) Smoothness

The local description of $\mathscr{M}_X^H(c_1, c_2)$ via the Kuranishi map yields immediately the following sufficient criterion for smoothness.

Criterion. $\mathscr{M}_X^H(c_1, c_2)$ is smooth at $[\mathscr{E}]$ if $h^2(X; \operatorname{End}_0\mathscr{E}) = 0$.

Example. $H^2(X; \operatorname{End}_0\mathscr{E})$ vanishes for every μ-stable bundle on surfaces with $\mathscr{K}_X \cong \mathscr{O}_K$ or $K_X \cdot H < 0$.

5) Dimension

The Kuranishi description of the germ $\left(\mathcal{M}_X^H(c_1,c_2),[\mathcal{E}]\right)$ shows that the dimension of $\mathcal{M}_X^H(c_1,c_2)$ at $[\mathcal{E}]$ is bounded from below:

$$\dim_{[\mathcal{E}]}\mathcal{M}_X^H(c_1,c_2)\geq h^1(X;\mathrm{End}_0\,\mathcal{E})-h^2(X;\mathrm{End}_0\,\mathcal{E})+h^1(X;\mathcal{O}_X)$$

as well as from above:

$$\dim_{[\mathcal{E}]}\mathcal{M}_X^H(c_1,c_2)\leq h^1(X;\mathrm{End}_0\,\mathcal{E})+h^1(X;\mathcal{O}_X)\ .$$

Using the Riemann-Roch formula for $\mathrm{End}_0\,\mathcal{E}$ the first inequality becomes in the case of a 2-bundle:

$$\dim_{[\mathcal{E}]}\mathcal{M}_X^H(c_1,c_2)\geq 4c_2(\mathcal{E})-c_1^2(\mathcal{E})-3\chi(\mathcal{O}_X)+q(X)\ .$$

The lower bound

$$4c_2(\mathcal{E})-c_1^2(\mathcal{E})-3\chi(\mathcal{O}_X)+q(X)$$

is the "expected dimension" of $\mathcal{M}_X^H(c_1,c_2)$.

6) Global Structure

The moduli spaces $\mathcal{M}_X^H(c_1,c_2)$ (for all H) are contained as open subsets in the moduli space $\mathcal{M}_X^s(c_1,c_2)$ of simple holomorphic structures. This last space is only locally Hausdorff. Two points $[\mathcal{E}],[\mathcal{E}']$ represented by simple bundles $\mathcal{E},\mathcal{E}'$ can not be separated if and only if there exist non-zero morphisms $\rho:\mathcal{E}\to\mathcal{E}'$, $\psi:\mathcal{E}'\to\mathcal{E}$ such that $\rho\circ\psi=0$ and $\psi\circ\rho=0$ ([L/O 1] p. 672). This is impossible for μ-stable bundles, i.e. $\mathcal{M}_X^H(c_1,c_2)$ is globally Hausdorff (this also follows from its construction).

Apart from the existence, in general little is known about the global structure of $\mathcal{M}_X^H(c_1,c_2)$.

In some cases results have been obtained by using special properties of the surface X ([T, M 5, M 6, M 7, BAR, H, LP, ST 2, MU, BR 1, BR 2, H/S, UM, O/V 1, O/V 2, S 2, L/O 2, F/M 3, FR, DR 1, DR 2, DR 3]).

In many situations the moduli spaces inherit characteristic properties of the surfaces as e.g. rationality, fibre structures or a symplectic structure.

There is however no general theorem to this effect.

We close this section with an easy example.

Example. Let X be the Godeaux surface, i.e. the quotient of the Fermat quintic in $\mathbb{P}_3$ by the standard action of $\mathbb{Z}/_5$ ([B/P/V], p. 170). The surface X is smooth of general type with $p_g(X)=0$, $q(X)=0$. Its canonical bundle $\mathcal{K}_X$ is ample since X contains no (-2)-curves (otherwise there would be (-2)-curves on the Fermat quintic, which is impossible since its canonical bundle is ample). The 2-canonical system $|2K_X|$ is a pencil whose base locus $Bs|2K_X|=\{z_1,\dots,z_4\}$ consists of 4 simple points ([P] p. 39). We fix a very ample pluricanonical divisor H, and claim that the moduli space $\mathcal{M}_X^H(K_X,1)$ is a nowhere reduced space with reduction

$$\mathcal{M}_X^H(K_X,1)_{\text{red}} \cong Bs|2K_X| .$$

The proof is very simple.

Using the Riemann-Roch theorem and the stability condition one finds that every μ-stable 2-bundle $\mathscr{E}$ with $c_1(\mathscr{E}) = c_1(\mathscr{K}_X)$, $c_2(\mathscr{E}) = 1$ has a global section leading to an extension

$$(\varepsilon) : 0 \longrightarrow \mathscr{K}_X^\vee \longrightarrow \mathscr{E}^\vee \longrightarrow \mathscr{I}_{Z/X} \longrightarrow 0$$

with $Z = \{z\}$ a simple point. As we have seen above, this point has the Cayley-Bacharach property with respect to $|2K_X|$, i.e. $z \in Bs|2K_X|$. The extension class $\varepsilon \in \operatorname{Ext}^1\left(\mathscr{I}_{\{z\}/X}, \mathscr{K}_X^\vee\right) \cong \mathbb{C}$ is uniquely determined up to multiplication by $\mathbb{C}^*$. Conversely, every point $z \in Bs|2K_X|$ and every class $(\varepsilon) \in \operatorname{Ext}^1\left(\mathscr{I}_{\{z\}/X}, \mathscr{K}_X^\vee\right) \setminus \{0\}$ leads to a stable bundle $\mathscr{E}$ as above. This yields an isomorphism $\mathcal{M}_X^H(K_X,1)_{\text{red}} \cong Bs|2K_X|$.

To prove that $\mathcal{M}_X^H(K_X,1)$ is nowhere reduced it suffices to show that $h^1(X;\operatorname{End}_0\mathscr{E}) \neq 0$. But

$$\begin{aligned} h^1(X;\operatorname{End}_0\mathscr{E}) &= h^2(X;\operatorname{End}_0\mathscr{E}) = h^2(X;\operatorname{End}\mathscr{E}) \\ &= h^0(X;\operatorname{End}\mathscr{E}\otimes\mathscr{K}_K) \geq h^0(X;\mathscr{E}) > 0 . \end{aligned}$$

It would be rather interesting to "understand" this non-reduced structure of $\mathcal{M}_X^H(K_X,1)$ from a geometric point of view.

II. Instantons

II.1 Yang-Mills Connections

The moduli spaces $\mathcal{M}_X^H(c_1,c_2)$ of μ-stable bundles over a projective surface X, which we described in the previous section, are the "integrable cases" of more general objects, the instanton moduli spaces.

Instantons – a notion which has its origin in the gauge theory of mathematical physics – can be defined on any oriented 4-dimensional Riemannian manifold. Their moduli spaces are the solution spaces of a certain system of non-linear semi-elliptic differential equations modulo the action of the corresponding gauge group.

It has been a completely new and far-reaching idea of Donaldson that these instanton spaces can be used to define invariants, which are independent of the Riemannian structure, but distinguish between different differentiable structures on the 4-manifold in question, seen as a topological manifold.

In this chapter we explain the construction of instanton moduli spaces and describe their main properties. The definition of Donaldson's invariants and their applications to the differential topology of algebraic surfaces will be given in the final chapters, after we have discussed in Chap. III an essential intermediate tool, namely Hermite-Einstein bundles.

Let X be a closed, connected, oriented differentiable 4-manifold, G a compact Lie group embedded in some orthogonal group. G will typically be one of the groups S^1, SU(2), U(2) or PU(2). Consider a principal bundle P over X with structure group G. The space of G-connections on P is an infinite-dimensional affine space

$$\mathscr{A}(P) = A_0 + A^1(\mathfrak{g}_P)$$

modelled after the vector space $A^1(\mathfrak{g}_P)$ of 1-forms with values in the adjoint bundle

$$\mathscr{G}_P = P \times_{\mathrm{Ad}(G)} \mathscr{G} \, .$$

A connection $A \in \mathscr{A}(P)$ on P induces, for every linear representation $\rho : G \to \mathrm{GL}(V)$, a differential operator

$$D_A : A^0(E) \longrightarrow A^1(E)$$

on the associated vector bundle $E = P \times_\rho V$. These operators satisfy the Leibniz rule

$$D_A(\varphi \cdot s) = d\varphi \otimes s + \varphi \cdot D_A(s) \, ,$$

and they can be extended to operators on E-valued p-forms in an obvious way ([K 2], p. 3). The curvature

$$F_A \in A^2(\mathfrak{g}_P)$$

of A, which is the 2-form representing the composition $D_A \circ D_A$, satisfies the Bianchi identity

$$D_A F_A \equiv 0 \, .$$

Let Aut(P) be the gauge group of P/X, i.e. the group of principal bundle automorphisms of P covering the identity on X. Then Aut(P) acts in a natural way on the space of connections $\mathscr{A}(P)$ such that the differential operator $D_{A \cdot f}$, corresponding to the transformed connection $A \cdot f$, $f \in \mathrm{Aut}(P)$, is given by

$$D_{A \cdot f} = f^{-1} \circ D_A \circ f \, .$$

The isotropy group $\mathrm{Aut}(P)_A$ of this action at a point $A \in \mathscr{A}(P)$ is the centraliser of the holonomy of A, which always contains the centre Z_G of G. A connection A is irreducible if its holonomy is dense in G, in which case $\mathrm{Aut}(P)_A$ reduces to Z_G. The quotient $\mathrm{Aut}^*(P) := \mathrm{Aut}(P)/Z_G$ acts on $\mathscr{A}(P)$ restricting to a free action on the subset $\mathscr{A}^*(P) \subset \mathscr{A}(P)$ of irreducible connections. Let

$$\mathscr{B}(P) := \mathscr{A}(P)/\mathrm{Aut}(P)$$

be the moduli space of gauge equivalence classes of connections on P and let

$$\mathscr{B}^*(P) := \mathscr{A}^*(P)/\mathrm{Aut}(P)$$

be the open subset corresponding to irreducible connections. The instanton moduli spaces will be defined as closed, finite-dimensional subsets in these infinite dimensional spaces; their definition depends on the choice of a metric

g on X. So let g be a fixed Riemannian metric on X. The Yang-Mills functional (with respect to P and g) is the functional

$$YM_X^g(P) : \mathcal{A}(P) \longrightarrow \mathbb{R}_{\geq 0}$$

which associates to a connection A the norm

$$\|F_A\|^2 := -\int_X \operatorname{tr}(F_A \wedge *F_A)$$

of its curvature, where $* = *_g$ is the Hodge star-operator defined by g and the orientation of X. Since $-\operatorname{tr} : \mathfrak{g} \otimes \mathfrak{g} \rightarrow \mathbb{R}$ is an Ad-invariant metric and $F_{A \cdot f} = f^{-1} \circ F_A \circ f$ for a gauge transformation f, the Yang-Mills functional descends to a well-defined functional on the moduli space $\mathcal{B}(P)$. Yang-Mills connections on P are defined as the critical points of the Yang-Mills functional. The Yang-Mills equations, i.e. the Euler-Lagrange equations of the corresponding variational problem, follow easily from the basic formula

$$F_{A+\alpha} = F_A + D_A(\alpha) + [\alpha \wedge \alpha]$$

for a connection $A \in \mathcal{A}(P)$ and a form $\alpha \in A^1(\mathfrak{g}_P)$.

In fact, setting $A_t = A + t\alpha$ one finds:

$$\begin{aligned} 0 = \frac{d}{dt} \|F_{A_t}\|^2 &= \frac{d}{dt} \left\|F_A + tD_A(\alpha) + t^2[\alpha \wedge \alpha]\right\|^2 \\ &= 2(F_A, D_A(\alpha)) = 2(D_A^* F_A, \alpha) , \end{aligned}$$

where $D_A^* = - * \circ D_A \circ *$ denotes the formal adjoint of D_A with respect to the Hodge inner product ([WEL], p. 173).

In other words, A is a Yang-Mills connection if and only if its curvature F_A satisfies the Yang-Mills equations:

$$D_A(*F_A) = 0 .$$

These equations form a system of second order partial differential equations in the variable A. A nice survey from this analytical point of view can be found in Bourguignon's DMV article ([BOU]). The equation $D_A(*F_A) = 0$, together with the Bianchi identity $D_A F_A \equiv 0$, expresses the fact that Yang-Mills connections are harmonic with respect to their own Laplacians. In this sense Yang-Mills theory is a non-linear version of Hodge theory. For abelian structure groups G it actually reduces to Hodge theory.

Example. Let $G = S^1$ and denote the unique g-harmonic representative of $c_1(P) \in H^2_{DR}(X)$ by ω. The connection $A \in \mathcal{A}(P)$ is Yang-Mills if its curvature is harmonic, i.e. if and only if $-\frac{1}{2\pi i} F_A = \omega$. Fix such a connection A_0. Since every other connection is of the form $A = A_0 + i\alpha$ for some 1-form α, we see that the set of all Yang-Mills connections forms an affine space

$$A_0 + iZ^1 ,$$

where $Z^1 = \{\alpha \in A^1 | d\alpha = 0\}$ is the vector space of closed 1-forms. The gauge group $\mathrm{Aut}(P)$ in this abelian case is simply the group $C^\infty(X; S^1)$ of C^∞-maps from X to S^1. Its identity component $\mathrm{Aut}^0(P)$ consists of maps which lift to $\mathbb{R}$ and it acts on $A_0 + iZ^1$ in the obvious way. The orbit space of this $\mathrm{Aut}^0(P)$-action is isomorphic to the first De Rham group $H^1_{\mathrm{DR}}(X)$ with $\mathrm{Aut}(P)/\mathrm{Aut}^0(P) \cong H^1(X;\mathbb{Z})$ acting naturally. The set of gauge equivalence classes of Yang-Mills connections can therefore be identified with the quotient

$$H^1_{\mathrm{DR}}(X)/H^1(X,\mathbb{Z}) .$$

So far we have not used the assumption that our base manifold X is of dimension 4. This 4-dimensional case has three important special features.

(i) *Conformal Invariance.* The Yang-Mills functional $YM^g_X(P)$ depends only on the conformal class $[g]$ of the metric since the integrand $-\mathrm{tr}(F_A \wedge *_g F_A)$ does not change if we replace g by a point-wise conformal metric $\lambda^2 \cdot g$. This can be translated into a scaling invariance, a property which is crucial for many results like e.g. Uhlenbeck's theorem about the removability of point singularities (compare Theorem II.3.4).

(ii) *Self Duality.* The Hodge operator $*$ is an involution on the bundle $\bigwedge^2 = \bigwedge^2 T^*_X$ of 2-forms on X, decomposing it into (± 1)-eigen bundles

$$\textstyle\bigwedge^2 = \bigwedge_+ \oplus \bigwedge_- ,$$

the bundles of self-dual and anti-self-dual forms. This splitting is point-wise orthogonal with respect to the Hodge inner product and leads to the decomposition of curvature forms

$$F_A = F_{A+} + F_{A-}$$

into self-dual and anti-self-dual components. These components are characterised by

$$*F_{A\pm} = \pm F_A .$$

(iii) *Topological Lower Bound.* By Chern-Weil theory ([BE], p. 192) the differential form $-\frac{1}{4\pi^2}\mathrm{tr}(F_A \wedge F_A)$ represents a characteristic class of the bundle P/X, i.e. the integral

$$i(P) := -\int_X \mathrm{tr}(F_A \wedge F_A) = \|F_{A_+}\|^2 - \|F_{A_-}\|^2$$

depends only on the bundle P and the volume of X, but not on the connection A. Using

$$\|F_A\|^2 = \|F_{A_+}\|^2 + \|F_{A_-}\|^2 = 2\,\|F_{A_+}\|^2 - i(P) = 2\,\|F_{A_-}\|^2 + i(P)$$

one finds the topological lower bound

$$YM_X^g(P)(A) \geq i(P)$$

for the Yang-Mills functional. Moreover, equality occurs if and only if F_A is self-dual or anti-self-dual.

II.2 (Anti-)Self-Duality. Instantons

Definition. *A connection A on P is self-dual (anti-self-dual) with respect to the metric g if*

$$*_g F_A = F_A \qquad (*_g F_A = -F_A) .$$

Self-dual (s.d.) and anti-self-dual (a.s.d.) connections are automatically Yang-Mills connections. As we have seen at the end of Sect. I, they give an absolute minimum for the Yang-Mills functional if they exist. An obvious necessary condition for a bundle P to admit s.d. (a.s.d.) connections is $i(P) \geq 0$ ($i(P) \leq 0$). This is usually not sufficient.

Example. Consider a $\mathrm{U}(r)$-bundle P over a Riemannian 4-manifold X. The exact sequence

$$1 \longrightarrow S^1 \longrightarrow \mathrm{U}(r) \longrightarrow \mathrm{PU}(r) \longrightarrow 0$$

induces a decomposition

$$u(r) = i\mathbb{R} \oplus \mathrm{pu}(r)$$

of the Lie algebra into a central and a trace free part. The curvature form of a connection $A \in \mathscr{A}(P)$ splits accordingly

$$F_A = \frac{1}{r}\mathrm{tr}(F_A) \cdot \mathrm{id} + F_A^0$$

into a central component representing $-\frac{2\pi i}{r}c_1(P)$ and a trace free component F_A^0, which represents the curvature $F_{\overline{A}}$ of the induced $\mathrm{PU}(r)$-connection $\overline{A}$ on P/S^1. For the Yang-Mills action $YM_X^g(P)(A) = \|F_A\|^2$ (relative to an arbitrary metric g) this yields $\|F_A\|^2 = \frac{1}{r}\|\mathrm{tr}(F_A)\|^2 + \|F_A^0\|^2$ with $\|F_A^0\|^2 = \|F_{\overline{A}}\|^2 = \|F_{\overline{A}_+}\|^2 + \|F_{\overline{A}_-}\|^2$. Substituting $\|F_{\overline{A}_-}\|^2 - \|F_{\overline{A}_+}\|^2 = -i(P/S^1) = (c_2(P) - \frac{r-1}{2r}c_1^2(P))8\pi^2 \mathrm{vol}_g(X)$ we find

$$\|F_A\|^2 = \frac{1}{r}\|\mathrm{tr}(F_A)\|^2 - i(P/S^1) + 2\|F_{\overline{A}_+}\|^2 = \frac{1}{r}\|\mathrm{tr}(F_A)\|^2 + i(P/S^1) + 2\|F_{\overline{A}_-}\|^2 .$$

Since the central and the trace free part of the curvature can be minimised independently, we see that the set of absolute minima of $YM_X^g(P)$ consists of connections $A \in \mathscr{A}(P)$ for which $\mathrm{tr}(F_A)$ is harmonic and $F_{\overline{A}}$ self-dual or anti-self-dual, depending on the sign of $c_2 - \frac{r-1}{2r}c_1^2$. Thus for P to admit s.d. or a.s.d. connections the harmonic representative of $c_1(P)$ has to be self-dual or anti-self-dual. If the intersection form of X is indefinite and $c_1(P) \neq 0$, this will not be the case for a generic metric g. For other groups like e.g. $\mathrm{SO}(4)$ there are further topological restrictions coming from other characteristic classes.

A g-instanton on a bundle P is now defined as a connection which is either self-dual or anti-self-dual. The gauge group $\mathrm{Aut}(P)$ leaves the subset

$$\mathscr{A}_{\pm}(P) = \left\{ A \in \mathscr{A}(P) \mid *_g F_A = \pm F_A \right\}$$

of instantons invariant.

Definition. *The moduli space of g-instantons on P is the orbit space*

$$\mathscr{M}_X^g(P) := \mathscr{A}_{\pm}(P)/\mathrm{Aut}(P)\ .$$

Let $\mathscr{M}_X^g(P)^* := \mathscr{M}_X^g(P) \cap \mathscr{B}^*(P)$ be the subspace of irreducible instantons. Then $\mathscr{M}_X^g(P)^*$ has a nice description as the zero set of a section $F_{\pm}$ in the infinite dimensional vector bundle

$$\begin{array}{c} \mathscr{A}^*(P) \times_{\mathrm{Aut}^*(P)} A^2_{\pm}(\mathfrak{g}_P) \\ \downarrow \quad \uparrow_{F_{\pm}} \\ \mathscr{B}^* \quad (P) \end{array}$$

over $\mathscr{B}^*(P)$; $F_{\pm}$ is induced by the $\mathrm{Aut}^*(P)$-equivariant map

$$\mathscr{A}^*(P) \longrightarrow A^2_{\pm}(\mathfrak{g}_P)$$

sending A to $F_{A_{\pm}}$. It is this description which makes it possible to prove smoothness results for $\mathscr{M}_X^g(P)^*$, for sufficiently general g, by Sard's theorem.

There are some cases in which the instanton moduli space admits a simple global description. The most important one is the case of the principal SU(2)-bundle P over S^4 arising from the quarternionic Hopf fibration. In this case one finds that the moduli space of self-dual connections $\mathscr{M}_{S^4}^g(P)$ is naturally isomorphic to $\mathrm{SO}(5,1)^+/\mathrm{SO}(5) = \mathbb{B}^5$ (see [A/H/S], p. 458). Other cases are described in [A/D/H/M].

In general it is only possible to describe $\mathscr{M}_X^g(P)$ locally, to calculate the "expected dimension" and construct "universal bundles". In the sections below we explain this for a principal G-bundle P/X with $i(P) \leq 0$.

II.3 Properties of Instantons

1) Local Structure

To any a.s.d. connection A on P there is associated a fundamental elliptic complex

$$(*) \qquad 0 \longrightarrow A^0(\mathfrak{g}_P) \xrightarrow{D_A} A^1(\mathfrak{g}_P) \xrightarrow{D_{A_+}} A^2_+(\mathfrak{g}_P) \longrightarrow 0$$

($D_{A_+} \circ D_A = F_{A_+} = 0$ since A is anti-self-dual) with finite dimensional cohomology groups H^i_A. These groups have the following significance: $A^0(\mathfrak{g}_P)$ is the tangent space $T_{\mathrm{Aut}(P)}(\mathrm{id})$ of the gauge group at the identity. Identifying $A^1(\mathfrak{g}_P)$

with the tangent space $T_{\mathscr{A}(P)}(A)$ of $\mathscr{A}(P)$ at A, one sees that the operator $D_A : A^0(\mathfrak{g}_P) \to A^1(\mathfrak{g}_P)$ is just the differential of the orbit map

$$A \cdot \mathrm{Aut}(P) : \mathrm{Aut}(P) \longrightarrow \mathscr{A}(P)$$

at the identity. The kernel H_A^0 of D_A is therefore the tangent space to the isotropy group $\mathrm{Aut}(P)_A$ at the identity, whereas $\mathrm{Im}(D_A)$ corresponds to the tangent space $T_{A\cdot\mathrm{Aut}(P)}(A)$ of the orbit ([L], p. 25).

Another connection $A + \alpha$, $\alpha \in A^1(\mathfrak{g}_P)$ is anti-self-dual if and only if it satisfies the a.s.d. equation $(F_{A+\alpha})_+ = 0$, i.e.

$$D_{A_+}(\alpha) + [\alpha \wedge \alpha]_+ = 0\ .$$

The linearisation $D_{A_+}(\alpha) = 0$ of this equation describes the space $T_{\mathscr{A}_-(P)}(A)$ of tangents to curves in $\mathscr{A}_-(P)$ through A. Thus $H_A^1 = \ker(D_{A_+})/\mathrm{Im}(D_A)$ becomes the (Zariski) tangent space to $\mathscr{M}_X^g(P)$ at $[A]$ ([L], p. 48). In fact, if $\mathrm{Aut}^*(P)_A = 1$ and $H_A^2 = 0$, then $\mathscr{M}_X^g(P)$ is a manifold near $[A]$ with $T_{\mathscr{M}_X^g(P)}([A]) \cong H_A^1$. In general we have ([A/H/S], p. 446), [L], p. 48):

Theorem II.3.1. *Let A be an anti-self-dual connection on P. There is a neighbourhood U of 0 in H_A^1 and an $\mathrm{Aut}^*(P)_A$-equivariant differential map*

$$K_{[A]} : U \longrightarrow H_A^2$$

with $K_{[A]}(0) = 0$, such that $(K_{[A]}^{-1}(0)/\mathrm{Aut}^*(P)_A, 0)$ describes the germ of $\mathscr{M}_X^g(P)$ at $[A]$.

The proof of this "Kuranishi description" of $\mathscr{M}_X^g(P)$ is an adaptation of the classical arguments of Kodaira-Spencer and Kuranishi. An exposition can be found in Lawson's notes ([L], p. 48).

Example. Let P/X be a non-flat principal PU(2)-bundle over a manifold X with $H^1(X;\mathbb{Z}/2) = 0$. A connection $A \in \mathscr{A}(P)$ is reducible if and only if $\mathrm{Aut}(P)_A$ is non-trivial, in which case one has

$$\mathrm{Aut}(P)_A \cong S^1$$

([F/S], p. 530). This means that the adjoint bundle $\mathfrak{g}_P$ and the connection D_A on $\mathfrak{g}_P$ split in the form $\mathfrak{g}_P = L \oplus \varepsilon$, $D_A = D_L \oplus d$, where L is a unitary line bundle with a S^1-connection D_L and ε the trivial real line bundle with the trivial connection d. Suppose that A is a.s.d. relative to a metric g (notice that this can only happen if the harmonic representative of $c_1(L)$ is a.s.d.). In this case the fundamental complex (*) splits in two parts

$$\begin{array}{ccccccccc}
0 & \longrightarrow & A^0(L) & \xrightarrow{D_L} & A^1(L) & \xrightarrow{D_{L+}} & A_+^2(L) & \longrightarrow & 0 \\
 & & \oplus & & \oplus & & \oplus & & \\
0 & \longrightarrow & A^0 & \longrightarrow & A^1 & \longrightarrow & A_+^2 & \longrightarrow & 0
\end{array}$$

Assume for simplicity that D_{L_+} is surjective, i.e. $H_A^2 = \mathbb{H}_+^2(X)$, the space of self-dual harmonic forms. The action of $\mathrm{Aut}(P)_A \cong S^1$ on the fundamental complex induces a complex structure on $H_L^1 := \ker D_{L_+}/\mathrm{Im} D_L$, but is trivial on the lower complex.

The Kuranishi map

$$K_{[A]} : H_L^1 \longrightarrow \mathbb{H}_+^2(X)$$

is S^1-equivariant, i.e. it factors through the cone over the complex projective space $\mathbb{P}(H_L^1)$:

$$\widehat{K}_{[A]} : c\mathbb{P}(H_L^1) \longrightarrow \mathbb{H}_+^2(X) \; .$$

The zero-set of $\widehat{K}_{[A]}$ near the cone point is then a local model for $\mathcal{M}_X^{\mathfrak{g}}(P)$ near the reducible connection $[A]$.

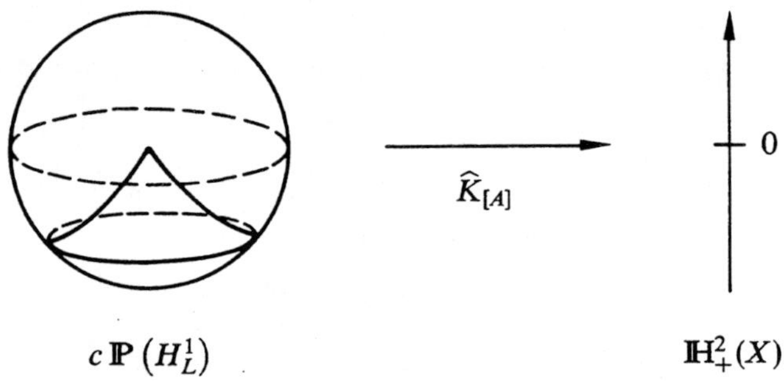

2) Dimension

The exact sequence of groups

$$1 \longrightarrow Z_G \longrightarrow G \longrightarrow \mathrm{Ad}(G) \longrightarrow 1$$

corresponds on the Lie algebra level to an equivariant splitting

$$\mathfrak{g} = z \oplus \overline{\mathfrak{g}}$$

into the centre z and the Lie algebra $\mathfrak{g}$ of the adjoint group $\mathrm{Ad}(G)$. For a given a.s.d. connection A on P this leads to a decomposition of the fundamental complex $(*)$

$$\begin{array}{ccccccccc} 0 & \longrightarrow & A^0(\mathfrak{g}_{\overline{P}}) & \xrightarrow{D_{\overline{A}}} & A^1(\mathfrak{g}_{\overline{P}}) & \xrightarrow{D_{\overline{A}_+}} & A_+^2(\mathfrak{g}_{\overline{P}} & \longrightarrow & 0 \\ & & \oplus & & \oplus & & \oplus & & \\ 0 & \longrightarrow & A^0 \otimes z & \xrightarrow{d \otimes \mathrm{id}} & A^1 \otimes z & \xrightarrow{d_+ \otimes \mathrm{id}} & A_+^2 \otimes z & \longrightarrow & 0 \end{array}$$

into the fundamental complex for the induced connection $\overline{A}$ on the $\mathrm{Ad}(G)$-bundle P/Z_G and a central part. The homology groups decompose accordingly and the second component of the Kuranishi map

$$K_{[A]} : U \longrightarrow H^2_{\overline{A}} \oplus \left(\mathbb{H}^2_+(X) \otimes z\right)$$

vanishes (its definition involves the Lie bracket). From these considerations we obtain the following formula for the virtual dimension of $\mathcal{M}^{\mathfrak{g}}_X(P)$ at a point $[A]$:

$$\dim_{[A]} \mathcal{M}^{\mathfrak{g}}_X(P) = \dim H^1_A - \dim H^2_{\overline{A}} - \dim H^0_{\overline{A}} .$$

Using the equation $\dim H^1_A = \dim H^1_{\overline{A}} + b_1(X) \cdot \dim Z_G$ and the Atiyah-Singer index theorem for the fundamental complex of $\overline{A}$ we find that

$$\dim_{[A]} \mathcal{M}^{\mathfrak{g}}_X(P) = -2p_1(\mathfrak{g}_P) - \dim \mathrm{Ad}(G)(1 - b_1(X) + b_+(X)) + b_1(X) \cdot \dim Z_G .$$

The right hand side of this formula is the "expected dimension" of the instanton moduli space.

Example. Consider a principal U(2)-bundle with Chern classes c_1, c_2 over an algebraic surface X. The formula for the expected dimension gives

$$\begin{aligned} &2(4c_2 - c_1^2) - 3(1 + b_+(X) - b_1(X)) + b_1(X) \\ &= 2(4c_2 - c_1^2) - 3(1 + 2p_g(X) - 2q(X)) + 2q(X) \\ &= 2(4c_2 - c_1^2 - 3\chi(\mathcal{O}_X) + q(X)) \end{aligned}$$

(compare [B/P/V], Chap. IV). We have met this expression already in I.3, at the end of 5).

3) Universal Bundles

Properly speaking, a universal bundle in the context of a.s.d.-connections would be something like a G-bundle $\mathfrak{P}$ on $\mathcal{M}^{\mathfrak{g}}_X(P)^* \times X$, provided with a connection B, such that, for every $[A] \in \mathcal{M}^{\mathfrak{g}}_X(P)^*$, the restriction of $\mathfrak{P}$ to $[A] \times X$ would be isomorphic to P, with the restriction of B being gauge equivalent to A. But in general such an object does not exist. However, it is still possible to construct a useful universal bundle in the following sense.

We take on $\mathcal{B}^*(P) \times X$ the bundle $\mathcal{A}^*(P) \times_{\mathrm{Aut}(P)} P/Z_G$, denoting it by $\widetilde{\mathbb{P}}/\mathcal{B}^*(P) \times X$. Its restriction to a fibre $[A] \times X$ is isomorphic to P/Z_G and comes with an a.s.d. connection gauge equivalent to A. But in general, this bundle can't be lifted to a G-bundle.

From this point on (for the rest of this section) we confine ourselves to the case $G = \mathrm{PU}(2)$ ($\cong \mathrm{SO}(3)$, the adjoint group for both U(2) and SU(2)) and suppose that $H^1(X;\mathbb{Z}/2) = 0$. Principal PU(2)-bundles P are classified by two characteristic classes, $w_2 \in H^2(X;\mathbb{Z}/2)$ and $p_1(P) \in H^4(X;\mathbb{Z})$.

P lifts to a U(2)-bundle $\widehat{P}$ if and only if w_2 is the reduction of an integral class (this is always the case if $H^1(X;\mathbb{Z}/2) = 0$) and it lifts to a SU(2)-bundle if and only if $w_2 = 0$. The Chern classes c_i of a U(2)-bundle $\widehat{P}$ with $P \cong \widehat{P}/S^1$ determine the characteristic classes of P by the formulas $w_2 = \overline{c}_1 \pmod 2$, $p_1 = c_1^2 - 4c_2$ ([H/H], p. 163).

The following fundamental theorem of K. Uhlenbeck describes some global properties of the instanton spaces $\mathcal{M}_X^g(P)$.

Theorem II.3.2 (Generic metrics). *Let P/X be a non-flat principal* PU(2)*-bundle over X. There is an open dense set of (generic) metrics g, such that $\mathcal{M}_X^g(P)$ has the following properties:*

i) $\mathcal{M}_X^g(P)^*$ is smooth,

ii) $\mathcal{M}_X^g(P) = \mathcal{M}_X^g(P)^*$ if $b_+(X) > 0$,

iii) $\mathcal{M}_X^g(P) \setminus \mathcal{M}_X^g(P)^*$ consists of a discrete set of points with neighbourhoods equivalent to cones on projective spaces if $b_+(X) = 0$.

This is essentially a transversality theorem ([F/U], p. 71, [F/U], p. 85), whose main ingredient is the Sard-Smale theorem ([F/U], p. 70). The reason for ii) is that the space $\mathbb{H}^2_-(X) \subset H^2_{\mathrm{DR}}(X)$ of a.s.d. 2-forms is a proper subspace which generically avoids a given set of integral classes $c_1(L)$ ([F/U], p. 73). The neighbourhood structure for a reducible point in case iii) follows from the example after Theorem II.3.1.

At the heart of vital compactness properties of instanton moduli spaces (in particular their compactifications) are two other theorems of Uhlenbeck, namely the bubble theorem ([U 2], p. 33) and the theorem on removable singularities ([U 1], p. 24).

Theorem II.3.3 (Bubble theorem). *Let $\{A_i\}$ be a sequence of anti-self-dual connections on P. Then either*

i) there is a subsequence $\{A_{i'}\}$ and a sequence $\{\widetilde{A}_{i'}\}$ of gauge equivalent connections converging to an anti-self-dual connection on P, or

ii) there exist finitely many points $x_1, \ldots, x_m \in X$, a subsequence $\{A_{i'}\}$ of $\{A_i\}$ and a sequence of gauge equivalent connections $\{\widetilde{A}_{i'}\}$ which converges to an anti-self-dual connection on the restriction $P|_{X\setminus\{x_1,\ldots,x_m\}}$ of P to $X \setminus \{x_1, \ldots, x_m\}$.

Theorem II.3.4 (Theorem on removable singularities). *Let A_0 be an anti-self-dual connection on a* PU(2)*-bundle P_0 over $X \setminus \{x_1, \ldots, x_m\}$ with $\|F_{A_0}\|^2 < \infty$. Then there exists a gauge transformation $f \in \mathrm{Aut}(P_0)$ such that f^*P_0 and f^*A_0 extend smoothly to a* PU(2)*-bundle P over X and an anti-self-dual connection A on P.*

This is actually a local result which applies more generally to Yang-Mills connections with finite action, i.e. connections A satisfying $D_A(*F_A) = 0$ and $-\int_{X\setminus\{x_1,\ldots,x_m\}} \mathrm{tr}(F_A \wedge *F_A) < \infty$.

As an important global consequence it shows that Yang-Mills connections on the euclidean space $\mathbb{R}^4$ with finite action extend via stereographic projection to Yang-Mills connections over S^4.

Using the preceding two theorems, it is not difficult to show ([F/S], p. 536):

Corollary II.3.5. *The space $\mathcal{M}^g_X(P)$ is always compact for* PU(2)*-bundles P for which either $-3 \leq p_1 \leq 0$ or $p_1 = -4$, and w_2 is not the pull-back of a class in $H^2\left(K(\pi_1(X),1);\mathbb{Z}/2\right)$.*

For instanton spaces $\mathcal{M}^g_X(P)$ with $p_1(P) \leq -4$ the same basic facts yield a natural compactification $\overline{\mathcal{M}}^g_X(P)$ of $\mathcal{M}^g_X(P)$ by "ideal connections". Intuitively this means: if a sequence $\{[A_i]\}$ in $\mathcal{M}^g_X(P)$ has no convergent subsequence, then a subsequence $\{[\widetilde{A}_{i'}]\}$ can be found, which "converges" to some point $([A'],\{x_1,\dots,x_m\}) \in \mathcal{M}^g_X(P') \times S^mX$. Here P' is a PU(2)-bundle with $w_2(P') = w_2(P)$ and $p_1(P') \geq p_1(P)$, S^mX the m-th symmetric product of X with $m = \frac{1}{4}(p_1(P') - p_1(P))$. These limiting "ideal connections" $([A'],\{x_1,\dots,x_m\})$ are the points in the compactification $\overline{\mathcal{M}}^g_X(P)$. More precisely, using the "right" topology on

$$\bigcup_{0\leq m\leq\left[\frac{-p_1}{4}\right]} \mathcal{M}^g_X(P') \times S^mX\,,$$

$\overline{\mathcal{M}}^g_X(P)$ can be defined as the closure of $\mathcal{M}^g_X(P)$ in this union ([D 3], p. 292).

This "Donaldson compactification" will play an important role in applications.

Two other results will also be needed. The first is due to Donaldson ([D 6], p. 36).

Theorem II.3.6 (Orientation). *The moduli spaces $\mathcal{M}^g_X(P)^*$ are orientable for a generic metric g. An orientation is given by the choice of an integral lift of w_2 and an orientation of $\mathbb{H}^2_+(X)$.*

Let $c \in H^2(X;\mathbb{Z})$ be a lift of w_2; another lift $c + 2l$, together with a fixed orientation of $\mathbb{H}^2_+(X)$, defines the same orientation if and only if $l^2 \equiv 0 \pmod 2$. Choosing the opposite orientation on $\mathbb{H}^2_+(X)$ reverses the orientation of $\mathcal{M}^g_X(P)^*$ ([D 5], p. 11).

So far we do not know on which bundles P/X instantons actually exist. For principal SU(2)-bundles Taubes ([TA], p. 518) has proved the following general existence theorem:

Theorem II.3.7 (Existence). *Let P/X be an* SU(2)*-bundle with second Chern class $c_2 \geq 0$ over an oriented, closed Riemannian 4-manifold (X,g). Then*

$$\mathcal{M}^g_X(P)^* \neq \emptyset$$

if

$$c_2 \geq \begin{cases} 4 & \text{for } b_+(X) = 2 \\ \max\left(\frac{4}{3}b_+(X), 1\right) & \text{otherwise} \end{cases} .$$

III. Hermite-Einstein Structures

III.1 Simple Bundles

Let E/X be a differentiable complex vector bundle over a smooth, compact complex surface X with $c_1(E)$ algebraic, i.e. $c_1(E) \in \mathrm{NS}(X)$. Let $\wedge^{p,q}$ be the bundle of (p,q)-forms and let $A^{p,q}(E)$ be the space of sections in $\wedge^{p,q} \otimes E$ over X. We shall need the differential-geometric construction of the moduli space $\mathcal{M}^s_X(E)$ of simple holomorphic structure on E [(2)]. It goes as follows.

A holomorphic bundle $\mathcal{E}$ differentiably equivalent to E, induces a canonical differential operator

$$\overline{\partial}_{\mathcal{E}} : A^0(E) \longrightarrow A^{0,1}(E)$$

satisfying the $\overline{\partial}$-Leibniz rule

$$\overline{\partial}_{\mathcal{E}}(\varphi \cdot s) = \overline{\partial}\varphi \otimes s + \varphi \cdot \overline{\partial}_{\mathcal{E}}(s) \ ,$$

such that (using the same symbol for the extension to $A^{p,q}(E)$)

$$\overline{\partial}^2_{\mathcal{E}} = 0 \ .$$

Conversely, let $\overline{\partial}_A$ be a semi-connection on E, i.e. a differential operator

$$\overline{\partial}_A : A^0(E) \longrightarrow A^{0,1}(E) \ ,$$

which satisfies the $\overline{\partial}$-Leibniz rule. If $\overline{\partial}_A$ is integrable in the sense that

$$\overline{\partial}^2_A = 0 \ ,$$

then there exists a holomorphic bundle $\mathcal{E}_A$, C^∞-equivalent to E, such that

$$\overline{\partial}_{\mathcal{E}_A} = \overline{\partial}_A \ .$$

This statement is the vector bundle version of the Newlander-Nirenberg theorem ([A/H/S], p. 439).

Let

$$\overline{\mathcal{A}}(E) = \overline{\partial}_{A_0} + A^{0,1}(\mathrm{End}\, E)$$

be the affine space of semi-connections and denote by

$$\mathcal{H}(E) := \left\{ \overline{\partial}_A \in \overline{\mathcal{A}}(E) \,|\, \overline{\partial}^2_A = 0 \right\}$$

the subset of holomorphic structures.

Two holomorphic structures $\overline{\partial}_A, \overline{\partial}_{A'}$ define holomorphically equivalent bundles $\mathcal{E}_A, \mathcal{E}_{A'}$ if and only if there exists an $f \in \mathrm{Aut}(E)$ such that

$$\overline{\partial}_{A'} \circ f = f \circ \overline{\partial}_A \ .$$

---

(2) Compare the footnote at p. 201.

Consider the subset

$$\mathscr{S}(E) := \{\overline{\partial}_A \in \mathscr{H}(E) \mid H^\circ(X; \mathrm{End}\mathscr{E}_A) = \mathbb{C}\mathrm{id}_E\}$$

of simple holomorphic structures on E. The natural action of the group

$$\mathrm{Aut}^*(E) = \mathrm{Aut}(E)/\mathbb{C}^* \cdot \mathrm{id}_E$$

on $\mathscr{H}(E)$ leaves $\mathscr{S}(E)$ invariant, restricting to a free action on this set. The orbit space

$$\mathscr{M}^s_X(E) := \mathscr{S}(E)/\mathrm{Aut}^*(E)$$

is the modul space of simple holomorphic structure on E.

Theorem III.1.1 ([K 2], p. 261, [L/O 1], p. 671). *$\mathscr{M}^s_X(E)$ is a locally Hausdorff complex space of finite dimension.*

The natural complex structure on $\mathscr{M}^s_X(E)$ is in general not reduced. Local models can be described as zero sets of (germs of) Kuranishi maps

$$K_{[\overline{\partial}_A]} : H^1(X; \mathrm{End}\mathscr{E}_A) \longrightarrow H^2(X; \mathrm{End}\mathscr{E}_A) .$$

III.2 Hermite-Einstein Bundles

From this point on we assume that X supports a Kähler metric g with Kähler form $\omega = \omega_g \in A^{1,1}_{\mathbb{R}}$.

Choose a Hermitian metric h on E and let P be the corresponding principal $\mathrm{U}(r)$-bundle. We want to describe the set of absolute minima of the Yang-Mills functional

$$YM^g_X(P) : \mathscr{A}(P) \longrightarrow \mathbb{R}$$

associated to these data.

From the first example in Sect. II.2 we know, that this set consists of connections A, for which $\mathrm{tr}(F_A)$ is harmonic and $F^0_A = F_A - \frac{1}{r}\mathrm{tr}(F_A) \cdot \mathrm{id}_E$ self-dual or anti-self-dual.

We treat the anti-self-dual case.

The choice of the unitary structure on E defines a map

$$\mathscr{A}(P) \longrightarrow \overline{\mathscr{A}}(E)$$

sending a Hermitian connection A to the $(0,1)$-component of the operator

$$d_A : A^0(E) \longrightarrow A^1(E) .$$

This map is bijective and identifies the set $\mathscr{H}(E)$ of holomorphic structures with the set

$$\mathscr{A}^{1,1}(P) := \{A \in \mathscr{A}(P) \mid F_A \in A^{1,1}(\mathrm{End}\, E)\}$$

of "Chern connections". The Kähler metric g on the other hand splits the bundle $\wedge^2_{\mathbb{C}}$ of $\mathbb{C}$-valued 2-forms into an orthogonal sum

$$\wedge^2_{\mathbb{C}} = \wedge_+ \otimes \mathbb{C} \oplus \wedge_- \otimes \mathbb{C}$$

with

$$\wedge_+ \otimes \mathbb{C} = \wedge^{2,0} \oplus \wedge^{0,2} \oplus \wedge^0_{\mathbb{C}} \cdot \omega$$

and

$$\wedge_- \otimes \mathbb{C} = \wedge^{1,1}_{\perp}$$

consisting of primitive forms, i.e. $(1,1)$-forms orthogonal to ω. Let

$$F_A = F_A^{2,0} + F_A^{0,2} + \widehat{F}_A \cdot \omega + F_{A\perp}^{1,1}$$

be the corresponding decomposition of curvature forms. The endomorphism $i\widehat{F}_A$, the component parallel to ω, is the Ricci curvature of A.

Definition. *A connection $A \in \mathscr{A}^{1,1}(P)$ is a Hermite-Einstein connection if its Ricci curvature is a constant multiple of the identity:*

$$i\widehat{F}_A = c \cdot \mathrm{id}_E \ .$$

In other words, A is Hermite-Einstein (H.-E.) if and only if the curvature

$$F_A = \frac{1}{r}\mathrm{tr}(F_A) \cdot \mathrm{id}_E + F_A^0$$

has a harmonic central part $\frac{1}{r}\mathrm{tr}(F_A) \cdot \mathrm{id}_E$ of type (1,1), and a trace-free component F_A^0, which is anti-self-dual.

Definition. *A Hermite-Einstein bundle over a Kähler surface is a vector bundle E which admits a H.-E. connection A for some Hermitian metric h.*

If this is the case, then we say that h is a Hermite-Einstein metric on the holomorphic bundle $\mathscr{E}_A$ associated to A.

This concept has been introduced by Kobayashi ([K 1]) as a generalisation of the notion of a Kähler-Einstein metric. Kobayashi showed that H.-E. bundles are not Bogomolov-unstable (K 1]), but it soon became clear that his concept was more closely related to stability in the sense of Mumford and Takemoto.

Let us first note some simple properties of H.-E. bundles.

a) Lübke's Inequality

Theorem III.2.1. *If E/X is a Hermite-Einstein bundle over X, then*

$$c_2(E) - \frac{r-1}{2r}c_1^2(E) \geq 0 \ ,$$

with equality if and only if $\mathbb{P}(E)$ admits a flat $\mathrm{PU}(r)$-structure.

The proof is very easy (cf. [LU 1], p. 139 for the original proof). Namely, the trace free part F_A^0 of a unitary connection $A \in \mathscr{A}(P)$ represents the curvature $F_{\overline{A}}$ of the induced PU(r)-connection $\overline{A}$ on P/S^1.

Now the Chern-Weil equality

$$\left\| F_{\overline{A}_-} \right\|^2 - \left\| F_{\overline{A}_+} \right\|^2 = \left(c_2(E) - \frac{r-1}{2r} c_1^2(E) \right) 8\pi^2 \operatorname{vol}_g(X)$$

immediately implies the above inequality for projectively anti-self-dual connections.

In fact, the Hermite-Einstein condition characterises exactly the absolute minima of $YM_X^g(P)$ for an U(r)-bundle P with $c_1(P) \in \mathrm{NS}(X)$ and $c_2(P) - \frac{r-1}{2r} c_1^2(P) \geq 0$ over a Kähler surface.

b) Kobayashi's Vanishing Theorem

The constant c appearing in the definition of H.-E. connections is determined by cohomological invariants of (X, g) and E:

$$c = \frac{\deg_\omega(E)}{r} \cdot \frac{\pi}{\operatorname{vol}_g(X)},$$

where the real number

$$\deg_\omega(E) = \langle c_1(E) \cup [\omega], [X] \rangle$$

is the ω-degree of E. The following basic vanishing theorem, due to Kobayashi (K 1], p. 10) explains its significance.

Theorem III.2.2. *Suppose that a bundle E/X admits a Hermite-Einstein connection A with constant c. If $c < 0$, then $H^0(X; \mathscr{E}_A) = 0$. If $c = 0$, then every holomorphic section of $\mathscr{E}_A$ is parallel.*

The proof follows immediately from the formula

$$\overline{\Delta}_A = \frac{1}{2} \left(\Delta_A - i\widehat{F}_A \right)$$

for the Laplacians of $\overline{\partial}_A$ and d_A.

A Hermite-Einstein connection is called *reducible* if its holonomy is contained in a subgroup of type U$(k) \times$ U$(r-k)$ for some $0 < k < r$. This means that the bundle splits as an orthogonal sum of Hermite-Einstein bundles with the same constant. Irreducible H.-E. connections A define simple holomorphic bundles $\mathscr{E}_A$ since by Kobayashi's vanishing theorem every holomorphic section of End $\mathscr{E}_A$ must be parallel.

III.3 Hermite-Einstein Bundles and Stability

ω-Stability

The ω-degree of a coherent torsion free sheaf $\mathscr{F}$ over X is defined as the ω-degree of its determinant line bundle ([K 2], p. 166).

Definition. *A holomorphic r-bundle $\mathscr{E}$ over a Kähler surface (X,g) is ω-stable (ω-semi-stable) if every coherent subsheaf $\mathscr{F} \subset \mathscr{E}$ with $0 < \mathrm{rk}(\mathscr{F}) < r$ satisfies*

$$\frac{\deg_\omega(\mathscr{F})}{\mathrm{rk}(\mathscr{F})} < \frac{\deg_\omega(\mathscr{E})}{r} \qquad \left(\frac{\deg_\omega(\mathscr{F})}{\mathrm{rk}(\mathscr{F})} \leq \frac{\deg_\omega(\mathscr{E})}{r}\right).$$

If (X,g) is a projective-algebraic surface, whose metric is induced by an embedding $X \subset \mathbb{P}_N$, then ω-stability is clearly equivalent to μ-stability with respect to a hyperplane section H of X.

Theorem III.3.1 (Kobayashi, Lübke). *Hermite-Einstein bundles are ω-semi-stable. They are either ω-stable or direct sums of ω-stable Hermite-Einstein bundles with the same degree/rank ratio.*

Here is the idea of the proof ([LU 2], p. 245, [K 2], p. 179):

A coherent subsheaf $\mathscr{F} \subset \mathscr{E}$ of rank s induces a non-trivial morphism

$$\det \mathscr{F} \longrightarrow \wedge^s \mathscr{E},$$

which can be interpreted as a non-zero section in the H.-E. bundle $\wedge^s\mathscr{E} \otimes (\det \mathscr{F})^\vee$. The constant of this bundle is

$$\left[\frac{s \deg_\omega(\mathscr{E})}{r} - \deg_\omega(\mathscr{F})\right] \cdot \frac{\pi}{\mathrm{vol}_g(X)},$$

which, according to Kobayashi's vanishing theorem, must be ≥ 0. This already proves ω-semi-stability. If equality holds, then $\mathscr{F}$ defines a parallel holomorphic subbundle with a holomorphic orthogonal complement.

The converse has been conjectured independently by Kobayashi and Hitchin.

Of course, ω-stable bundles and Hermite-Einstein metrics can be defined on compact Kähler manifolds of arbitrary dimensions. The proof of the Kobayashi-Lübke theorem remains valid in this situation and one has the general

Conjecture (Kobayashi, Hitchin). *Every ω-stable vector bundle over a compact Kähler manifold admits a Hermite-Einstein metric.*

Notice that in the case of a 1-dimensional base, i.e. for bundles over Riemann surfaces, the conjecture already follows from a theorem of Narasimhan-Seshadri, who showed that stable bundles over Riemann surfaces correspond to irreducible $PU(r)$-representations of the fundamental group ([N/S], p. 560). A direct proof of this result has been proposed by Donaldson ([D 1], p. 269). Later Donaldson established the Kobayashi-Hitchin conjecture for algebraic surfaces ([D 2], p. 1).

Theorem III.3.2 (Donaldson). *A holomorphic vector bundle over a projective surface $X \subset \mathbb{P}_N$ is μ-stable with respect to a hyperplane section, if and only if it admits an irreducible Hermite-Einstein connection with respect to the induced Hodge metric.*

Afterwards Donaldson extended his proof to projective manifolds of arbitrary dimensions. The general (Kähler) case has been treated by Yau and Uhlenbeck ([U/Y]). The Bourbaki article of Margerin ([MAG]) contains an outline of the proofs.

We want to interpret these results in terms of moduli spaces. Let

$$\mathscr{A}^{1,1}_{\mathrm{H.-E.}}(P) := \left\{ A \in \mathscr{A}^{1,1}(P) \mid i\widehat{F}_A = c \cdot \mathrm{id}_E \right\}$$

be the set of Hermite-Einstein connections on the $U(r)$-bundle P. The unitary gauge group $\mathrm{Aut}(P)$ acts on $\mathscr{A}^{1,1}_{\mathrm{H.-E.}}(P)$, leaving the subset $\mathscr{A}^{1,1}_{\mathrm{H.-E.}}(P)^*$ of irreducible H.-E. connections invariant. Thus the unitary structure on E induces a well-defined map

$$\Phi_h : \mathscr{A}^{1,1}_{\mathrm{H.-E.}}(P)^*/\mathrm{Aut}(P) \longrightarrow \mathscr{M}^s_X(E)$$

sending $[A]$ to $[\bar{\partial}_A]$.

Φ_h is an open embedding whose image does not depend on the choice of the Hermitian metric h on E ([K 2], p. 266, [L/O 1], p. 673). We denote the image by

$$\mathscr{M}_{\mathrm{H.-E.}}(E)^* .$$

This is the moduli space of irreducible Hermite-Einstein structures on E.

Now let $X \subset \mathbb{P}_N$ be a projective surface with Hodge metric g and hyperplane section H. Then Donaldson's solution of the Kobayashi-Hitchin conjecture together with the theorem of Kobayashi-Lübke imply the existence of a diffeomorphism

$$\mathscr{M}_{\mathrm{H.-E.}}(E)^* \cong \mathscr{M}^H_X(E) .$$

Actually, one also needs a result of Miyajima ([MI], p. 314), showing that the moduli space $\mathscr{M}^s_X(E)$ of simple holomorphic structures on E is complex-analytically equivalent to the complex space underlying the corresponding algebraic moduli space of algebraic simple bundles. The latter space has been constructed by Altman and Kleiman ([A/K]), p. 99).

As a consequence of this equality we have the following "algebraic interpretation" of $PU(2)$-instanton spaces ([O/V 2], p. 608).

Corollary III.3.3. *Let E/X be a differentiable vector bundle of rank 2 over a projective surface X. Suppose that $H^1(X;\mathbb{Z}/2) = 0$ and $c_1(E) \in \mathrm{NS}(X)$. Then* *every choice of a Hermitian metric h on E with corresponding* U(2)*-bundle P induces a diffeomorphism*

$$\mathcal{M}_X^g(P/S^1)^* \cong \mathcal{M}_X^H(E) .$$

Moreover, if $\mathcal{M}_X^H(E)$ is smooth, then the natural orientation of $\mathcal{M}_X^g(P/S^1)^*$ coincides with the complex orientation of $\mathcal{M}_X^H(E)$ ([D 6], p. 38). And if $\mathcal{M}_X^H(E)$ is a fine moduli space with a universal bundle $\mathfrak{E}$, then the projectivised bundle $\mathbb{P}(\mathfrak{E})$ is equivalent to the universal PU(2)-bundle $\widetilde{\mathbb{P}}$ over $\mathcal{M}_X^g(P/S^1)^* \times X$.

A modified version of the preceding result also holds without the assumption on $H^1(X;\mathbb{Z}/2)$.

IV. The Donaldson Invariants

IV.1 General Principles

In this section we describe the construction of differential-topological invariants by Donaldson and others.

The following topological considerations provide the general principle underlying Donaldson's method. Fix a non-flat principal bundle P with structure group SU(2) or PU(2) over a closed oriented, smooth 4-manifold X with $b_+(X) \equiv 1 \pmod 2$. We shall assume that $\pi_1(X) = 1$ (this is not necessary, but simplifies many arguments).

The first Pontrjagin class $p_1(\widetilde{\mathbb{P}}) \in H^4(\mathcal{B}^*(P) \times X;\mathbb{Z})$ of the universal bundle $\widetilde{\mathbb{P}}/\mathcal{B}^*(P) \times X$ induces a homomorphism

$$\mu : H_2(X;\mathbb{Z}) \longrightarrow H^2(\mathcal{B}^*(P);\mathbb{Z}) ,$$

obtained by decomposing $p_1(\widetilde{\mathbb{P}})$ into its Künneth components; in other words, μ is defined by the usual slant product ([SP], p. 351).

$$/ : H^4(\mathcal{B}^*(P) \times X;\mathbb{Z}) \otimes H_2(X;\mathbb{Z}) \longrightarrow H^2(\mathcal{B}^*(P);\mathbb{Z})$$

by setting $\mu(c) := p_1(\widetilde{\mathbb{P}})/c$.

Let g be a generic metric on X. Then $\mathcal{M}_X^g(P)^* \subset \mathcal{B}^*(P)$ is smooth by Theorem II.3.2. We also choose an orientation $\mathfrak{o}$ for $\mathcal{M}_X^g(P)^*$, and assume that it defines a fundamental class

$$[\mathcal{M}_X^g(P)^*] \in H_{2d}(\mathcal{B}^*(P);\mathbb{Z}) ,$$

where $d = -p_1(P) - \frac{3}{2}(1 + b_+(X))$. Then we can evaluate

$$\langle \mu(c_1) \cup \ldots \cup \mu(c_d), [\mathcal{M}_X^g(P)^*] \rangle$$

and in this way construct a polynomial function

$$\gamma_{X,\mathfrak{o}}^g(P) : S^d H_2(X;\mathbb{Z}) \longrightarrow \mathbb{Z} .$$

There are special circumstances under which this construction actually works, giving well-defined C^∞-invariants, independent of the metric g ([KOT, O/V 2]).

In general however, one faces two basic difficulties:

(i) The moduli spaces $\mathcal{M}^g_X(P)^*$ are not compact in most cases, and therefore do not define fundamental classes in the usual sense.

(ii) The definition of $\mathcal{M}^g_X(P)^*$ depends on the choice of a Riemannian structure on X, and not only on the differential topology.

Both difficulties can be overcome, but it is necessary to distinguish between several cases depending on the invariants $b_+(X)$, $w_2(P)$ and $p_1(P)$.

In fact, there is a certain "stable range" of the form

$$-p_1(P) > 3(1 + b_1(X)) + 2 + \delta(w_2) ,$$

in which the expression $\langle \mu(c_1) \cup \ldots \cup \mu(c_d), \left[\mathcal{M}^g_X(P)^*\right]\rangle$ is always well-defined; outside of this range the definition of $\gamma^g_{X,\mathfrak{o}}(P)$ has to be modified.

The second problem leads to a basic distinction between the special case $b_+(X) = 1$ and the general situation $b_+(X) > 1$.

For $b_+(X) > 1$ the polynomials $\gamma^g_{X,\mathfrak{o}}(P)$ are actually independent of the metric and therefore invariants of the differential structure. In the case $b_+(X) = 1$ however, they do depend on g. This dependence leads to a more complicated type of C^∞-invariant.

IV.2 The Definition of the Polynomials

The idea for the definition of $\langle \mu(c_1) \cup \ldots \cup \mu(c_d), \left[\mathcal{M}^g_X(P)^*\right]\rangle$ is very simple ([D 5], p. 14). The classes $\mu(c_i) \in H^2(\mathcal{B}^*(P);\mathbb{Z})$ are represented geometrically by zero-sets V_{c_i} of the corresponding complex line bundles, such that all intersections

$$V_{c_{i_1}} \cap \ldots \cap V_{c_{i_k}} \cap \mathcal{M}^g_X(P)^*$$

are transverse of codimension $2k$ in $\mathcal{M}^g_X(P)^*$.

Now, if

$$V_{c_1} \cap \ldots \cap V_{c_d} \cap \mathcal{M}^g_X(P)^*$$

is finite, then, using the orientation of $\mathcal{M}^g_X(P)^*$ and the natural orientation of the normal bundles of the V_{c_i}'s to attach a sign ± 1 to the points in $V_{c_1} \cap \ldots \cap V_{c_d} \cap \mathcal{M}^g_X(P)^*$, we define $\langle \mu(c_1) \cup \ldots \cup \mu(c_d), \left[\mathcal{M}^g_X(P)^*\right]\rangle$ as the algebraic sum of these (oriented) points.

To carry out this idea, a condition is required which guarantees the compactness of $V_{c_1} \cap \ldots \cap V_{c_d} \cap \mathcal{M}^g_X(P)^*$. This condition implies that the method only works in a certain range. Donaldson obtained such a criterion using a "dimension counting argument" (D 5], p. 13).

A less geometric way to arrive at his condition is the following ([D 8]): the complex line bundles $L_{c_i}/\mathcal{M}^g_X(P)^*$, which are defined by the restrictions of the classes $\mu(c_i)$ to $\mathcal{M}^g_X(P)^*$, extend naturally to line bundles $\overline{L}_{c_i}$ over the Donaldson compactification $\overline{\mathcal{M}}^g_X(P)$. This is a stratified space with $\mathcal{M}^g_X(P)^*$ as open stratum. If the lower strata of $\overline{\mathcal{M}}^g_X(P)$ have codimension at least 2, then $\overline{\mathcal{M}}^g_X(P)$ carries a fundamental class, so that

$$\langle \mu(\overline{L}_{c_1}) \cup \ldots \cup \mu(\overline{L}_{c_d}), \left[\overline{\mathcal{M}}^g_X(P)\right] \rangle$$

is well-defined; in other words, we need that

$$\text{codim } \left(\overline{\mathcal{M}}^g_X(P) \setminus \mathcal{M}^g_X(P)^*\right) \geq 2 .$$

This condition defines the stable range.

Example. Consider a PU(2)-bundle P/X with characteristic classes w_2, p_1 and fix again a generic metric g on X. The dimension of a stratum

$$\overline{\mathcal{M}}^g_X(P) \cap \mathcal{M}^g_X(P') \times S^m X$$

corresponding to a non-flat bundle P' with $p_1(P') > p_1$ (and $m = \frac{1}{4}(p_1(P') - p_1)$) is at most

$$-2p_1(P') - 3(1 + b_+(X)) + 4m ;$$

its codimension is therefore at least 4. If P' is flat however, $\mathcal{M}^g_X(P')$ is isomorphic to the space

$$\text{Hom}^{W_2}(\pi_1(X), \text{PU}(2))/\text{conj}.$$

of conjugacy classes of representations of $\pi_1(X)$ in PU(2) with second Stiefel-Whitney class w_2. For a general fundamental group this is a compact space whose dimension $\delta(\pi_1(X), w_2)$ may be large ([B/O]).

For $\pi_1(X) = 1$ it is empty if $w_2 \neq 0$ and consists of a single point $[\theta]$ if $w_2 = 0$. The condition $\dim(\overline{\mathcal{M}}^g_X(P) \setminus \mathcal{M}^g_X(P)^*) \geq 2$ is therefore empty if $w_2 \neq 0$ and equivalent to

$$-p_1 \geq 3(1 + b_+(X)) + 2$$

if $w_2 = 0$.

Outside of the stable range there is another way to make sense of the expression

$$\langle \mu(c_1) \cup \ldots \cup \mu(c_d), \left[\mathcal{M}^g_X(P)^*\right] \rangle .$$

The idea is to compactify $\mathcal{M}^g_X(P)^*$ by "truncating" it and then introduce a compactifying tail at infinity. This has to be done in a way which works for all generic metrics. The method has been used by Donaldson in the construction of his Γ-invariant ([D 4]).

Example. Let P/X be a SU(2)-bundle with $c_2(P) = 1$ over a (1-connected) manifold X with $b_+(X) = 1$ and $b_-(X) > 0$.

The instanton space $\mathcal{M}^g_X(P)^*$ will be either empty or smooth of dimension 2 for a generic metric g, but usually not compact. Its end can be described in the following way:

Let $\omega_g \in \mathbb{H}^2_+(X) \setminus \{0\}$ be a non-zero self-dual 2-form and denote by $Z(\omega_g) \subset X$ the zero-set of the corresponding section in Λ_+.

If g is generic, then $Z(\omega_g)$ will be a smooth curve, which is independent of the choice of ω_g since $b_+(X) = 1$. Taubes' construction in the proof of his existence Theorem II.3.5 ([TA], p. 520) now shows that the end of $\mathcal{M}^g_X(P)^*$ is modelled after $Z(\omega_g) \times (0,1)$. More precisely: there is a map

$$\tau : X \times (0,1) \longrightarrow \mathcal{B}^*(P)$$

mapping $Z(\omega_g) \times (0,1)$ diffeomorphically onto an open subset of $\mathcal{M}^g_X(P)^*$, such that the complement $\mathcal{M}^g_X(P)^* \setminus \tau(Z(\omega_g) \times (0,1))$ is compact.

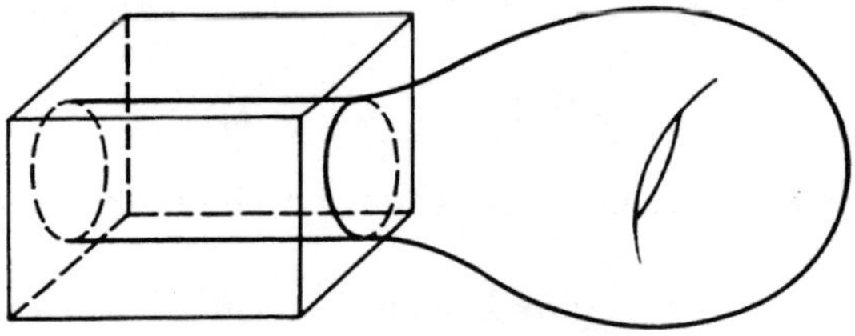

$\tau(X \times (0,1))$ $\qquad$ $\mathcal{M}^g_X(P)^* \setminus \tau(Z(\omega_g) \times (0,1))$

If we set $\tau_1 = \tau|X \times 1$, then the truncated moduli space

$$\widetilde{\mathcal{M}}^g_X(P) := \mathcal{M}^g_X(P)^* \setminus \tau(Z(\omega_g) \times (0,1))$$

is a compact surface with boundary $\tau_1(Z(\omega_g))$, so that the choice of an orientation defines a relative fundamental class

$$\left[\widetilde{\mathcal{M}}^g_X(P)\right] \in H_2(\mathcal{B}^*(P), \tau_1(Z(\omega_g)); \mathbb{Z}) .$$

Let

$$e \in H_2(X, Z(\omega_g); \mathbb{Z})$$

be the Lefschetz dual of the Euler class of the quotient bundle $\Lambda_+ / \Lambda^0 \cdot \omega_g$ over $X \setminus Z(\omega_g)$. A simple homology diagram argument shows that the corrected class $2\left[\widetilde{\mathcal{M}}^g_X(P)\right] - \tau_{1*}(e)$ uniquely determines an element

$$[\mathcal{M}^g] \in H_2(\mathcal{B}^*(P); \mathbb{Z}) .$$

Evaluation of $\mu(C)$ on $[\mathcal{M}^g]$ yields a well-defined linear form

$$\gamma^g_{X,\mathfrak{o}}(P) : H_2(X; \mathbb{Z}) \longrightarrow \mathbb{Z} ,$$

i.e. an element of $H^2(X, \mathbb{Z})$ since $\pi_1(X) = 1$.

IV.3 The Dependence on the Metric

Now suppose that we have polynomials

$$\gamma^g_{X,\mathfrak{o}}(P) : S^d H_2(X;\mathbb{Z}) \longrightarrow \mathbb{Z}$$

for any generic metric g.

In order to construct C^∞-invariants one has to study the dependence of the functions $\gamma^g_{X,\mathfrak{o}}(P)$ on the metric. This can be done by analysing the behaviour of instanton spaces $\mathcal{M}^{g_t}_X(P)$ in 1-parameter families of metrics. If $b_+(X) > 1$, then any path of metrics $(g_t)_{t\in[0,1]}$, connecting generic metrics g_0 and g_1, can be approximated arbitrarily close by a path $\widetilde{g}_t$ with the same end points, such that, for a non-flat bundle P, every

$$\mathcal{M}^{\widetilde{g}_t}_X(P)\ , \qquad t \in [0,1]$$

is contained in $\mathcal{B}^*(P)$. The new path can even be chosen in such a way that

$$\mathcal{M}^{\widetilde{g}}_X(P) := \bigcup_{t\in[0,1]} \mathcal{M}^{\widetilde{g}_t}_X(P) \times \{t\} \subset \mathcal{B}^*(P) \times [0,1]$$

is a manifold with boundary

$$\partial\mathcal{M}^{\widetilde{g}}_X(P) = \left(\mathcal{M}^{g_0}_X(P)\right) \amalg \left(-\mathcal{M}^{g_1}_X(P)\right) .$$

The same can be achieved simultaneously for all non-flat bundles P' which occur in the Donaldson compactification. This implies that $\gamma^{g_0}_{X,\mathfrak{o}}(P) = \gamma^{g_1}_{X,\mathfrak{o}}(P)$ in the stable range, if $b_+(X) > 1$ ([D 5], p. 8 *and* p. 15).

In Sect. II (Theorem II.3.6) we have seen that an orientation $\mathfrak{o}$ of an instanton space $\mathcal{M}^g_X(P)^*$ is determined by the choice of an orientation of $\mathbb{H}^2_+(X)$ and the choice of an integral lift c of $w_2(P)$.

Definition. *A H_+-orientation of X is an orientation $\mathfrak{o}_+$ of a maximal positive subspace $H_+ \subset H^2_{DR}(X)$ (maximal with respect to the intersection form).*

Since the space of all such subspaces is contractible, the choice of H_+ is not essential.

Theorem IV.3.1 (Donaldson). *Let X be a 1-connected, closed, oriented differentiable 4-manifold with $b_+(X) = 2p+1 > 1$. Let P be a non-flat principal* PU(2)*-bundle in the stable range. Then, associated to any choice of an H_+-orientation $\mathfrak{o}_+$ and an integral lift c of $w_2(P)$, there is a polynomial*

$$\gamma_{X,c,\mathfrak{o}_+}(P) : S^d H_2(X;\mathbb{Z}) \longrightarrow \mathbb{Z}$$

of degree $d = -p_1(P) - 3(1+p)$ with the following properties:

i) $\gamma_{X,c,-\mathfrak{o}_+}(P) = -\gamma_{X,c,\mathfrak{o}_+}(P)$;

ii) $\gamma_{X,c+2l,\mathfrak{o}_+}(P) = \varepsilon(l)\gamma_{X,c,\mathfrak{o}_+}(P)$ *where*

$$\varepsilon(l) = \begin{cases} 1 & \text{if } \bar{l}^2 = 0 \\ -1 & \text{if } \bar{l}^2 \neq 0 \, ; \end{cases}$$

iii) If $f : X' \to X$ *is an orientation-preserving diffeomorphism, then*

$$\gamma_{X',f^*c,f^*\mathfrak{o}_+}(f^*P) = f^*\gamma_{X,c,\mathfrak{o}_+}(P) \, .$$

If $w_2(P) = 0$, *i.e. if* P *is induced by a* SU(2)*-bundle* $\widehat{P}_k$ *with* $p_1(P) = -4c_2(\widehat{P}_k) = -4k$, *then the obvious choice* $c = 0$ *leads to invariants*

$$\gamma^k_{X,\mathfrak{o}_+} : S^d H_2(X;\mathbb{Z}) \longrightarrow \mathbb{Z}$$

of degree $d = 4k - 3(1+p)$, *indexed by the integers* $k > \frac{3}{2}(1+p)$ *if* $b_+(X) > 1$ ([D 5], p. 18).

IV.4 The Case $b_+ = 1$

In the special case $b_+(X) = 1$ new features occur, due to the appearance of reducible connections in generic 1-parameter families $\mathcal{M}^{\tilde{g}_t}_X(P')$ of instanton spaces.

A transversality argument, similar to the one used above, shows that in this case there will usually exist finitely many values of t, for which $\mathcal{M}^{\tilde{g}_t}_X(P')$ is not entirely contained in $\mathcal{B}^*(P')$. This happens at every t for which a topological reduction of P' admits a $\tilde{g}_t$-anti-self-dual connection; in other words, it happens if for some t there exists an integral lift c' of $w_2(P')$ with $(c')^2 = p_1(P')$, which is contained in the subspace $\mathbb{H}^2_-(X)$ of $\tilde{g}_t$-anti-self-dual forms. If this phenomenon occurs at a time t_0 for a bundle P', which is needed in the Donaldson compactification of $\mathcal{M}^{g}_X(P)^*$, then the polynomials $\gamma^{\tilde{g}_t}_{X,\mathfrak{o}}(P)$ will, in general, be different for $t < t_0$ and $t > t_0$ ([D 4, D 7, MO 2]).

To construct C^∞-invariants in the case $b_+(X) = 1$, we can therefore proceed as follows ([MO 2], p. 12). Consider a manifold X as above with $b_+(X) = 1$. The positive cone

$$\Omega_X := \left\{\alpha \in H^2_{\mathrm{DR}}(X) \mid \alpha \cdot \alpha > 0\right\}$$

consists of two components, Ω_+ and Ω_-, with $\Omega_- = -\Omega_+$, and choosing an H_+-orientation becomes the same as choosing one of these components

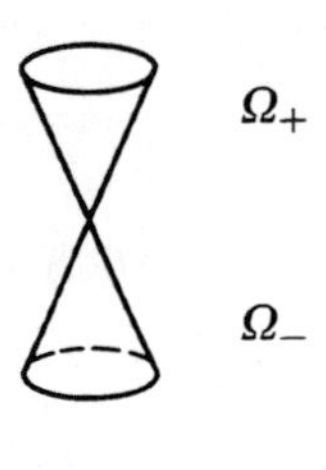

Fix a non-flat bundle P over X, such that there are well-defined polynomials

$$\gamma^g_{X,c,\mathfrak{o}_+}(P) : S^d H_2(X;\mathbb{Z}) \longrightarrow \mathbb{Z}$$

associated to generic metrics g and choices of H_+-orientations $\mathfrak{o}_+$ and integral lifts c.

For every integral class $e \in H^2(X;\mathbb{Z})$ with $\overline{e} = w_2(P)$ and $p_1(P) \leq e^2 < 0$ let

$$W_e := \{\alpha \in \Omega_X \mid \alpha \cdot e = 0\}$$

be the corresponding wall in Ω_X. Denote the set of connected components of

$$\Omega_X \setminus \bigcup_{\substack{\overline{e}=w_2 \\ p_1 \leq e^2 < 0}} W_e$$

by $\mathscr{C}_X(P)$; this is the set of chambers.

Now we associate the polynomial $\gamma^g_{X,c,\mathfrak{o}_+}(P)$ (of degree $d = -p_1(P) - 3$) to a chamber containing the intersection of $H^2_+(X)$ (relative to g) with the chosen component of Ω_X. The "function"

$$\Gamma_{X,c}(P) : \mathscr{C}_X(P) \dashrightarrow S^d H_2(X;\mathbb{Z})^*$$

has to be extended to all of $\mathscr{C}_X$ by way of a universal formula (i.e. a formula which is independent of the C^∞-structure) for the difference $\gamma^{g'}_{X,c,\mathfrak{o}_+}(P) - \gamma^g_{X,c,\mathfrak{o}_+}(P)$ of polynomials corresponding to adjacent chambers.

There are four cases in which all this has been carried out.

a) Donaldson's Γ-Invariant

This invariant corresponds to a SU(2)-bundle P with $c_2(P) = 1$ and the lift $c = 0$, over an indefinite 4-manifold X ([D 4], p. 143). The set $\mathscr{C}_X$ of chambers consists of the components of

$$\Omega_X \setminus \bigcup_{e^2=-1} W_e \, .$$

Theorem IV.4.1 (Donaldson). *There is a map*

$$\Gamma_X : \mathscr{C}_X \longrightarrow H^2(X;\mathbb{Z})$$

with the following properties:

i) $\Gamma_X(-C) = -\Gamma_X(C)$;

ii) If $C_+, C_- \in \Omega_+$, *then*

$$\Gamma_X(C_+) = \Gamma_X(C_-) + 2\sum e \, ,$$

where the summation is over all $e \in H^2(X;\mathbb{Z})$ with $e^2 = -1$ and $C_- \cdot e < 0 < C_+ \cdot e$.

iii) $\Gamma_{X'} \circ f^* = f^* \circ \Gamma_X$ for an orientation-preserving diffeomorphism $f : X' \to X$.

Notice that this is an invariant in the unstable range; its definition uses the "truncation method".

The Γ-invariant was the first invariant to be constructed and it had extremely important applications. It can be extended to the non-simply-connected case, provided ii) is properly modified ([D 4], p. 165).

b) Mong's Polynomials

Mong considers polynomials associated to SU(2)-bundles P_k with $c_2(P_k) = k > 1$, i.e. bundles in the stable range. He obtains "functions"

$$\Gamma_X^k : \mathscr{C}_X^k \dashrightarrow S^{4k-3}H_2(X;\mathbb{Z})^*$$

on the set of chambers $\mathscr{C}_X^k$ of

$$\Omega_X \setminus \bigcup_{-k \le e^2 < 0} W_e \,.$$

Γ_X^k has properties analogous to the properties i) and iii) of the Γ-invariant; a universal formula for the difference between the values on different chambers, corresponding to property ii) of the Γ-invariant, is in general not known. Mong computes Γ_X^2 for $X = \mathbb{P}_2$, $X = \mathbb{P}_2 \# \mathbb{P}_2$ and for X homeomorphic to $\mathbb{P}_1 \times \mathbb{P}_1$ ([MO 1], p. 2, p. 6, [MO 4], p. 2, [MO 2], p. 3, p. 15).

There are two particularly simple invariants, both corresponding to PU(2)-bundles P with $w_2(P) \neq 0$, for which the set of chambers $\mathscr{C}_X(P)$ consists only of the two components Ω_+ and Ω_- of Ω_X.

c) Kotschick's φ-Invariant

Let X be a manifold with $w_2(X) \neq 0$, such that for integral lifts k_X of $w_2(X)$ we have $k_X^2 \not\equiv -3 \pmod 8$. A PU(2)-bundle P with $w_2(P) = w_2(X)$ and $p_1(P) = -3$ yields an invariant

$$\varphi_{X,k_X} : \{\Omega_+, \Omega_-\} \longrightarrow \mathbb{Z}$$

of degree 0, counting the points of a 0-dimensional instanton space ([KOT], p. 595).

d) The $\overline{\Gamma}$-Invariant

The $\overline{\Gamma}$-invariant is the PU(2)-analogue of Donaldson's Γ-invariant ([O/V 2], p. 603).

Theorem IV.4.2. *Let P be an irreducible* PU(2)*-bundle with $w_2(P) \neq 0$ and $p_1(P) = -4$. To every integral lift c of $w_2(P)$ there is an associated map*

$$\overline{\Gamma}_{X,c} : \{\Omega_+, \Omega_-\} \longrightarrow H^2(X;\mathbb{Z})$$

with the following properties:

i) $\overline{\Gamma}_{X,c}(\Omega_-) = -\overline{\Gamma}_{X,c}(\Omega_+)$;

ii) $\overline{\Gamma}_{X,c+2l} = \varepsilon(l)\overline{\Gamma}_{X,c}$, *where*

$$\varepsilon(l) = \begin{cases} 1 & \text{if } \bar{l}^2 = 0 \\ -1 & \text{if } \bar{l}^2 \neq 0\,; \end{cases}$$

iii) $\overline{\Gamma}_{X',f^*c} \circ f^* = f^* \circ \overline{\Gamma}_{X,c}$ *for an orientation-preserving diffeomorphism* $f : X' \to X$.

Compared to Donaldson's Γ-invariant, the $\overline{\Gamma}$-invariant has some advantages. First, its definition uses a moduli space which is a priori compact, as we have seen in II.3. Also, the chamber structure is trivial in this case, since P is irreducible, so there is no dependence on a metric. Finally, the $\overline{\Gamma}$-invariant is often much easier to compute than the Γ-invariant ([O/V 1], p. 359, [O/V 2], p. 608).

All these various types of invariants are shown in the following table.

	$w_2 = 0$, unstable	$w_2 = 0$, stable	$w_2 \neq 0$
$b_+ = 1$	Γ_X	Γ_X^k, $k \geq 2$	$\Gamma_{X,c}(P)$
$b_+ > 1$	?	γ_{X,o_+}^k, $k > \frac{3}{4}(1+b_+)$	$\gamma_{X,c,o_+}(P)$

Mong has some partial results for invariants in the box ?, whereas Kotschick works on the $\Gamma_{X,c}(P)$-invariants ([D 7]).

V. Applications

V.1 The Topological Structure of Algebraic Surfaces

In this final chapter we describe the most important applications of the Donaldson invariants to the differential topology of algebraic surfaces. Mostly, we shall only consider the 1-connected case.

In order to do so, we have to recall the topological classification of 1-connected 4-manifolds. Let

$$S_X : H^2(X;\mathbb{Z})/\text{Tors} \times H^2(X;\mathbb{Z})/\text{Tors} \longrightarrow \mathbb{Z}$$

be the intersection form of a compact oriented 4-manifold. Then S_X is a symmetric bilinear form, which is unimodular by Poincaré duality. It can be odd (= of Type I) or even (= of Type II).

Freedman's celebrated classification theorem ([F], p. 368) can be phrased in the following way.

Theorem V.1.1. *The natural map*

$$\{\textit{oriented homeomorphism types of 1-connected topological 4-manifold}\}$$
$$\downarrow$$
$$\{\mathbb{Z}\textit{-equivalence classes of unimodular, symmetric bilinear forms}\}$$

is 1-*to*-1 *for even forms and* 2-*to*-1 *for odd forms.*

In other words, every unimodular, symmetric bilinear form S can be realised as the intersection form of a 1-connected manifold X. It is unique if S is even. If S is odd, there are two different topological manifolds $X, \widetilde{X}$ with S_X and $S_{\widetilde{X}}$ isomorphic to S. They are distinguished by the fact that exactly one of the 5-manifolds $X \times S^1$ and $\widetilde{X} \times S^1$ is smoothable. In particular, once it is known that a 1-connected topological 4-manifold is smoothable, then its (oriented) homeomorphism type completely determined by its intersection form.

Intersection forms of 1-connected algebraic surfaces are always indefinite, with the single exception of $\mathbb{P}_2$ (see [Y]). Indefinite forms are characterised by rank, signature and parity ([SER], p. 54). In our case the first two invariants are simply $b_2(X)$ and $b_+(X) - b_-(X)$, whereas the parity corresponds to $w_2(X)$ being 0 or $\neq 0$.

Corollary V.1.2. *The homeomorphism type of a* 1-*connected algebraic surface is determined by its Chern numbers* c_1^2, c_2 *and the parity of its canonical class.*

Using standard notation ([B/P/V], p. 14) we have $S_{\mathbb{P}_2} = \langle 1 \rangle$, $S_{\mathbb{P}_1 \times \mathbb{P}_1} = H$ and $S_K = -2E_8 + 3H$ for the intersection forms of $\mathbb{P}_2, \mathbb{P}_1 \times \mathbb{P}_1$, where K is a $K3$-surface.

Every other 1-connected algebraic surface X is homeomorphic to a connected sum of a number of copies of these three surfaces, possibly with their orientation reversed.

This is clear if S_X is of Type I, whereas in the even case the Miyaoka-Yau inequality is needed to prove this fact.

Example. Let $X^k \subset \mathbb{P}_1 \times \mathbb{P}_2$ be the zero-set of a *generic* bihomogeneous polynomial $f \in \mathbb{C}[X_0, X_1, Y_0, Y_1, Y_2]$ of bidegree $(k+1, 3)$. Then X^k is a smooth surface, which is 1-connected by Lefschetz.

The projection of $\mathbb{P}_1 \times \mathbb{P}_2$ onto the first factor shows that X^k is a relatively minimal elliptic surface over $\mathbb{P}_1$ with plane elliptic curves as fibres. Choose two relatively prime numbers $p, q \in \mathbb{N}$ and perform logarithmic transformations of multiplicities p and q over 0 and ∞ ([B/P/V], p. 164). The result is a relatively minimal elliptic surface

$$\begin{array}{c} X^k_{p,q} \\ \downarrow \\ \mathbb{P}_1 \end{array}$$

over $\mathbb{P}_1$ with two multiple fibres F_p and F_q. If F denotes a general fibre, then $pF_p \sim qF_q \sim F$ in $H_2(X;\mathbb{Z})$. Furthermore, $\pi_1(X^k_{p,q}) = 1$ ([UE], p. 639), $p_g(X^k_{p,q}) = k$, and $c_2(X^k_{p,q}) = 12(1+k)$. The "canonical bundle formula" ([B/P/V], p. 161) yields

$$K_{X^k_{p,q}} \sim (k-1)F + (p-1)F_p + (q-1)F_q \ ,$$

i.e. the order of divisibility $d(X^k_{p,q})$ of the canonical class is

$$d(X^k_{p,q}) = (k+1)pq - (p+q) \ .$$

This implies that $S_{X^k_{p,q}}$ is of Type II if k, p and q are odd, and of Type I otherwise.

Summarising we find ($\approx$ means homeomorphic)

$$X^k_{p,q} \approx \begin{cases} \#_{\frac{k+1}{2}} K \# \#_{\frac{k-1}{2}} \mathbb{P}_1 \times \mathbb{P}_1 & \text{if } k \equiv p \equiv q \equiv 1 \pmod 2 \\ \#_{2k+1} \mathbb{P}_2 \# \#_{10k+9} \overline{\mathbb{P}}_2 & \text{otherwise.} \end{cases}$$

The surface $X^0_{p,q}$ is rational if $\min(p,q) = 1$; $X^1_{1,1}$ is a $K3$ surface, but all other $X^k_{p,q}$'s are properly elliptic surfaces.

Using the Enriques-Kodaira classification ([B/P/V], p. 188), one can show that every 1-connected algebraic surface X, which is not of general type, is a deformation of one of the following surfaces:

i) $\mathbb{P}_1 \times \mathbb{P}_1$;

ii) $\widehat{\mathbb{P}}_2(y_1, \ldots, y_m)$, $m \geq 0$;

iii) $\widehat{X}^k_{p,q}(x_1, \ldots, x_n)$, $n \geq 0$, $\min(p,q) > 1$ if $k = 0$.

The surfaces $X^0_{p,q}$ with $\min(p,q) > 1$ are known as Dolgachev surfaces.

With these preliminaries out of the way, we can come to the applications of Donaldson's invariants to the differential topology of algebraic surfaces.

There are four major types of results which we want to discuss.

a) The failure of the h-cobordism conjecture for smooth 4-manifolds and the existence of infinitely many "exotic" C^∞-structures on some topological 4-manifolds. (Sect. V.2.)

b) The essential indecomposability of 1-connected algebraic surfaces. (Sect. V.3.)

c) The C^∞-invariance of the canonical classes of certain surfaces. (Sect. V.4.)

d) The relation between deformation types and C^∞-structures of algebraic surfaces. (Sect. V.5.)

Other applications, like restrictions on the realisability of homology classes by embedded 2-spheres, will not be discussed. For these and further applications we refer to the survey article of Friedman and Morgan ([F/M 1]).

V.2 Many Distinct C^∞-Structures

The h-cobordism conjecture asks whether two closed, 1-connected, oriented manifolds are homeomorphic (diffeomorphic) if they are h-cobordant. This question has an affirmative answer for smooth manifolds of dimension bigger than 4 and for topological manifolds of dimension equal to 4. The first result, due to Smale ([SM]) is one of the cornerstones of the classification theory of higher dimensional differentiable manifolds and essentially reduces this classification problem to questions in homotopy theory ([MA]). The 4-dimensional topological case is equivalent to Theorem V.1.1. It was a big breakthrough, when in 1985 Donaldson showed ([D 4], p. 142) that the h-cobordism conjecture for differentiable 4-manifolds is false. More precisely, using his Γ-invariant, Donaldson proved that the Dolgachev surface $X^0_{2,3}$ is not diffeomorphic to $\widehat{\mathbb{P}}_2(y_1,\dots,y_9)$, the projective plane blown up in nine points. For both surfaces, he computed the values of the Γ-invariant on chambers containing the Kähler classes of a suitable Hodge metric. To do so, he interpreted the corresponding instanton spaces as algebraic moduli spaces of μ-stable vector bundles and then used algebraic geometry to determine their structure. We shall illustrate below this method of computation with the simpler $\overline{\Gamma}$-invariant.

The natural question arising from this example, was: what can be said about the C^∞-structures of the other Dolgachev surfaces? Also using the Γ-invariant Friedman-Morgan and the authors proved ([F/M 2], p. 298, [0/V 1], p. 370)

Theorem V.2.1. *The Dolgachev surfaces induce countably many different C^∞-structures on the topological manifold X underlying $\widehat{\mathbb{P}}_2(y_1,\dots,y_9)$. They are all different from the differentiable structure determined by the rational structure on X.*

This theorem exhibits another remarkable difference between the 4-dimensional and the higher-dimensional cases. For on a 1-connected topological manifold of dimension ≥ 5 there exist only finitely many C^∞-structures; a fact, which is essentially due to the homotopy theoretic nature of the classification problem and so ultimately depends on the h-cobordism theorem.

Thus the result above shows that the h-cobordism conjecture for smooth 4-manifolds fails in a very essential way.

Friedman and Morgan further extended the preceding theorem by showing that it remains valid after blowing up. As a corollary they obtained ([F/M 2], p. 299)

Corollary V.2.2. *The topological manifold $\mathbb{P}_2 \# m\overline{\mathbb{P}}_2$ admits infinitely many distinct C^∞-structures for every $m \geq 9$.*

As to the case $m \leq 8$, there is essentially only one example known of a non-rational algebraic surface, homeomorphic to $\widehat{\mathbb{P}}_2(y_1, \ldots, y_m)$, namely the surface constructed by Barlow ([BA], p. 297).

Example (Barlow's surface). Let $Q \subset \mathbb{P}_4$ be the singular quintic surface given by the equations

$$\sum z_i = 0\,, \quad \sum z_i^5 - \frac{5}{4}\left(\sum z_i^3\right)\left(\sum z_i^2\right) = 0\,.$$

This quintic has 20 nodes, namely the points in the orbit of $(2 : 2 : 2 : -3-\sqrt{-7} : 3+\sqrt{7})$ under the obvious action of the alternating group A_5, but no other singularities. Q is the canonical image of a non-singular surface Q' with the canonical map

$$\varphi_{|\mathscr{K}_{Q'}} : Q' \longrightarrow Q$$

being the orbit map of a "canonical" involution

$$\imath : Q' \longrightarrow Q'\,.$$

From the A_5-action on Q and the canonical involution $\imath$ one can construct an action of the dihedral group D_{10} on Q' such that the quotient Q'/D_{10} is smooth, except for 4 nodes $p_1, \ldots, p_4$. The blowing-up

$$\tau : Y \longrightarrow Q'/D_{10}$$

of these nodes yields a smooth, minimal surface Y of general type with $\pi_1(Y) = 1$ and Chern classes $c_1(Y)^2 = 1$, $c_2(Y) = 11$. This is Barlow's surface, which by Freedman's theorem is topologically equivalent to the projective plane $\widehat{\mathbb{P}}_2(y_1, \ldots, y_8)$ blown up in eight points.

At present Y is (up to deformations) the only known example of a minimal surface of general type which is homeomorphic to a rational surface. Thus it is certainly of interest to decide whether or not Y is diffeomorphic to a rational surface, i.e. $\widehat{\mathbb{P}}_2(y_1, \ldots, y_8)$. More generally, one can ask if any blow-up of Y can be diffeomorphic to a rational surface or to (a blow-up of) a Dolgachev surface $X^0_{p,q}$.

Theorem V.2.3 ([O/V 2], p. 613). *The Barlow surface blown up in one point is neither diffeomorphic to $\mathbb{P}_2$ blown up in nine points, nor to any Dolgachev surface.*

Our result implies of course that the Barlow surface itself is not diffeomorphic to the projective plane blown up in eight points. This result has been obtained independently by Kotschick ([KOT], p. 591), who used the invariant described in IV.4.c).

We give an outline of the proof, which is based on the $\overline{\Gamma}$-invariant. For details we refer to [O/V 2], p. 608.

Let X be either the surface $\widehat{\mathbb{P}}_2(y_1, \dots, y_9)$, one of the surface $X^0_{p,q}$ with $\min(p,q) > 1$, or the surface $\widehat{Y}(y)$. Over X there is a unique irreducible PU(2)-bundle P with $w_2(P) = w_2(X) \neq 0$ and $p_1(P) = -4$. Consider the $\overline{\Gamma}$-invariant

$$\overline{\Gamma}_{X,k_X} : \{\Omega_+, \Omega_-\} \longrightarrow H^2(X;\mathbb{Z})$$

corresponding to this bundle and the lift $k_X = c_1(\mathcal{K}_X)$ (see IV.4, d).

For each surface we choose a suitable embedding $X \subset \mathbb{P}_N$, with associated Hodge metric g and hyperplane section H. Then we have the algebraic interpretation

$$\mathcal{M}^g_X(P)^* \cong \mathcal{M}^H_X(k_X, 1)$$

of the PU(2)-instanton space $\mathcal{M}^g_X(P)^*$ as the moduli space of μ-stable vector bundles $\mathcal{E}$ (relative to the polarisation H) with Chern classes $c_1(\mathcal{E}) = k_X$, $c_2(\mathcal{E}) = 1$.

These algebraic moduli spaces will turn out to be disjoint unions of complete curves $\mathcal{M}_i$, whose natural complex structure in general will be non-reduced. Nevertheless, by Maruyama's criterion (I.2.2) there exist universal bundles $\mathfrak{E}$ over $\mathcal{M}^H_X(k_X, 1) \times X$. Consequently, the value of $\overline{\Gamma}_{X,k_X}$ on the chamber Ω_+ containing $[H]$ is given by

$$\overline{\Gamma}_{X,k_X}(\Omega_+) = \sum_i m_i p_1(\mathfrak{E}_i) / \left[\widehat{(\mathcal{M}_i)_{\text{red}}} \right] .$$

Here $\widehat{(\mathcal{M}_i)_{\text{red}}}$ denotes the normalisation of the reduction of the i-th curve $\mathcal{M}_i$ in $\mathcal{M}^H_X(k_X, 1)$, m_i the multiplicity of $\mathcal{M}_i$ and $\mathfrak{E}_i$ the "restriction" of $\mathfrak{E}$ to $\widehat{(\mathcal{M}_i)_{\text{red}}}$. So the problem is to determine explicitly the moduli spaces $\mathcal{M}^H_X(k_X, 1)$ and their universal bundles for all three types of surfaces.

In the case $X = \widehat{\mathbb{P}}_2(y_1, \dots, y_9)$ this moduli space is empty, as is easily seen. Hence $\overline{\Gamma}_{X,k_X} = 0$ in this case.

In the case where X is one of the surfaces $X^0_{p,q}$, the moduli space, considered as a set, is found by using Riemann-Roch, Cayley-Bacharach (Theorem I.2.1), the definition of stability and the canonical bundle formula for elliptic surfaces. The result is that for $X^0_{p,q}$ the normalisation of the reduction of $\mathcal{M}^H_X(k_X, 1)$ is isomorphic to the disjoint union of a (strictly positive) number of copies of F_p and a number of copies of F_q (in most cases the space $\mathcal{M}^H_X(k_X, 1)$ itself is not reduced). Then, using the Serre construction, a universal bundle is found in an explicit way. Once all this has been carried out, it is easy to write down the $\overline{\Gamma}$-invariant

$$\overline{\Gamma}_{X,k_X}(\Omega_+) = -4 m_{p,q} l ,$$

with $m_{p,q} \in \mathbb{N}$ and $l \in H^2(X;\mathbb{Z})$ primitive, $l^2 = 0$.

The number $m_{p,q}$ satisfies the inequality $m_{p,q} \geq q x_{p,q} + p y_{p,q}$, where

$$x_{p,q} := \left\{ (x,y) \in \mathbb{N}^2 \;\middle|\; \frac{2x+1}{p} + \frac{2y+1}{q} < 1 < \frac{2x+3}{p} + \frac{2y+1}{q}\,,\ p > 2x+2 \right\}$$

$$y_{p,q} := \left\{ (x,y) \in \mathbb{N}^2 \;\middle|\; \frac{2x+1}{p} + \frac{2y+1}{q} < 1 < \frac{2x+1}{p} + \frac{2y+3}{q}\,,\ q > 2y+2 \right\}.$$

From the explicit description of Barlow's surface it is not hard to see that the 2-canonical system $|2K_Y|$ has four isolated fixed points but no fixed curves ([O/V 2], p. 610, [KOT], p. 600). It turns out that this set $\mathrm{Bs}|2K_Y|$ of base points determines the structure of the relevant moduli space of bundles over $\widehat{Y}(y)$. To begin with, we fix a polarisation H on Y of the form $H = H^0 + mK_Y$ with $H \cdot K_Y \equiv 1 (\mathrm{mod}\, 2)$ for a very ample divisor H^0 and a sufficiently large integer m. Let

$$\sigma : (\widehat{Y}(y), E) \longrightarrow (Y, y)$$

be the blowing up of a sufficiently general point $y \in Y$. Then $\widehat{H} := n(\sigma^* H) - E$, with n large and even, is a hyperplane section of a suitable embedding of $\widehat{Y}(y)$. We compute $\mathscr{M}^{\widehat{H}}_{\widehat{Y}(y)}(k_{\widehat{Y}(y)}, 1)$ in terms of a universal bundle over $\mathscr{M}^H_Y(k_Y, 1) \times Y$. Thus we first determine $\mathscr{M}^H_Y(k_Y, 1)$.

Consider a stable bundle $\mathscr{E}$ over Y with $c_1(\mathscr{E}) = k_Y$, $c_2(\mathscr{E}) = 1$. Using the Riemann-Roch formula and the stability condition, one easily shows that $\mathscr{E}$ can be uniquely represented by an extension

$$0 \longrightarrow \mathscr{O}_Y \longrightarrow \mathscr{E} \longrightarrow \mathscr{J}_z \otimes \mathscr{K}_Y \longrightarrow 0\,,$$

where the point z must have the Cayley-Bacharach property with respect to $|2K_Y|$, i.e. $z \in \mathrm{Bs}|2K_Y|$. Conversely, every such non-trivial extension defines a point in $\mathscr{M}^H_Y(k_Y, 1)$. Thus we have an identification

$$\mathscr{M}^H_Y(k_Y, 1)_{\mathrm{red}} \cong \mathrm{Bs}|2K_Y|$$

($\mathscr{M}^H_Y(k_Y, 1)$ is non-reduced at every point).

Let $\mathfrak{E}$ be a universal bundle over $\mathscr{M}^H_Y(k_Y, 1) \times Y$; it exists by Maruyama's criterion (I.3.2). We denote its restriction to $\mathscr{M}^H_Y(k_Y, 1) \times \{y\}$ by $\mathfrak{E}(y)$. Then we prove

$$\mathscr{M}^{\widehat{H}}_{\widehat{Y}(y)}(k_{\widehat{Y}(y)}, 1)_{\mathrm{red}} \cong \mathbb{P}(\mathfrak{E}(y))_{\mathrm{red}}\,.$$

This description of $\mathscr{M}^{\widehat{H}}_{\widehat{Y}(y)}(k_{\widehat{Y}(y)}, 1)$ yields

$$\overline{\Gamma}_{\widehat{Y}(y), k_{\widehat{Y}(y)}}(\Omega_+) = -4m[E]\,,$$

where $[E] \in H^2(\widehat{Y}(y); \mathbb{Z})$ is the dual class of the exceptional curve and $m \geq 1$ an integer which takes the multiplicities of $\mathscr{M}^{\widehat{H}}_{\widehat{Y}(y)}(k_{\widehat{Y}(y)}, 1)$ into account.

So we find for the value of the $\overline{\Gamma}$-invariant on Ω_+ with respect to k_X the values 0, a class $\neq 0$ but with square 0 and a class with square $\neq 0$. From the properties of the $\overline{\Gamma}$-invariant our theorem now follows.

The inequality $m_{p,q} \geq qx_{p,q} + py_{p,q}$ implies the existence of infinitely many different differentiable structures coming from Dolgachev surfaces.

If we summarise the preceding computations in a table

X	$\mathrm{kod}(X)$	$P_2(X)$	$\mathrm{Bs}\lvert 2K_X\rvert$	$\mathcal{M}_X^H(k_X, 1)_{\mathrm{red}}$	$\overline{\Gamma}_{X,k_X}(\Omega_+)$
$\widehat{\mathbb{P}}_2(y_1,\ldots,y_9)$	$-\infty$	0	$\emptyset$	$\emptyset$	0
$X^0_{p,q}, \min(p,q) > 1$	1	1	$(p-2)F_p + (q-2)F_q$	$\coprod_{x_{p,q}} F_p \amalg \coprod_{y_{p,q}} F_q$	$-4m_{p,q}l$
$\widehat{Y}(y)$	2	2	$\{z_1,\ldots,z_4\} \cup E$	$\coprod_{i=1}^{4} \{z_i\} \times E$	$-4m[E]$

we see that the $\overline{\Gamma}$-invariant is very closely related to properties of the 2-canonical system $\lvert 2K_X\rvert$ of each surface. Because of Castelnuovo's criterion, this is the linear system which determines whether or not a 1-connected algebraic surface is rational.

V.3 Indecomposability of Surfaces

As we have seen in Sect. V.1, every 1-connected algebraic surface is *homeomorphic* to a connected sum of simple pieces, namely $\mathbb{P}_2, \mathbb{P}_1 \times \mathbb{P}_1$ and K ($K3$ surface), possibly with reversed orientation. But from the point of view of differentiable equivalence, the situation is once again totally different.

First, some background. A good starting point is the following theorem of Wall ([W 2], p. 141).

Theorem V.3.1. *Let X and X' be 1-connected compact, oriented, differentiable 4-manifolds. If X and X' are homeomorphic, then there exists a number $r \geq 0$, such that $X\#r(\mathbb{P}_1 \times \mathbb{P}_1)$ and $X'\#r(\mathbb{P}_1 \times \mathbb{P}_1)$ are diffeomorphic.*

In other words, X and X' are stably diffeomorphic. Using the standard diffeomorphism $\mathbb{P}_1 \times \mathbb{P}_1\#\mathbb{P}_2 \cong 2\mathbb{P}_2\#\overline{\mathbb{P}}_2$, we see that for every pair of homeomorphic 4-manifolds X, X' as above there exist numbers $k, l \geq 0$, such that

$$X\#k\mathbb{P}_2\#l\overline{\mathbb{P}}_2 \cong X'\#k\mathbb{P}_2\#l\overline{\mathbb{P}}_2 \ .$$

Applying this consequence of Wall's theorem to the connected sums $X\#\mathbb{P}_2\#\overline{\mathbb{P}}_2$ and $(b_+(X)+1)\mathbb{P}_2\#(b_-(X)+1)\overline{\mathbb{P}}_2$, one also finds integers $a, b \geq 0$ with

$$X\#a\mathbb{P}_2\#b\overline{\mathbb{P}}_2 \cong \alpha\mathbb{P}_2\#\beta\overline{\mathbb{P}}_2$$

($\alpha = a + b_+(X)$, $\beta = b + b_-(X)$) for every given 1-connected differentiable 4-manifold X.

Example. Let K be a $K3$ surface. For every pair of relatively prime positive integers (p, q) there exist diffeomorphisms:

i) $X^k_{p,q}\#\mathbb{P}_1 \times \mathbb{P}_1 \cong \frac{k+1}{2}(K\#\mathbb{P}_1 \times \mathbb{P}_1)$ if $k \equiv p \equiv q \equiv 1 (\mathrm{mod}\, 2)$,

ii) $X^k_{p,q}\#\mathbb{P}_2 \cong 2(k+1)\mathbb{P}_2\#(10k+9)\overline{\mathbb{P}}_2$ otherwise.

This and similar results can be found in Mandelbaum's survey article ([MA], p. 43).

A smooth 4-manifold X, for which $X\#\mathbb{P}_2$ decomposes completely as a sum of $\mathbb{P}_2$'s and $\overline{\mathbb{P}}_2$'s, i.e.

$$X\#\mathbb{P}_2 \cong \alpha\mathbb{P}_2\#\beta\overline{\mathbb{P}}_2 \ ,$$

is called "almost completely decomposable". Mandelbaum conjectures that all 1-connected algebraic surfaces have this property. Together with Moishezon he proved the conjecture for all complete intersections and for all 1-connected elliptic surfaces. Their results show that the differential topology of 4-manifolds simplifies drastically if one leaves the algebraic category by taking smooth connected sums.

Of course, there is a trivial case in which an algebraic surface X can be decomposed, namely if X is non-minimal. For if X is obtained from $X_{\min}$ by blowing up k points, then

$$X \cong X_{\min}\#k\overline{\mathbb{P}}_2 \ .$$

In particular, if $X_{\min}$ is one of the basic building blocks $\mathbb{P}_2, \mathbb{P}_1 \times \mathbb{P}_1$ or K, then we obtain in this way a complete decomposition. But all other efforts to decompose algebraic surfaces themselves, however, were in vain. This fact led Mandelbaum and Moishezon to the conjecture that, apart from trivial cases, algebraic surfaces are never completely reducible in the differentiable category. The following as astonishing as admirable result of Donaldson implies this conjecture for surfaces with $b_+ > 1$ and explains the phenomenon ([D 5], p. 3).

Theorem V.3.2. *If a 1-connected algebraic surface X is diffeomorphic to a connected sum $X_1\#X_2$, then one of the 4-manifolds X_i has a negative definite intersection form.*

Example. There is no algebraic surface diffeomorphic to a connected sum

$$k\mathbb{P}_2\#l\overline{\mathbb{P}}^2$$

if $k \geq 2$, but there are many algebraic surfaces homeomorphic to such a sum, e.g. all complete intersections of odd and sufficiently high degree.

Donaldson conjectures that minimal models of 1-connected algebraic surfaces are indecomposable as differentiable manifolds.

The following result is an application of Donaldson's indecomposability theorem (compare [D 8]).

Corollary V.3.3. *Every 1-connected algebraic surface X with $b_+(X) > 1$ admits an "exotic" C^∞-structure, i.e. a differentiable structure different from the one induced by the algebraic structure.*

Indeed, given such a surface X, there exist a homeomorphism (see Sect. V.1)

$$X \approx \begin{cases} a(\pm K)\#b\mathbb{P}_1 \times \mathbb{P}_1 & \text{if } S_X \text{ is even} \\ \alpha\mathbb{P}_2\#\beta\overline{\mathbb{P}}_2 \ \ (\alpha > 1) & \text{if } S_X \text{ is odd.} \end{cases}$$

The manifold on the right induces an exotic structure except if $a = 1$ and $b = 0$. In this case X is either a honest $K3$ surface or a "homotopy $K3$ surface" $X^1_{p,q}$, $p \equiv q \equiv 1 (\mathrm{mod}\, 2)$, $p \cdot q \neq 1$ and one needs a result of Freedman and Morgan, to be explained in V.5.

Donaldson's indecomposability theorem is an immediate consequence of two other results which describe fundamental properties of his polynomial invariants ([D 5], p. 4).

Theorem V.3.4 (Vanishing theorem). *If X is a smooth connected sum $X \cong X_1\#X_2$ of 1-connected 4-manifolds X_i with each $b_+(X_i) > 0$, then all polynomial invariants γ^k_{X,o_+} vanish.*

Theorem V.3.5 (Positivity theorem). *If X is a 1-connected algebraic surface with $b_+(X) > 1$, and H a hyperplane section of X, then*

$$\gamma^k_{X,o_+}([H], \ldots, [H]) > 0$$

for sufficiently large k.

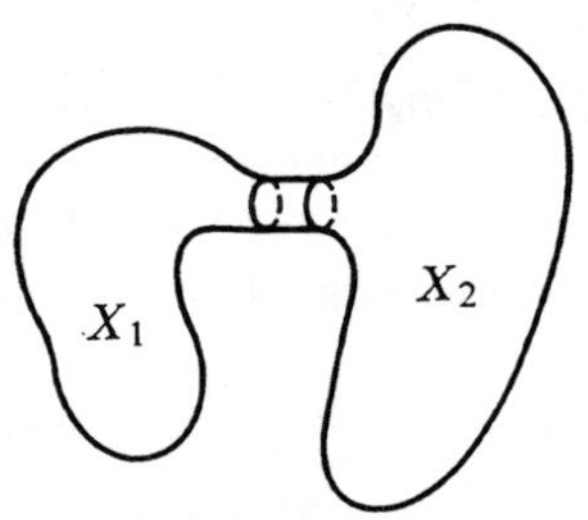

To prove the vanishing theorem Donaldson chooses metrics g_λ with a small "neck" of "radius" approximately λ and studies the limiting behaviour of the instanton spaces $\mathcal{M}^{g_\lambda}_X\left(\widehat{P}_k\right)^*$ as λ tends to zero. In this way he relates a k-instanton on X to a $(k - l)$-instanton on X_1 and an l-instanton on X_2, together with a "glueing parameter". ([D 5], p. 26.)

The proof of the positivity theorem, on the other hand, depends on the algebraic interpretation of instanton spaces over algebraic surfaces as moduli spaces of stable vector bundles.

Consider a Hodge metric g corresponding to an embedding $X \subset \mathbb{P}_N$ and a hyperplane section H. The instanton spaces $\mathcal{M}_X^g(\widehat{P}_k)^*$ can then be identified with the moduli spaces $\mathcal{M}_X^H(k)$ of μ-stable vector bundles $\mathcal{E}$ over X with $c_1(\mathcal{E}) = 0$ and $c_2(\mathcal{E}) = k$. Donaldson shows that the restriction of the class $\mu([H]) \in H^2(\mathcal{B}(P_k)^*;\mathbb{Z})$ to $\mathcal{M}_X^g(\widehat{P}_k)^*$ defines a polarisation of $\mathcal{M}_X^H(k)$ via this identification ([D 5], p. 57).

Intuitively this polarisation is given by the "divisor" of bundles $\mathcal{E}$ in $\mathcal{M}_X^H(k)$ whose restriction to H is "special".

The degree of $\mathcal{M}_X^H(k)$ with respect to this polarisation is certainly positive for large k, since $\mathcal{M}_X^H(k)$ is quasi-projective and non-empty by Taubes' existence theorem (or similar algebraic arguments).

Donaldson now shows that computing this degree is an admissable way of calculating $\gamma_{X,\mathfrak{o}_+}^k([H],\ldots,[H])$ if k is large enough.

This poses a problem since the Hodge metric g is, in general, not one of the generic metrics which were used in the definition of $\gamma_{X,\mathfrak{o}_+}^k$. The key point is an estimate of the dimensions of the subspace $\Sigma_k \subset \mathcal{M}_X^H(k)$ consisting of bundles $\mathcal{E}$ with $h^2(X;\mathrm{End}_0\mathcal{E}) \neq 0$, showing that $\mathcal{M}_X^H(k) \setminus \Sigma_k$ is non-empty (and smooth of the expected dimension) for $k \gg 0$ ([D 5], p. 58).

V.4 Surfaces with Large Groups of Diffeomorphisms

The naturality of Donaldson's polynomial invariants with respect to orientation-preserving diffeomorphisms suggests that the possibilities for these polynomials should be very limited for 4-manifolds with many symmetries. Friedman and Morgan have shown this to be true for algebraic surfaces which admit many non-trivial self-diffeomorphisms coming from algebraic geometry (F/M/M], p. 284).

Let X be a 1-connected algebraic surface with canonical class $k_X \in H_2(X;\mathbb{Z})^*$ and consider S_X as an element $q_X \in S^2H_2(X;\mathbb{Z})^*$. Let $\mathrm{O}(q_X)$ be the group of automorphisms of $H_2(X;\mathbb{Z})$ which preserve q_X, and $\mathrm{SO}(q_X)$ the subgroup of $\mathrm{O}(q_X)$ of automorphisms with determinant 1. Furthermore, let

$$\mathrm{SO}(q_X)_{k_X} := \left\{\alpha \in \mathrm{SO}(q_X) \mid \alpha(k_X) = k_X\right\}$$

be the stabiliser of the canonical class. All these are algebraic group schemes, defined over $\mathbb{Z}$. We denote their groups of integral or complex points by $\mathrm{O}_{\mathbb{Z}}(q_X)$ or $\mathrm{O}_{\mathbb{C}}(q_X)$, and similarly for $\mathrm{SO}(q_X)$ and $\mathrm{SO}(q_X)_{k_X}$ ([F/M/M], p. 284).

Let $\mathrm{Diff}^+(X)$ be the group of orientation preserving diffeomorphisms of X,

$$\psi : \mathrm{Diff}^+(X) \longrightarrow \mathrm{O}_{\mathbb{Z}}(q_X)$$

the natural map sending f to f^*; ψ factors through the group of components of $\mathrm{Diff}^+(X)$ inducing a map

$$\psi_0 : \pi_0(\mathrm{Diff}^+(X)) \longrightarrow \mathrm{O}_{\mathbb{Z}}(q_X)\ .$$

Now consider a deformation of X, i.e. a smooth proper holomorphic map

$$\pi : \mathfrak{X} \longrightarrow S$$

of connected complex spaces with fibre $\pi^{-1}(s_0) \cong X$ for a base point $s_0 \in S$. Every other fibre $\pi^{-1}(s)$ is diffeomorphic to X by way of a diffeomorphism which preserves the natural orientations and the canonical classes. The monodromy representation

$$\rho : \pi_1(S, s_0) \longrightarrow \pi_0(\mathrm{Diff}^+(X))$$

induces a map

$$\psi_0 \circ \rho : \pi_1(S, s_0) \longrightarrow \mathrm{O}_{\mathbb{Z}}(q_X)k_X \, .$$

Definition (Friedman-Morgan). *The family* $\pi : \mathfrak{X} \to S$ *has big monodromy if* $\mathrm{Im}(\psi_0 \circ \rho) \cap \mathrm{SO}_{\mathbb{Z}}(q_X)k_X$ *is Zariski-dense in* $\mathrm{SO}_{\mathbb{C}}(q_X)k_X$. *A surface has big monodromy if it is a fibre of a family with big monodromy.*

Example. Smooth complete intersections and minimal, 1-connected elliptic surfaces have big monodromy.

This follows from results of Ebeling ([E 1]).

The next theorem describes the effect of having big monodromy on Donaldson's polynomial invariants.

Theorem V.4.1 (Friedman-Morgan). *Let X be a 1-connected algebraic surface with $p_g(X) > 0$ and big monodromy. Then for all sufficiently large k, $\gamma^k_{X,\mathfrak{o}_+}$ is a polynomial in q_X and k_X:*

$$\gamma^k_{X,\mathfrak{o}_+} = \sum_{i=0}^{[\frac{d}{2}]} a_i q_X^i k_X^{d-2i} \, ,$$

$d = 4k - 3(1 + p_g(X))$, $a_i \in \mathbb{C}$.

Example. Let K be a $K3$ surface. Then

$$\gamma^k_{K,\mathfrak{o}_+} = a_{2k-3} q_X^{2k-3}$$

for sufficiently large k.

This theorem has an important corollary for surfaces with $p_g \equiv 0 (\mathrm{mod}\, 2)$. Under this hypothesis $d = 4k - 3(1 + p_g(X))$ is odd, so the canonical class divides $\gamma^k_{X,\mathfrak{o}_+}$, being its only linear factor up to multiples ([F/M/M], p. 685).

Corollary V.4.2. *Let $f : X \to X'$ be an orientation-preserving diffeomorphism of 1-connected surfaces with $p_g(X)$, $p_g(X') > 0$ and big monodromy. Suppose that $p_g(X) \equiv p_g(X') \equiv 0 (\mathrm{mod}\, 2)$. Then*

i) $f^* k_{X'} = \lambda(f) k_X$ *for some* $\lambda(f) \in \mathbb{Q}$;

ii) *if* $X = X'$ *or if* $k_X^2 \neq 0$, *then* $\lambda(f) = \pm 1$.

This corollary makes it possible to produce many examples of homeomorphic but not diffeomorphic surfaces.

For suppose that X and X' are minimal, 1-connected surfaces of general type with $p_g(X), p_g(X') > 0$, $p_g(X) \equiv p_g(X') \equiv 0 (\mathrm{mod}\ 2)$ and big monodromy. If X and X' are diffeomorphic, then the divisibilities of the canonical classes k_X and $k_{X'}$ in integral cohomology must be equal, since $k_X^2 \neq 0$.

Using this criterion, Moishezon constructed an infinite set of pairs (X_n, X_n') of homeomorphic but not diffeomorphic surfaces of general type ([F/M/M], p. 285). His surfaces are abelian covers of $\mathbb{P}_1 \times \mathbb{P}_1$. Salvetti applied similar ideas to certain ramified coverings of $\mathbb{P}_2$ and proved that for any given positive integer n there exist n surfaces $X_1, \ldots, X_n$ of general type which are all homeomorphic to each other but pairwise not diffeomorphic ([SA], p. 157).

Of course, an obvious class of surfaces for which these phenomena should be studied, is the class of complete intersections. In this case everything can be expressed in terms of the multi-degrees of the surfaces, and the problem is reduced to a question of number theory. If $X(d_1, \ldots, d_r) \subset \mathbb{P}_{r+2}$ denotes a complete intersection surface of multi-degree $(d_1, \ldots, d_r)$, then we have the following

Example. The surfaces $X(10, 7, 7, 6, 3, 3) \subset \mathbb{P}_8$ and $X(9, 5, 3, 3, 3, 3, 3, 2, 2) \subset \mathbb{P}_{11}$ are homeomorphic but not diffeomorphic.

This example is due to Ebeling, who used a computer to find it ([E 2], p. 4).

It would be surprising if the analogue of Salvetti's result would not hold for complete intersections.

V.5 Deformation Types and C^∞-Structures

The applications of the gauge-theoretic invariants to the differential topology of algebraic surfaces, which we have described so far, indicate that the algebraic classification of these surfaces may be closely related to the C^∞-classification of their underlying 4-manifolds. How closely ? Of course, one has to compare algebraic surfaces up to deformation with smooth oriented 4-manifolds up to orientation-preserving diffeomorphisms.

In this direction Friedman and Morgan have announced the following important result ([F/M 4], p. 2):

Theorem V.5.1. *The natural map*

$$\{\textit{deformation types of algebraic surfaces}\}$$
$$\downarrow$$
$$\{\textit{oriented } C^\infty\textit{-structures on 4-manifolds}\}$$

is finite-to-one.

The proof of their theorem consists of three distinct steps.

i) Reduction to 1-Connected Elliptic Surfaces ([F/M 1], p. 10)

It follows from the Enriques-Kodaira classification that in the case of surfaces of Kodaira dimension $-\infty$ or 0 deformation type and C^∞-type coincide because of obvious topological reasons. Surfaces of Kodaira dimension 2, with a fixed underlying topological structure, are parametrised by a finite union of quasi-projective varieties ([B/P/V], p. 206), so that on a given topological 4-manifold only finitely many C^∞-structures can be induced by surfaces of general type.

This reduces the problem already to surfaces of Kodaira dimension 1, i.e. honestly elliptic surfaces.

Now the theory of elliptic surfaces shows that the theorem is true for all elliptic surfaces, if it holds for (not necessarily relatively minimal) 1-connected elliptic surfaces ([F/M 1], p. 11).

Since every such surface is a deformation of one of the surfaces $\widehat{X}^{p_g}_{p,q}(x_1,\dots,x_r)$ (see Sect. V.2) with $r \geq 0$ and p,q relatively prime, it is enough to consider these standard models. Recall that the surfaces $X^{p_g}_{p,q}$ are honestly elliptic except for the $X^0_{p,q}$ with $\min(p,q) = 1$, which are rational, and the $K3$ surface $X^1_{1,1}$. The cases $p_g = 0$ and $p_g > 0$ have to be dealt with separately since different types of invariants are needed to distinguish the C^∞-structures.

ii) The Dolgachev Surfaces $X^0_{p,q}$

Using Donaldson's Γ-invariant and computations of moduli spaces for stable bundles on Dolgachev surfaces and their blow-ups, Friedman-Morgan showed ([F/M 2], p. 298, p. 299, p. 301):

Theorem V.5.2. *There exists a function $n(p,q)$ from unordered pairs of relatively prime integers $p,q > 1$ to the natural numbers, satisfying $n(p,q) \geq pq-p-q$ such that, if two blown-up Dolgachev surfaces $\widehat{X}^0_{p,q}(x_1,\dots,x_r)$ and $\widehat{X}^0_{p',q'}(x_1,\dots,x_{r'})$ are diffeomorphic, then $r = r'$ and $n(p,q) = n(p',q')$.*

This result implies Theorem V.5.1 for blow-ups of Dolgachev surfaces.

iii) The Elliptic Surfaces $X^{p_g}_{p,q}$, with $p_g > 0$

For these surfaces X Donaldson's polynomial invariantes $\gamma^k_{X,\mathfrak{o}_+}$ (see Theorem IV.3.1) are defined for sufficiently large k. According to Theorem V.4.1 they can be written in the form

$$\gamma^k_{X,\mathfrak{o}_+} = \sum_{i=0}^{[\frac{d}{2}]} a_i q^i_X l^{d-2i}_X ,$$

where $d = 4k - 3(1+p_g)$, and l_X is a primitive integral class which is a positive rational multiple of the canonical class k_X. Analysing again the structure of the relevant moduli spaces of stable vector bundles Friedman and Morgan determine the leading coefficient of $\gamma^k_{X^{p_g}_{p,q},\mathfrak{o}_+}$ assuming $k \geq 2p_g + 3$ they show:

$$a_i = \begin{cases} 0 & \text{if } i > 2k - 2p_g - 1 \\ \frac{d\,!}{2^{i+1}i\,!}(pq)^{p_g} & \text{for } i = 2k - 2p_g - 1 \end{cases}$$

([F/M 4], p. 3). Using this computation and a general formula relating the polynomial invariants γ^k_{X,o_+} of a 4-manifold X to the corresponding invariants of the "blow-up" $X \# \overline{\mathbb{P}}_2$, Friedman and Morgan prove ([F/M 4], p. 2):

Theorem V.5.3. *If two blown up elliptic surfaces $\widehat{X}^{p_g}_{p,q}(x_1, \ldots, x_r)$ and $\widehat{X}^{p_g}_{p',q'}(x_1, \ldots, x_{r'})$ with $p_g > 0$ are diffeomorphic, then $r = r'$ and $pq = p'q'$.*

This result completes the proof of Theorem V.5.1. For $p_g = p = q = 1$ the preceding theorem yields the following

Example. The homotopy $K3$ surfaces $X^1_{p,q}$, $p \equiv q \equiv 1 (\text{mod } 2)$, $p \cdot q \neq 1$ are not diffeomorphic to a "honest" $K3$ surface $K \cong X^1_{1,1}$.

This example is needed for the proof of the Corollary V.3.3.

A remarkable topological consequence of the previous two theorems is the existence of infinitely many distinct C^∞-structures on the manifolds

$$(2k+1)\mathbb{P}_2 \# (10k+9+r)\overline{\mathbb{P}}_2$$

for every $k, r \geq 0$ and on

$$lK \# (l-1)\mathbb{P}_1 \times \mathbb{P}_1$$

for every $l \geq 1$.

On the base of their theorem Friedman and Morgan ask, whether the natural map from deformation types of algebraic surfaces to C^∞-structures on the underlying 4-manifolds is actually one-to-one ([F/M 1], p. 12). Without being so ambitious, one could try to prove that certain algebraic invariants, which depend only on the deformation type, like e.g. this plurigenera P_m, are invariants of the C^∞-structure.

Proving the C^∞-invariance of the bigenus P_2 would certainly be a nice start.

As an explicit, we mention the following conjecture ([V 2], p. 15), which may be a good guide for future investigations.

Conjecture. The Kodaira dimension of algebraic surfaces is a C^∞-invariant.

There is some evidence in favour of this conjecture; for example, it has been established for surfaces of the homotopy types of Enriques- and $K3$ surfaces ([O], p. 415, [F/M 4], p. 5).

References

[A/K] Altman, A.B., Kleiman, S.L.: Compactifying the Picard scheme. Adv. Math. **35** (1980) 50–112

[A/D/H/M] Atiyah, M.F., Drinfeld, V.G., Hitchin, N.J., Manin, Y.I.: Constructions of instantons. Phys. Lett. **65A** (1978) 185–187

[A/H/S] Atiyah, M.F., Hitchin, N.J., Singer, I.M.: Self duality in four dimensional Riemannian geometry. Proc. Roy. Soc. London **A 362** (1978) 425–461

[B] Bănică, C.: Topologisch triviale holomorphe Vektorbündel auf $\mathbb{P}^n$. Crelle's J. **344** (1983) 102–119

[B/L] Bănică, C., Le Potier, J.: Sur l'existence des fibrés vectoriels holomorphes sur les surfaces non-algèbriques. J. reine angew. Math. **378** (1987) 1–31

[BA] Barlow, R.: A simply connected surface of general type with $p_g = 0$. Invent. math. **79** (1985) 293–301

[BAR] Barth, W.: Moduli of vector bundles on the projective plane. Invent. Math. **42** (1977) 63–91

[B/P/V] Barth, W., Peters, C., Van de Ven, A.: Compact complex surfaces. (Ergebnisse der Mathematik und ihrer Grenzgebiete, 3. Folge, Band 4.) Springer, Berlin Heidelberg New York Tokyo 1984

[B/O] Bauer, S., Okonek, C.: The algebraic geometry of representation spaces associated to Seifert fibered homology 3-spheres. Preprint 27, MPI Bonn 1989

[BE] Besse, A.: Géométrie riemannienne en dimension 4. Séminaire Arthur Besse 1978/79. "Textes mathématiques", 3. Cedic, Paris 1981

[BOU] Bourguignon, J.-P.: Analytical problems arising in geometry: examples from Yang-Mills theory. Jahresber. d. Dt. Math. Verein. **87** (1985) 67–89

[BR 1] Brosius, J.E.: Rank-2 vector bundles on a ruled surface I. Math. Ann. **265** (1983) 155–168

[BR 2] Brosius, J.E.: Rank-2 vector bundles on a ruled surface II. Math. Ann. **266** (1983) 199–214

[BRU] Brussee, R.: Stable bundles on blown up surfaces. Preprint, Leiden 1989

[DO] Dolgachev, I.: Algebraic surfaces with $q = p_g = 0$. In: Algebraic surfaces. C.I.M.E., Liguori Napoli (1981), pp. 97–215

[D 1] Donaldson, S.K.: A new proof of a theorem of Narasimhan and Seshadri. J. Diff. Geom. **18** (1983) 269–277

[D 2] Donaldson, S.K.: Anti-self-dual Yang-Mills connections over complex algebraic surfaces and stable vector bundles. Proc. London Math. Soc. **50** (1985) 1–26

[D 3] Donaldson, S.K.: Connections, cohomology and the intersection forms of 4-manifolds. J. Diff. Geom. **24** (1986) 275–341

[D 4] Donaldson, S.K.: Irrationality and the h-cobordism conjecture. J. Diff. Geom. **26** (1987) 141–168

[D 5] Donaldson, S.K.: Polynomial invariants for smooth 4-manifolds. Topology (to appear)

[D 6] Donaldson, S.K.: The orientation of Yang-Mills moduli spaces and 4-manifold topology. J. Diff. Geom. (to appear)

[D 7] Donaldson, S.K.: Letter to Okonek, December 1988

[D 8] Donaldson, S.K.: Talk at Durham, July 1989

[DR 1] Drezet, J.: Fibrés exceptionnels et variétés de modules de faisceaux semistables sur $\mathbb{P}_2$. Crelle's J. **380** (1987) 14–58

[DR 2] Drezet, J.: Cohomologie des variétés de modules de hauteur nulle. Math. Ann. **281** (1988) 43–85

[DR 3] Drezet, J.: Groupe de Picard des variétés de modules de faisceaux semi-stables sur $\mathbb{P}_2(\mathbb{C})$. Ann. Inst. Fourier, Grenoble **38** (3) (1988) 105–168

[E 1] Ebeling, W.: An arithmetic characterization of the symmetric monodromy groups of singularities. Invent. math. **77** (1984) 85–99

[E 2] Ebeling, W.: An example of two homeomorphic, nondiffeomorphic complete intersection surfaces. Invent. math. **99** (1990) 651–654

[E/F] Elencwajg, G., Forster, O.: Vector bundles on manifolds without divisors and a theorem on deformations. Ann. Inst. Fourier, Grenoble **32** (1982) 25–51

[F/S] Fintushel, R., Stern, R.: SO(3)-connections and the topology of 4-manifolds. J. Diff. Geom. **20** (1984) 523–539

[F/U] Freed, D., Uhlenbeck, K.K.: Instantons and four manifolds. M.S.R.I. publ. no. 1. Springer, New York Berlin Heidelberg Tokyo 1984

[F] Freedman, M.: The topology of 4-manifolds. J. Diff. Geom. **17** (1982) 357–454

[FR] Friedman, R.: Rank two vector bundles over regular elliptic surfaces. Invent. math. **96** (1989) 283–332

[F/M/M] Friedman, R., Moishezon, B., Morgan, J.W.: On the C^∞ invariance of the canonical classes of certain algebraic surfaces. Bull. A.M.S. **17** (2) (1987) 283–286

[F/M 1] Friedman, R., Morgan, J.W.: Algebraic surfaces and 4-manifolds: some conjectures and speculations. Bull. A.M.S. **18** (1) (1988) 1–15

[F/M 2] Friedman, R., Morgan, J.W.: On the diffeomorphism type of certain algebraic surfaces I. J. Diff. Geom. **27** (1988) 297–369

[F/M 3] Friedman, R., Morgan, J.W.: On the diffeomorphism type of certain algebraic surfaces II. J. Diff. Geom. **27** (1988) 371–398

[F/M 4] Friedman, R., Morgan, J.W.: Complex versus differentiable classification of algebraic surfaces. Preprint, New York 1988

[F/S] Fujiki, A., Schuhmacher, G.: The moduli space of Hermite-Einstein bundles on a compact Kähler manifold. Proc. Japan Acad. **63** (1987) 69–72

[G] Gieseker, D.: On the moduli of vector bundles on an algebraic surface. Ann. Math. **106** (1977) 45–60

[GO] Gompf, R.E.: On sums of algebraic surfaces. Preprint, Austin 1988

[G/H] Griffiths, Ph., Harris, J.: Residues and zero-cycles on algebraic varieties. Ann. Math. **108** (1978) 461–505

[GR] Grothendieck, A.: Techniques de construction et théorèmes d'existence en géométrie algébrique, IV: Les schémas de Hilbert. Sém. Bourbaki no. 221 (1961)

[H/H] Hirzebruch, F., Hopf, H.: Felder von Flächenelementen in 4-dimensionalen Mannigfaltigkeiten. Math. Ann. **136** (1956) 156–172

[H/S] Hoppe, H.J., Spindler, H.: Modulräume stabile 2-Bündel auf Regelflächen. Math. Ann. **249** (1980) 127–140

[H] Hulek, K.: Stable rank-2 vector bundles on $\mathbb{P}_2$ with c_1 odd. Math. Ann. **242** (1979) 241–266

[K 1] Kobayashi, S.: First Chern class and holomorphic tensor fields. Nagoya Math. J. **77** (1980) 5–11

[K 2] Kobayashi, S.: Differential geometry of complex vector bundles. Iwanami Shoten and Princeton University Press 1987

[K/O] Kosarew, S., Okonek, C.: Global moduli spaces and simple holomorphic bundles. Publ. R.I.M.S., Kyoto Univ. **25** (1989) 1–19

[KOT] Kotschick, D.: On manifolds homeomorphic $C\mathbb{P}^2\#8C\overline{\mathbb{P}}^2$. Invent. math. **95** (1989) 591–600

[L] Lawson, H.B.: The theory of gauge fields in four dimensions. Regional Conf. Series, A.M.S. 58. Providence, Rhode Island 1985

[LP] Le Potier, J.: Fibrès stables de rang 2 sur $\mathbb{P}_2(\mathbb{C})$. Math. Ann. **241** (1979) 217–256

[LU 1] Lübke, M.: Chernklassen von Hermite-Einstein Vektorbündeln. Math. Ann. **260** (1982) 133–141

[LU/ 2] Lübke, M.: Stability of Einstein-Hermitian vector bundles. Manuscr. Math. **42** (1983) 245–257

[L/O 1] Lübke, M., Okonek, C.: Moduli spaces of simple bundles and Hermitian-Einstein connections. Math. Ann. **267** (1987) 663–674

[L/O 2] Lübke, M., Okonek, C.: Stable bundles on regular elliptic surfaces. Crelle's J. **378** (1987) 32–45

[MA] Mandelbaum, R.: Four-dimensional topology: an introduction. Bull. A.M.S. **2** (1) (1980) 1–159

[MAG] Margerin, C.: Fibrés stables et metriques d'Hermite-Einstein. Sém. Bourbaki no. 683 (1987)

[M 1] Maruyama, M.: Stable bundles on an algebraic surface. Nagoya Math. J. **58** (1975) 25–68

[M 2] Maruyama, M.: Moduli of stable sheaves I. J. Math. Kyoto Univ. **17** (1977) 91–126

[M 3] Maruyama, M.: Moduli of stable sheaves II. J. Math. Kyoto Univ. **18** (1978) 557–614

[M 4] Maruyama, M.: Elementary transformations in the theory of algebraic vector bundles. In: Aroca, J.M., Buchweitz, R., Giusti, M., Merle, M. (eds.): Algebraic Geometry, Proc. La Rábida (Lecture Notes in Mathematics, vol. 961). Springer, Berlin Heidelberg New York 1982, pp. 241–266

[M 5] Maruyama, M.: Moduli of stable sheaves – generalities and the curve of jumping lines of vector bundles on $\mathbb{P}^2$. Advanced Studies of Pure Math., I, Alg. Var. and Anal. Var. 1–27, Kinokuniya and North-Holland 1983

[M 6] Maruyama, M.: The equations of plane curves and the moduli spaces of vector bundles on $\mathbb{P}^2$. In: Algebraic and topological theories, to the memory of T. Miyata, Tokyo 1985, pp. 430–466

[M 7] Maruyama, M.: Vector bundles on $\mathbb{P}^2$ and torsion sheaves on the dual plane. In: Vector bundles on Algebraic Varieties. Proc. Bombay 1984. Oxford Univ. Press 1987, pp. 275–339

[MI] Miyajima, K.: Kuranshi family of vector bundles and algebraic description of the moduli space of Einstein-Hermitian connections. Publ. R.I.M.S., vol. 25, Kyoto Univ., 1989, pp. 301–320

[MO 1] Mong, K.-C.: Some polynomials on $\mathbb{P}_2(\mathbb{C})\#\overline{\mathbb{P}}_2(\mathbb{C})$. Preprint 32, M.P.I. Bonn 1989

[MO 2] Mong, K.-C.: On some possible formulation of differential invariants for 4-manifolds. Preprint 34, M.P.I. Bonn 1989

[MO 3] Mong, K.-C.: Moduli spaces of stable 2-bundles and polarizations. Preprint 36, M.P.I. Bonn 1989

[MO 4] Mong, K.-C.: Polynomial invariants for 4-manifolds of type $(1, n)$ and a calculation for $S^2 \times S^2$. Preprint 37, M.P.I. Bonn 1989

[MU] Mukai, S.: Symplectic structure of the moduli space of sheaves on an abelian or $K3$ surface. Invent. math. **77** (1984) 101–116

[N/S] Narasimhan, M.S., Seshadri, C.S.: Stable and unitary vector bundles on compact Riemann surfaces. Ann. Math. **82** (1965) 540–567

[N] Norton, V.A.: Analytic moduli of complex vector bundles. Indiana Univ. Math. J. **28** (1979) 365–387

[O] Okonek, C.: Fake Enriques surfaces. Topology **23** (4) (1988) 415–427

[O/S/S] Okonek, C., Schneider, M., Spindler, H.: Vector bundles over complex projective spaces. Progress in Math. 3. Birkhäuser, Boston Basel Stuttgart 1980

[O/V 1] Okonek, C., Van de Ven, A.: Stable bundles and differentiable structures on certain elliptic surfaces. Invent. math. **86** (1986) 357–370

[O/V 2] Okonek, C., Van de Ven, A.: Γ-type-invariants associated to PU(2)-bundles and the differentiable structure of Barlow's surface. Invent. math. **95** (1989) 601–614

[P] Peters, C.A.M.: On two types of surfaces of general type with vanishing geometric genus. Invent. math. **32** (1976) 33–47

[SA] Salvetti, M.: On the number of non-equivalent differentiable structures on 4-manifolds. Manuscr. Math. **63** (1989) 157–171

[S 1] Schwarzenberger, R.L.E.: Vector bundles on algebraic surfaces. Proc. London Math. Soc. (3) **11** (1961) 601–623

[S 2] Schwarzenberger, R.L.E.: Vector bundles on the projective plane. Proc. London Math. Soc. (3) **11** (1961) 623–640

[SE] Sedlacek, S.: A direct method for minimizing the Yang-Mills functional over 4-manifolds. Commun. Math. Phys. **86** (1982) 515–528

[SER] Serre, J.-P.: A course in arithmetic. (Graduate Texts in Mathematics, vol. 7). Springer, New York Heidelberg Berlin 1973

[SM] Smale, S.: Generalized Poincaré's conjecture in dimensions > 4. Ann. Math. **74** (1961) 391–466

[SP] Spanier, E.H.: Algebraic topology. McGraw-Hill 1966

[ST 1] Strømme, S.A.: Deforming vector bundles on the projective plane. Math. Ann. **263** (1983) 385–397

[ST 2] Strømme, S.A.: Ample divisors on fine moduli spaces on the projective plane. Math. Z. **187** (1984) 405–423

[T] Takemoto, F.: Stable vector bundles on algebraic surfaces. Nagoya Math. J. **47** (1972) 29–48

[TA] Taubes, C.H.: Self-dual connections on 4-manifolds with indefinite intersection matrix. J. Diff. Geom. **19** (1984) 517–560

[UE] Ue, M.: On the diffeomorphism types of elliptic surfaces with multiple fibres. Invent. math. **84** (1986) 633–643

[U 1] Uhlenbeck, K.K.: Removable singularities in Yang-Mills fields. Commun. Math. Phys. **83** (1982) 11–30

[U 2] Uhlenbeck, K.K.: Connections with L^p bounds on curvature. Commun. Math. Phys. **83** (1982) 31–42

[U/Y] Uhlenbeck, K.K., Yau, S.-T.: On the existence of Hermitian-Yang-Mills connections in stable vector bundles. Commun. Pure Appl. Math. **39** (1986) 257–293

[UM] Umemura, H.: Stable vector bundles with numerically trivial chern classes over a hyperelliptic surface. Nagoya Math. J. **59** (1975) 107–134

[V 1] Van de Ven, A.: Twenty years of classifying algebraic vector bundles. In: Journées de géométrie algébrique. Sijthoff and Noordhoff, Alphen aan den Rijn 1980

[V 2] Van de Ven, A.: On the differentiable structure of certain algebraic surfaces. Sém. Bourbaki no. 667 (1986)

[W 1] Wall, C.T.C.: Diffeomorphisms of 4-manifolds. J. London Math. Soc. **39** (1964) 131–140

[W 2] Wall, C.T.C.: On simply-connected 4-manifolds. J. London Math. Soc. **39** (1964) 141–149

[WE] Wehler, J.: Moduli space and versal deformation of stable vector bundles. Rev. Roumaine Math. Pures Appl. **30** (1985) 69–78

[WEL] Wells, R.O.: Differential analysis on complex manifolds. (Graduate Texts in Mathematics, vol. 65). Springer, New York Heidelberg Berlin 1980

[WU] Wu, W.: Sur les espaces fibrés. Publ. Inst. Univ. Strasbourg, XI. Paris 1952

[Y] Yau, S.-T.: Calabi's conjecture and some new results in algebraic geometry. Proc. Nat. Acad. Sci. USA **74** (1977) 1789–1799

The Theory of Teichmüller Spaces
A View Towards Moduli Spaces of Kähler Manifolds

Georg Schumacher

Contents

Introduction

Over the last five decades, beautiful results have been proved in the subject of Teichmüller theory. Recently this area has been influenced by the spirit of analytic and algebraic geometry as well as complex differential geometry. Deformation theory of compact complex manifolds was created in a seemingly independent way. Its methods are significantly different and, as opposed to its classical counterpart, deformation theory only provides a local solution of the classification problem. A (coarse) moduli space, i.e. a global parameter space for complex structures exists only under certain assumptions. The aim of this article is to discuss some aspects of Teichmüller theory and their relationships to recent results on moduli of compact complex manifolds.

In his paper of 1857 "Theorie der Abel'schen Functionen" Riemann counted the number of parameters of isomorphism classes of algebraic equations in two variables i.e. classes of compact Riemann surfaces up to biholomorphic equivalence.

Some eighty years later Teichmüller realized that these parameters are real coordinates on a cell. Its points are quadratic holomorphic differentials on a surface which has been choosen as base point. The distinguished surface is related to the other points of the Teichmüller space by quasiconformal mappings. This relationship yielded a natural metric on the parameter space of Teichmüller. His last paper in this area was devoted to the complex structure of this space. The foundation of modern Teichmüller theory is due to Ahlfors and Bers. In particular they endowed the Teichmüller space with a complex structure and constructed a universal holomorphic family of Riemann surfaces.

The Teichmüller metric can be understood from the complex point of view from Royden's result which shows that it is the hyperbolic metric. Royden also uses this result to prove that the group of biholomorphic mappings of the Teichmüller space is just the Teichmüller modular group.

André Weil examined the Teichmüller family from the point of view of deformation theory. He proposed a certain metric on the moduli space and conjectured it to be Kähler. Petersson had already considered the corresponding inner product in the context of automorphic forms. In the hindsight this Petersson-Weil metric seems to have been the turning point, namely for the introduction of methods of complex differential geometry into the study of moduli space of compact Riemann surfaces. After proving the Kähler property, Ahlfors showed in 1961 that the holomorphic sectional and Ricci curvatures were negative. In 1974 Royden conjectured a precise upper bound for the holomorphic sectional curvatures of the Petersson-Weil metric. This, along with the negativity of the sectional curvature was established independently by Wolpert and Tromba in 1986. Its strong negativity in the sense of Siu was proved in [SCH 1986].

Another characterization of the Teichmüller space is the space of metrics of constant curvature -1 modulo diffeomorphisms. This led to an approach based on harmonic maps. It was initiated by Gerstenhaber and Rauch with more recent developments due to Fischer, Tromba, M. Wolf, and Jost. One of the main results is that the energy functional on a space of harmonic maps yields a potential for the Petersson-Weil metric.

If one is willing to disregard questions involving universal families, then one may consider the space of isomorphism classes of compact Riemann surfaces, i.e. the quotient of the Teichmüller space by the Teichmüller modular group. This is referred to as the moduli space. It carries an additional structure of a quasi-projective variety.

The compactification of the moduli space can be looked upon from different viewpoints: one is the algebraic viewpoint due to Mumford, another analytic approach is by Kleinian groups. Both involve deformations of singular and punctured Riemann surfaces. It turns out that the completion of the moduli space with respect to the Petersson-Weil metric is its (usual) compactification. This followed from estimates proven by H. Mazur. Estimates for the curvature tensor near the compactifying divisor are contained in [SCH 1986].

In terms of Fenchel-Nielsen coordinates, the Petersson-Weil form has the surprisingly simple shape of the standard symplectic form – a result shown by Wolpert in 1985. Comparing the different differentiable structures on the compactified moduli space which are induced by the usual holomorphic and the Fenchel-Nielsen coordinates, Wolpert proved that the Petersson-Weil class can be extended to the compactification. In fact, the Petersson-Weil form possesses local continuous potentials near the boundary. An essential result in this regard is a fiber-integral formula for the Petersson-Weil form. His result that the Petersson-Weil class on the compactified moduli space is (up to a numerical constant) the Chern class of an ample line bundle unifies the approach from algebraic geometry and Kähler geometry. This ample line bundle is closely related to the bundle that arises as the determinant of a certain direct image sheaf (the Knudsen-Mumford approach).

Based upon the powerful methods of deformation theory one can construct even in the case of non-algebraic Kähler manifolds a moduli space. For this, the notion of a polarization of a projective variety, which is closely related to its realization as a submanifold of some projective space, had to be replaced by the assignment of a Kähler class. In higher dimensions there occur additional complications. In the first place there no longer exists a global family. Even if the base space of a holomorphic family is treated as an analytic space-germ, there is in general no natural action of the automorphism group of the distinguished fiber on the base. This is contrary to the classical situation, where the Teichmüller modular group (more precisely the isotropy group of a point) comes from the automorphism group of the corresponding Riemann surface. Here the Teichmüller family yields a local universal deformation, and locally the moduli space is the quotient of this group. A general compact complex manifold only possesses a "versal" rather than universal deformation. This fact is related to the occurence of locally analytic subsets of the base, where all fibers are isomorphic. An identification of such points always yields a non-hausdorff topology. In fact there exist very simple examples of this kind. Even if the manifolds in question possess universal deformations, this is not sufficient for the existence of a moduli space.

A construction of the moduli space of polarized Kähler manifolds of a certain type or with additional structure can often be based on a simple criterion: Assuming the existence of versal deformations which is an obvious necessary prerequesit a coarse moduli space exists, if roughly speaking given any two families of complex manifolds and given any two sequences of pairwise isomorphic fibers there are subsequences of isomorphisms which converge to an isomorphism of the limit fibers [SCH 1983/1984]. The assumptions of the criterion were first verified in a situation which is closely related to the classical one, where biholomorphic mappings of compact Riemann surfaces are exactly the (orientation-preserving) isometries with respect to metrics of constant curvature. If in general the polarization can be realized by a distinguished Kähler metric with the above relationship between isomorphisms in the analytic category and isometries, the assumptions are automatically satisfied. In this way a moduli space for polarized Kähler manifolds with vanishing first real Chern class was constructed [SCH 1983]. This uses the unique Ricci-flat metrics according to Yau's solution of the Calabi problem. More generally, later the moduli space of polarized non-uniruled Kähler manifolds was constructed by Fujiki in [FU 1984] and in [SCH 1984].

For Riemann surfaces the Petersson inner product for quadratic holomorphic differentials yields an inner product on the cotangent space of the Teichmüller space. In the general setting by means of the Kodaira-Spencer map, this product on the tangent space is just the L^2-inner product of harmonic representatives of first cohomology classes with values in the tangent sheaf. According to Koiso, one may treat canonically polarized manifolds and compact polarized Kähler manifolds with vanishing first real Chern class in an analogous way by using Kähler-Einstein metrics. The variation of such metrics in a family gives in fact rise to a Kähler metric on the smooth part of the base.

Work on moduli spaces in higher dimensions uncovered a number of natural questions. First, could the generalized Petersson-Weil form be represented as a fiber integral? This would yield the Kähler property even at singular points, and also the existence of a Kähler potential (a consequence of a theorem of Varouchas). Second, can one find a hermitian line bundle on the moduli space of canonically polarized manifolds say, with curvature equal to the Petersson-Weil metric? Third, is there a natural Kähler metric on the moduli space of compact polarized Kähler manifolds? These questions were treated in [F-S 1988], where for example a fiber integral formula was proved. To obtain the hermitian line bundle on the moduli space of canonically polarized manifolds the direct images of powers of the relative canonical bundles had to be replaced by determinant bundles of certain virtual bundles related to the canonical one equipped with the Quillen metric. Its curvature is the Petersson-Weil form. The link with the fiber-integral formula for the generalized Petersson-Weil form was furnished by the generalized Riemann-Roch formula (for Chern forms rather than classes) by Bismut, Gilet and Soulé which also holds for singular base spaces.

The third question is treated in the following manner: Equip a compact polarized Kähler manifold with an extremal Kähler form (which represents the polarization). The general framework for such considerations is given by the above existence theorem for a moduli space of compact polarized Kähler manifolds which are not uni-ruled. The latter condition guarantees that in particular the components of the automorphism groups are compact. On such manifolds, extremal metrics are just those of constant scalar curvature. The general criterion for the existence of a moduli space applies.

However, in the case of moduli spaces of extremal Kähler manifolds the L^2-inner product of harmonic representatives of Kodaira-Spencer classes does not give rise to a Kähler metric on the moduli space (at least this is very unlikely). What one needs is a close relationship between the variation of the complex structures in a family of compact Kähler manifolds on one hand (i.e. a representative of the Kodaira-Spencer class), and on the other hand the variation of the distinguished Kähler metrics on these fibers which is described by a certain symmetric tensor). The key for the solution lies in the Berger-Ebin formula for the variation of the metric tensor in a family of Riemannian metrics with constant scalar curvature and Calabi's investigations. In the Kähler case all terms except for one of this expression vanish. This yields a forth order elliptic equation. This equation generalizes and replaces harmonicity.

The generalized Petersson-Weil metric is now defined through the L^2-inner product of such "harmonic" representatives of Kodaira-Spencer classes. The Kähler property can be directly verified, and just as in the Kähler-Einstein case a fiber-integral formula yields the Kähler property even at singular points of the moduli space. One should mention that computations are done with respect to the base of a universal deformation: the moduli space is locally the quotient of such by a finite group of automorphisms. The Petersson-Weil form descends to the quotient, but its potential may only be continuous. For integer-valued polarizations the moduli space is equipped with a hermitian line bundle whose curvature is the Petersson-Weil form. This is achieved in the following way: (1) classify extremal manifolds with isomorphism classes of ample line bundles, (2) construct a hermitian bundle and a Petersson-Weil form on the respective moduli space, (3) descend to the desired moduli space by fiber integration and determinant bundle.

A consequence is that all (connected) compact complex subspaces of the moduli space of rationally polarized, extremal Kähler manifolds (including the cases of canonically polarized manifolds and of Kähler-Einstein manifolds with positive curvature) are projective.

Chapter A

Teichmüller Theory

1. Teichmüller's Approach

Oswald Teichmüller's most influential article "Extremale quasikonforme Abbildungen" [TE 1939] contains essentially what nowadays is called "Teichmüller-theory" – the classification of Riemann surfaces by means of a parameter space.

He chose a somehow indirect approach: The classification of Riemann surfaces up to biholomorphic maps leads to the so-called "(coarse) moduli space" which turned out to contain singularities. He introduced a refined equivalence relation instead. First a Riemann surface X_0 had to be fixed. Variable surfaces X had to be equipped with an orientation-preserving homeomorphism f from X_0 to X_1. Two such *"marked" surfaces* are identified, if a biholomorphic map from X_1 to X_2 takes such a homeomorphism f_1 to a homoeomorphism homotopic to f_2.

He was able to show that the set of such equivalence classes carries the differentiable structure of a cell contained in a euklidean space, whose dimension had already been computed by Riemann in 1857. From the beginning of his construction this set was equipped with an intrinsic metric which he also related to a Finsler metric. Both the space and the metric bear his name. The Teichmüller space of compact Riemann surfaces of genus p is denoted by $\mathscr{T}_p$.

At the end the moduli space of Riemann surfaces was determined as the quotient of the Teichmüller space by a discrete group of automorphisms, the Teichmüller modular group.

Teichmüller reduced the classification problem to constructing a distinguished homeomorphism from the reference surface to a given one – this homeomorphism was to be the so-called *extremal quasiconformal map.* Considering a complex structure on a Riemann surface as arising from a Riemannian metric with an arbitrary scalar factor, he introduced quadratic differentials as links between extremal quasiconformal maps and conformal structures.

In "Veränderliche Riemannsche Flächen" [TE 1944] Teichmüller developed ideas to equip the Teichmüller-space with a complex structure and a holomorphic family of Riemann surfaces.

One should mention that he also regarded at the same time non-orientable surfaces for which certain results also hold.

1.1 Quasiconformal Mappings

Until Teichmüller realized the power of the notion of quasiconformal mappings, these had been more of a side-issue of geometric function theory, a generalization of holomorphic functions. At some place in [TE 1939] he shortly reviews the known aspects. He remarks that quasiconformal mappings should be used to study holomorphic mappings between Riemann surfaces $\mathscr{A}$ and $\mathscr{B}$, but not like approximating these by quasiconformal ones. One should rather consider quasiconformal maps from a fixed surface $\mathscr{C}$ to these.

Notion of Quasiconformal Mappings and the Problem of Grötzsch. Quasiconformal mappings had been introduced by Grötzsch in 1928. He showed a Picard type theorem but also solved an elementary problem about extremal mappings which ultimately lead to the Teichmüller theorems.

For any differentiable map $f : G \to G'$, $w = f(z)$ its *dilatation at a point z* is defined as

$$K_f(z) = \frac{|f_z(z)| + |f_{\bar{z}}(z)|}{|f_z(z)| - |f_{\bar{z}}(z)|} \tag{1.1}$$

and plainly its dilatation

$$K[f] = \sup K_f(z) \ . \tag{1.2}$$

Grötzsch showed that for any diffeomorphism f of rectangles in the complex plane, taking vertices to vertices in a fixed order, the dilatation $K[f]$ is bounded from below by a positive constant K_0 which is only attained for affine maps which are the *extremal quasiconformal* maps for this problem.

The idea of his proof is now mostly called "length-area" argument. He considers the image of a horizontal line from one edge to the other. Its length, as a function of the vertical coordinate, is integrated. This integral satisfies an obvious inequality in terms of the lenghts of the edges. On the other hand this area integral can be estimated from above by an integral involving the jacobian and the dilatation.

Differentiable orientation preserving homeomorphisms $w = f(z)$ of domains in the complex plane or between Riemann surfaces with bounded dilatation are called *quasiconformal*, and, more generally, if absolutely continous on lines and differentiable almost everywhere (with bounded dilatation).

Teichmüller allowed bordered Riemann surfaces with boundary curves and isolated boundary points ("Hauptbereiche") – we will restrict ourselves later to the case, where no boundary curves, just punctures, are present. The non-oriented case will be disregarded here.

Examples of Extremal Quasiconformal Mappings. Before his article [TE 1939] appeared, extremal quasiconformal mappings had been determined in some particular cases. Teichmüller considered the analogue of the problem of Grötzsch for annuli in the complex plane and arrived at a similar result: The lower bound for the dilatation of a map between annuli is the quotient of the respective moduli and attained for an affine map. (Here the modulus of $\{z \in \mathbb{C}; r < |z| < R\}$ is by definition $M = \log(R/r)$, which is the conformal invariant of the annulus).

As for compact Riemann surfaces, he looked at quasiconformal mappings of compact tori. The classifying space of marked compex tori was known to be the upper half plane H: Any quasiconformal map from a torus with modulus $\omega \in H$ to a torus with period $\omega' \in H$ has a dilatation bounded from below by

$\Im(\omega')/\Im(\omega)$. Provided the real parts of ω and ω' are equal, this number is the dilatation of an affine map. (In general the lowest dilatation is different but the extremal maps are in any case affine.)

As an example of a punctured Riemann surface he considered surfaces X of genus zero with four points removed. Because of the marking these points are ordered and can be chosen as 0, 1, ∞ and λ. Any such X is related to a two-sheeted covering of $\mathbb{P}^1$ by an elliptic curve with these points as branch points. One can directly establish a biholomorphic mapping from $\mathbb{C} \setminus \{0, 1\}$ to the upper half plane sending λ to an appropriately chosen modulus of the elliptic curve. In terms of the elliptic curves extremal quasiconformal mappings are induced by affine mappings of the complex plane.

In all of these examples the dilatation of an extremal quasiconformal mapping in the sense of (1.1) is *constant*.

The Teichmüller Metric. The analysis of the above (and more) examples seems to be the cause for Teichmüller to define a metric on the Teichmüller space. He poses the following question: Given a fixed Riemann surface and an upper bound C for the dilatation which are the quasiconformal maps, i.e. marked Riemann surface to lie within that bound?

Remarkable is the discussion of the torus case. If the period of the reference surface is ω, the estimate $(1/C)\Im(\omega) \leq \Im(\omega') \leq C\Im(\omega)$ yields that the range for ω' is an intersection of domains, whose boundary consists of two circular arcs. It is computed to be the non-euklidean disc with center ω and radius $\log C$.

The Teichmüller distance of two marked Riemann surfaces is in fact defined in terms of the logarithm of the dilatation.

$$(1.3) \qquad \tau(P, Q) = (1/2) \log \inf\{K[f]; f : P \to Q \text{ quasiconformal}\}$$

From this formula and the above discussion it follows immediately that the Teichmüller metric on the space of marked elliptic curves is equal to the hyperbolic metric on the upper half plane.

This principle proved even sound in the case of annuli, where the classifying space consists of all positive real numbers. The Teichmüller distance of two moduli M and N is here

$$\tau(M, N) = |\log M - \log N| \,.$$

For $\mathbb{P}^1$ with four punctures the Teichmüller space had been related to $\mathbb{C}\setminus\{0, 1\}$. Teichmüller showed, using the result about tori that the above expression (1.3) gives the metric of constant curvature -1 on $\mathbb{C} \setminus \{0, 1\}$. In all of these examples the construction implies that (1.3) actually defines a distance function.

1.2 Teichmüller Deformations

Teichmüller was able to put his hands on the desired extremal quasiconformal mappings. In this way he could identify the Teichmüller space with the open unit ball in the space of quadratic holomorphic differentials with respect to a suitable norm.

Conformal Structures Induced by Holomorphic Quadratic Differentials. A quasiconformal map can be visualized locally by the associated horizontal and vertical foliations – the images of the lines $\Re(z) = \text{const}$ and $\Im(z) = \text{const}$. Teichmüller observed that any quasiconformal map can be recovered up to a conformal map from these foliations and the dilatation. He first argued heuristically that an extremal quasiconformal map ought to have constant dilatation. On the other hand he explicitly constructed extremal mappings from quadratic differentials and prescribed dilatation $K > 1$.

Let $\phi(z)dz^2$ be a holomorphic quadratic differential given in local coordinates and p_0 a point. Then

$$\zeta = \int_{p_0}^{p} \sqrt{\phi dz^2} \tag{1.4}$$

defines a local coordinate about p_0 if $\phi(p_0) \neq 0$ and the sets

$$\zeta^{-1}\{z; \Im(z) = \text{const}\}\ ,\quad \zeta^{-1}\{z; \Re(z) = \text{const}\}$$

are horizontal and vertical foliations. At zeroes of ϕdz^2 a certain number of horizontal and vertical leaves resp. emanate from the singularity.

So far, the complex structure has not been changed – for a given constant $K > 1$ with respect to these coordinates an affine transformation is performed with constant dilatation K: On the complement of the zeroes of ϕdz^2 one defines

$$\zeta' = K \cdot \Re(\zeta) + i \cdot \Im(\zeta) \tag{1.5}$$

This is in fact a complex structure of a Riemann surface with isolated points removed – a function is by definition holomorphic, provided its derivative with respect to $\overline{\zeta'}$ vanishes. Thus the zeroes of ϕ are neglible and can be inserted. The existence of these points is a reason why mappings must be allowed which are not everywhere differentiable.

This construction looks like a very local argument. However, the conformal structure arises from a Riemannian metric which is of interest only up to a scalar factor. The corresponding line element is say $|d\zeta'|^2$, in Teichmüller's notation

$$ds^2 = \lambda(|d\zeta^2| + c \cdot \Re(d\zeta^2)) \tag{1.6}$$

where $\lambda = (K^2+1)/2$ and $c = (K^2-1)/(K^2+1)$. By definition $d\zeta^2$ is the given holomorphic differential.

An equivalent version of (1.6) is

$$ds^2 = \lambda' |d\zeta + k\overline{d\zeta}|^2 \tag{1.7}$$

where $k = (K-1)/(K+1)$ and $\lambda' = (K+1)/2$. The latter equation yields

$$ds^2 = \lambda' |\phi|^{1/2} \left| dz + k \frac{\overline{\phi}}{|\phi|} \overline{dz} \right|^2 . \tag{1.8}$$

The conformal structure associated to a holomorphic quadratic differential $\phi = \phi(z)\,dz^2$ and a number k, $0 < k < 1$ is called a *Teichmüller deformation.* The corresponding quasiconformal map is set-theoretically the identity which also defines the marking.

Now the space of quadratic differentials is equipped with a norm e.g. $\|\phi\| = \int |\phi|\,|dz|^2$. For any ϕ one sets $k := \|\phi\|$. Then the set of all Teichmüller deformations can be identified with the open unit ball in the space of quadratic differentials. It turned out that this is exactly the *Teichmüller space.* Apparently it is a cell. The Teichmüller space of closed Riemann surfaces of genus p is denoted by $\mathcal{T}_p$.

Teichmüller's Theorems. Let X_0 the Riemann surface chosen as reference point, k a number between 0 and 1, and $\phi\,dz^2$ a holomorphic quadratic differential on X_0. Denote by X_1 the associated Teichmüller deformation. The *Teichmüller uniqueness theorem* states that the identity from X_0 to X_1 is an extremal quasiconformal map: Any diffeomorphism $f : X_0 \to X_1$ which is homotopic to the identity with $K[f] < \infty$ satisfies $K[f] \geq K = (1+k)/(1-k)$. Moreover equality holds only if f is the identity.

The core of Teichmüller's and any other proof is the length-area principle of Grötzsch. However, it required an investigation of the metric induced by $|\phi|\,|dz|^2$, in particular the existence of geodesics.

For closed Riemann surfaces Teichmüller actually showed in [TE 1943]

Theorem (Teichmüller's existence theorem). *Any complex structure on a compact Riemann surface, viewed upon as a marked Riemann surface, is a Teichmüller deformation.*

The theorem means that for any fixed Riemann surface X_0 and any diffeomorphism g from X_0 to a surface X_1 there exists a Teichmüller deformation of X_0 together with a biholomorphic map to X_1 which is homotopic to g.

Teichmüller's considerations included an analysis of the topology of $\mathcal{T}_p$ related to the representation of closed Riemann surfaces in terms of Fuchsian groups.

A proof of the existence theorem is nowadays mostly worded in terms of Beltrami differentials. These are implicitely contained in (1.8). The conformal structure of a Teichmüller deformation is defined by $|dz + k\frac{\overline{\phi}}{|\phi|}\overline{dz}|^2$ and

$$k\frac{\overline{\phi}}{|\phi|}\frac{\overline{dz}}{dz}$$

is called *Teichmüller differential.*

This is a particular case of a *Beltrami differential* $\mu\,\frac{\overline{dz}}{dz}$ (cf. Sect. 2). A solution of the Beltrami equation $f_{\overline{z}} = \mu f_z$ is a quasiconformal map, provided μ is bounded. The Beltrami equation was studied extensively by Ahlfors and Bers. In a sense it can be used as a hinge of Teichmüller theory.

Teichmüller's theorems state that extremal quasiconformal maps always exist and correspond exactly to Teichmüller differentials.

1.3 The Moduli Space

Once the Teichmüller space is constructed the moduli space is easy to identify: Let P and Q be points representing marked Riemann surfaces X_j equipped with homeomorphisms $f_j : X_0 \to X_j$, $j = 1, 2$. Denote by p the genus of X_0. Any isomorphism of X_1 and X_2 (in the abstract sense) certainly defines a homotopy class of a homoeomorphism of X_0. The group Γ_p of such homotopy classes is called Teichmüller modular group Γ_p. It can essentially be identified with the group $\mathrm{Sp}(2p, \mathbb{Z})$ of homomorphisms of $H^1(X_0, \mathbb{Z})$ which are compatible with the intersection pairing (cf. [GR 1961]). It acts now on $\mathscr{T}_p$ in a way that its orbits consist exactly of the equivalence classes of marked Riemann surfaces in the above sense, because it just operates on the markings. The quotient $\mathscr{T}_p/\Gamma_p$ is the moduli space $\mathscr{M}_p$. Once the Teichmüller space is equipped with a complex structure and a holomorphic family of compact Riemann surfaces, it will be clear that the action of Γ_p is holomorphic (cf. Sect. 2). A similar notion is used in the case of punctured Riemann surfaces.

2. The Analytic Structure on the Teichmüller Space

We already mentioned that in one of his last papers [TE 1944] Teichmüller considered the problem of imposing a natural complex structure on $\mathscr{T}_p$. He introduced the notion of an analytic family of compact Riemann surfaces ("analytische Schar Riemannscher Flächen") given by a holomorphic map of complex manifolds. He remarks that different so far not related approaches yielded the same result for the dimension of $\mathscr{T}_p$ seemingly by miracle – his aim was to put rather an intrinsic structure on it. He gave a description of the total space in terms of branched coverings with analytic dependence of the branch points using the Riemann-Roch theorem.

Rigorous proofs were later given by Ahlfors [AH 1960/61a] and Bers [BE 1960]. At the end one was able to understand the relationship between holomorphic families (defined by smooth, proper holomorphic maps), real analytic families of Fuchsian groups, and the space of Teichmüller differentials.

For any holomorphic family of Riemann surfaces it was natural to use the period map from the base to the generalized Siegel upper half-plane as essential means to introduce a complex structure on the Teichmüller space, with underlying metric and in particular topological structure being already fixed (cf. [AH 1960]).

The subsequent development was influenced by deformation theory. A. Weil had pointed out in [WE 1958] the strong relationship of Teichmüller theory and the theory of deformations of compact complex manifolds in the sense of Kodaira and Spencer. The variation of solutions of the Beltrami equation related to infinitesimal deformations provided new aspects.

2.1 Real Analytic Theory

Fricke Coordinates. The basic idea originating from Fricke [F-K 1926] was to represent Riemann surfaces of genus $p > 1$ as quotients of the upper half plane by a discrete sub-group of the Möbius group $\mathrm{PSL}(2,\mathbb{R}) = \mathrm{SL}(2,\mathbb{R})/\{\pm 1\}$ and to pick a set of generators, whose entries can serve as real coordinates. It was used in [AH 1960] and further developed by Keen [KE 1971]. In particular, the Teichmüller space is provided with a real analytic structure (cf. [AB 1980]).

Let X be a compact Riemann surface and $\alpha_1, \ldots, \alpha_{2p}$ a fixed set of generators of the fundamental group $\pi_1(X)$ with relations being generated by

$$(2.1) \qquad \prod_{i \text{ odd}} [\alpha_i, \alpha_{i+1}] = 1 \,,$$

where $[\ ,\ ]$ denotes the commutator. The geometric intersection number for $i \leq j$ is

$$(2.2) \qquad (\alpha_i, \alpha_j) = \begin{cases} \delta_{i+1,j} & \text{for } i \text{ odd} \\ 0 & \text{for } i \text{ even} \end{cases}$$

Such a set of generators defines in this context a *marking*, and the set of isomorphism classes of compact, marked Riemann surfaces of genus p is called the *Fricke space* $\mathcal{F}_p$.

Theorem. *The Fricke space $\mathcal{F}_p$ can be identified with an open subset of $\mathbb{R}^{6g-g}$. The natural map from the Teichmüller space $\mathcal{T}_p$ to $\mathcal{F}_p$ is a homeomorphism – in particular $\mathcal{T}_p$ carries the structure of a real analytic manifold.*

In order to establish such a relationship, one assigns to the marked Riemann surface a representation $\rho : \pi_1(X) \to \mathrm{Aut}(H) \simeq \mathrm{PSL}(2,\mathbb{R})$, where H denotes the upper half plane. The representation becomes unique under normalization. One can choose from different normalizations. One is $A_1(1) = 1$, $A_2(-1) = -1$ and $A_2(\sqrt{-1}) = \sqrt{-1}$ for $A_j = \rho(\alpha_j)$ (cf. [AH 1960]); another choice is A_{2p} with 0 as a repelling fixed point, ∞ as an attractive fixed point and $A_{2p-1}(1) = 1$ (cf. [AB 1980]). For those $A_j(z) = (a_j z + b_j)/(c_j z + d_j)$ not affected by the normalizing conditions one sets $a_j d_j - b_j c_j = 1$, and $a_j \geq 0$, and $b_j > 0$ if $a_j = 0$. This yields the $6g - 6$ real so-called Fricke coordinates which are with respect to the latter normalization just $(a_1, b_1, c_1, \ldots, a_{2p-2}, b_{2p-2}, c_{2p-2})$. In fact, the equation (2.1) in terms of matrices together with the determinant condition yield the components for $j = 2g - 1$ and $j = 2g$. Thus the Fricke coordinates provide an injection of the Teichmüller space into $\mathbb{R}^{6p-6}$. By a description of transition functions a real analytic structure is imposed on $\mathcal{T}_p$ in [AH 1960] based upon the former, slightly different normalization. Moreover, a Fricke space of punctured Riemann surfaces can also be constructed.

The fact that the topologies of the Teichmüller and Fricke spaces match is a consequence of the Teichmüller existence theorem (with parameters) (cf. [AB 1980]).

In a direct way the introduction of a complex structure on $\mathcal{F}_p$ was accomplished by Kyoji Saito in [SAI 1988].

Fenchel-Nielsen Coordinates. Natural invariants of a marked Riemann surface are the lengths of closed geodesics. Also punctured Riemann surfaces are allowed. (As usual the Riemann sphere with fewer than three punctures and closed tori have to be excluded). The genus p of a punctured Riemann surface is by definition the genus of its closure.

The idea is to pick sufficiently many elements α_i of the homotopy group, represent these by closed geodesics and consider the lengths $\ell(\alpha_i)$ with respect to the hyperbolic metric. On the other hand any such α corresponds to an element γ of the Möbius group, and the quotient of the upper half plane by the cyclic group generated by γ is an annulus or punctured disk so that $\ell(\alpha)$ can be expressed in terms of γ as a real analytic function. L. Keen showed in [KE 1971] that a set of $6p-6+n$ of such geodesic length functions provide real analytic coordinates on the Fricke space, i.e. these are real analytic coordinates on $\mathscr{T}_p$.

The approach of Fenchel and Nielsen [F-N] is different although it also relies in a very explicit way on the hyperbolic structure of a compact Riemann surface X. A surface X of genus p with n punctures is cut along a simple geodesic loop α. A sequence of such cuts along closed geodesics α_j which do not intersect, decomposes X into building blocks which are conformally equivalent to so-called "pants". These are disks with two disjoint disks removed. The removed disks may also degenerate to points in order to allow punctures. The number of these pants is $2p-2+n$. As a matter of fact the conformal structure of the pants is determined by three non-negative numbers namely the geodesic lengths $\ell(\alpha_j)$ of the boundary components. The $\ell(\alpha_j)$ different from zero yield in all $3p-3+n$ invariants.

Conversely, given a set of pants fitting in the above combinatorial pattern with (non-complete) hyperbolic metrics and matching geodesic lengths of the boundaries, the identification of boundaries turns the union into a Riemann surface $\tilde{X}$. The ambiguity of its complex structure is due to a possible displacement δ_i of the endpoints of the α_i. This gives rise to another set of $3p-3+p$ real numbers $\theta_i = 2\pi\delta_i/\ell(\alpha_i)$. In terms of the complex structure on the Riemann surface the θ_i are unique up to multiples of 2π. The angles θ_i together with the inverse numbers $\ell(\alpha_j)^{-1}$ define a map $\mathscr{T}_p \to (S^1)^{3p-3+n} \times (\mathbb{R}^+)^{3p-3+n}$ which is real analytic by a similar argument as above. Its lift $\mathscr{T}_p \to \mathbb{R}^{3p-3+n} \times (\mathbb{R}^+)^{3p-3+n}$ is a real analytic diffeomorphism whose components are called *Fenchel-Nielsen coordinates.* These have an obvious description in terms of quasiconformal mappings. The gluing with respect to an angle different from a multiple of 2π, carries geodesics transversal to a cut to broken geodesics. Multiples of 2π induce isomorphic complex structures, i.e. elements of the modular group. The related mappings of the Riemann surface are called *Dehn twists.*

2.2 The Tangent Space of $\mathscr{T}_g$ and Its Complex Structure

We follow Ahlfors' arguments and use his notations (cf. [AH 1961a,b]). Let $X = H/\Gamma$ be a compact Riemann surface of genus $p > 1$ say and μ a Beltrami differential on H which is Γ-invariant, i.e. a complex measurable function on H

$$(\mu \circ A) \cdot \overline{A'} = \mu \cdot A' \tag{2.3}$$

for all $A \in \Gamma$ with $\|\mu\|_\infty < 1$. The space of Γ-invariant differentials in $L^\infty(H)$ is denoted by $B(\Gamma)$ and the open unit ball in it by $B(\Gamma)_1$.

For $\mu \in B(\Gamma)_1$, the normalized solution $f^\mu(z)$ of the Beltrami-equation

$$\bar{\partial} f(z) = \mu(z) \partial f(z) \tag{2.4}$$

satisfies by definition $f^\mu(0) = 0$, $f^\mu(1)$, and $f^\mu(\infty) = \infty$.

The Γ-invariance of the Beltrami differential yields a transformation formula

$$f^\mu \circ A = A^\mu \circ f^\mu \,, \tag{2.5}$$

where $A^\mu \in \mathrm{PSL}(2, \mathbb{R})$ can be associated to A in a natural way – it yields a group Γ^μ related to the complex structure on X induced by μ – a link with Fricke's approach. From the group-theoretic description in Sect. 2.1 the Teichmüller space could be considered as the quotient $B_1(\Gamma)/R$, where the equivalence relation R identifies Beltrami differentials μ_1 and μ_2, provided $A^{\mu_1} = A^{\mu_2}$ for all $A \in \Gamma$.

The Tangent Space of $\mathscr{T}_g$. In order to introduce local parameters on $\mathscr{T}_g$ in a neighborhood of the reference point μ, it is sufficient to consider real analytic Beltrami differentials which are real analytic with respect to parameters. It is convenient to set $f(v) = f^v$. Given fixed Beltrami differentials $v_1, \ldots, v_k$, for small real $t_1, \ldots, t_k$ the function $f(\mu + t_1 v_1 + \ldots + t_k v_k)$ is differentiable, even real analytic, with respect to these parameters: The induced natural map which assigns to $(t_1, \ldots, t_k)$ the corresponding point in the Fricke space $\mathscr{F}_p$ is in fact real analytic. How could one recover the tangent space? Certainly it is necessary to introduce partial derivatives in the direction of a Beltrami differential:

$$\dot{f}^\mu[v] := \frac{\partial}{\partial t} f(\mu + tv)\Big|_{t=0} \tag{2.6}$$

One may assume now that the reference structure on X is the given one, i.e. $\mu = 0$. Then the normalization of f and (2.4) imply

$$\bar{\partial} \dot{f}[v] = v \tag{2.7}$$

It is obvious how to replace the real parameters by complex ones $s_1, \ldots, s_k$ and obtain from (2.7) for complex s

$$\bar{\partial}\Big(\frac{\partial}{\partial s} f(sv)\Big|_{s=0}\Big) = v \tag{2.8}$$

and

$$\bar{\partial}\Big(\frac{\partial}{\partial \bar{s}} f(sv)\Big|_{s=0}\Big) = 0 \tag{2.9}$$

This equation tells us that $\Phi[v] := \big(\frac{\partial}{\partial \bar{s}} f(sv)\big|_{s=0}\big)$ is holomorphic. It turned out from explicit integral representations that the range of all such Φ consists of all holomorphic functions with a certain growth condition, namely $\Phi(z) = O((1 -$

$|z|^2)^{-2}$) (with normalization $\Phi(1) = 0$, $\Phi(-1) = 0$, and $\Phi(\sqrt{-1}) = 0$). Turning to Γ-invariant Beltrami differentials, one observes that the third derivatives of such functions $\phi[\nu] = \Phi[\nu]'''$ in fact give rise to quadratic holomorphic differentials on the Riemann surface X chosen as reference point. Such holomorphic functions Φ are referred to as *generalized abelian integrals.* On the other hand given a quadratic holomorphic differential on X with pull-back ϕ to the unit disk U, the growth condition follows from the invariance under Γ. Iterated integration yields a *unique* function Φ, vanishing at 1, -1 and $\sqrt{-1}$ which satisfies $\Phi''' = \phi$. On the other hand any quadratic Γ-invariant holomorphic differential ϕ gives rise to an invariant Beltrami differential

$$\nu = -\frac{1}{4}(1 - |z|^2)^2\overline{\phi(z)} \tag{2.10}$$

from which one can recover ϕ as $\Phi[\nu]'''$. This means that the space $Q(\Gamma)$ of invariant quadratic differentials can be identified with the quotient $B(\Gamma)$ of all invariant Beltrami differentials (which are say differentiable) by the subspace $N \subset B(\Gamma)$ of those, for which Φ vanishes. Its complex dimension $n = 3p - 3$ is well known.

Now the tangent space can be identified with $Q(\Gamma)$: Let a set of Beltrami differentials $\nu_1, \dots, \nu_n$ correspond to a basis of $Q(\Gamma)$ over $\mathbb{C}$. For $s \in \mathbb{C}^n$ close to the origin the $f(\nu)$, $\nu = s_1\nu_1 + \dots + s_n\nu_n$ yield generators $A_j \in \mathrm{Aut(H)}$ of Γ^ν which depend on the parameter s (cf. Sect. 2.1). These are directly related with the Fricke coordinates. It is explicitly shown that the partial derivative of all A_j in a direction ν only vanishes, if $\dot{f}[\nu] = 0$. But this means together with the above facts that the tangent map at the origin is an $\mathbb{R}$-linear isomorphism. With respect to the obvious structures of complex vector spaces it is anti-linear.

Lemma [AH 1961a]. *The tangent space of the Teichmüller space at the reference point can be identified with $B(\Gamma)/N$. The latter carries a natural complex structure and is $\mathbb{C}$-anti-linear isomorphic to the space of holomorphic quadratic differentials.*

The Complex Structure on $\mathcal{T}_p$ and the Universal Family. The above local, complex coordinate functions $s_1, \dots, s_n$ are referred to as *Bers coordinates.* Bers showed in [BE 1960] starting from (2.5) that the choice of different reference points yields holomorphic transition functions. Bers coordinates play an essential role in geometric investigations. A main result is the existence of a holomorphic fiber space $\mathcal{X}_p \to \mathcal{T}_p$ whose fibers are the Riemann surfaces determined by the Teichmüller points. The complex structure on $\mathcal{T}_p$ was inherited from the (infinite dimensional) complex vector space $B(\Gamma)$ – the natural map from $B_1(\Gamma)$ to the Teichmüller space shall be denoted by $\mu \mapsto \theta^\mu$ with reference point $\theta = \theta^0$. Now the product $U \times \mathcal{T}_p$ is equipped with a holomorphic structure as follows: A complex function F on a neighborhood of a point (z_0, θ) is by definition holomorphic, if $F(\theta^\mu, f^\mu(z_0))$ is holomorphic for all μ in a neighborhood of zero in $B(\Gamma)$, a condition which descends to the quotient structure and is invariant under the choice of the reference point.

The family $\mathscr{X} = \mathscr{X}_p \to \mathscr{T}_p$ of marked Riemann surfaces has the following universal property: For any family $\mathscr{Y} \to S$ of marked Riemann surfaces (with the same topological type) the (unique) map $\phi : S \to \mathscr{T}_p$ such that the fibers of s and $\phi(s)$ are isomorphic marked Riemann surfaces is holomorphic. Moreover it can be lifted to a map $\Phi : \mathscr{Y} \to \mathscr{T}_p$, whose restrictions to fibers are isomorphisms.

Action of the Modular Group. In the first section we have already seen that a change of markings yields a homeomorphism of the Teichmüller space. For $\gamma \in \Gamma$ and $t \in \mathscr{T}_p$, the point $t' = \gamma(t)$ represents the Riemann surface $\mathscr{X}_t$ with marking induced by γ. Thus γ induces an isomorphism ϕ_γ of $\mathscr{T}_p$ which sends t to t' and lifts to an isomorphism of total spaces. In particular, the isotropy group Γ_t of t say acts holomorphically on the Riemann surface $\mathscr{X}_t$ corresponding to t. On the other hand automorphisms of Riemann surfaces change the marking – the natural representation of $\mathrm{Aut}(\mathscr{X}_t)$ has Γ_t as image. It is injective for $p > 2$ and has kernel $\mathbb{Z}_2$ for $p = 2$. The moduli space $\mathscr{M}_p$ now carries the structure of a normal complex space, since locally it can be identified with an open subset of $\mathbb{C}^{3p-3}$ divided by the finite group of automorphisms Γ_p.

A Deformation Theoretic Viewpoint. In which way does this set-up fit into deformation theory? One may observe that the notion v is just a short-hand term for

$$v = v(z)\frac{\partial}{\partial z}\overline{dz}\,,$$

a ($\bar\partial$-closed) $(0,1)$-form with values in the holomorphic tangent bundle of the upper half-plane. The transformation law (2.3) just means that v descends to the Riemann surface X (with respect to its complex structure). So it determines by Dolbeault cohomology a class in $H^1(X, \Theta_X)$, where Θ_X denotes the sheaf of holomorphic vector fields. For such a cohomology class represented by a Beltrami differential v, and a solution f of the Beltrami-equation, (2.8) and (2.9) mean that v represents the Kodaira-Spencer class $\rho(\partial/\partial s)$ associated to the complex tangent vector $\frac{\partial}{\partial s}$. The corresponding map $\rho : T_0(\mathscr{T}_p) \to H^1(X, \Theta_X)$ is called Kodaira-Spencer map. Now a Hodge theoretic point of view immediately implies:

Lemma. *Let $X = D/\Gamma$ be a Riemann surface of genus $p > 1$ equipped with the hyperbolic metric $g = (1 - |z|^2)^{-2}|dz|^2$. Then the harmonic representative of a Kodaira-Spencer class is of the form*

$$v(z)\frac{\partial}{\partial z}\overline{dz} = \frac{\overline{\phi}}{g}\frac{\partial}{\partial z}\overline{dz}\,,$$

where $\phi(z)dz^2$ is a quadratic holomorphic differential.

Conceptually the above explicit $\mathbb{C}$-anti-linear map from the tangent space of $\mathscr{T}_p$ to the space of holomorphic quadratic differentials may be replaced by a duality of complex vector spaces:

For $\mu\frac{\partial}{\partial z}\overline{dz}$ and ϕdz^2, the natural pairing is

$$(\mu\frac{\partial}{\partial z}\overline{dz}, \phi dz^2) \mapsto \int_X \mu\phi dz\,\overline{dz}\,. \tag{2.11}$$

One can verify directly that the pairing vanishes identically for all Beltrami differentials which are $\overline{\partial}$-coboundaries and that it yields a duality

$$H^1(X, \Theta_X) \times \Gamma(X, \Omega_X^{\otimes 2t}) \to \mathbb{C} \tag{2.12}$$

We have seen that the harmonic representative of the Kodaira-Spencer class can be computed from the solution of the Beltrami equation.

Another approach is by the variation of the hyperbolic metrics. This is a general phenomenon for families of Kähler-Einstein manifolds, which we discuss in Sect. 9.1 in more detail. Here we give the result which follows from [F-S 1988]:

Proposition. *Let $\mathscr{X} \to S$ be a family of compact Riemann surfaces, and $g(z,s)dz\,\overline{dz}$ the family of hyperbolic metrics (of constant curvature -1). Let $\partial/\partial s$ be a tangent vector of S at a point $s_0 \in S$. Then the harmonic representative of $\rho(\partial/\partial s)$ is*

$$\left[-\frac{\partial}{\partial \overline{z}}\left(\frac{\partial^2 \log g}{\partial s \partial \overline{z}}/2g\right)\right]\frac{\partial}{\partial z}\overline{dz}\,. \tag{2.13}$$

The Cotangent Space of $\mathscr{T}_g$. A remarkable property of the duality (2.12) is its relation with the period map $\mathscr{T}_p \to \mathscr{H}_p$, where $\mathscr{H}_p$ denotes the Siegel upper half plane. The second symmetric power $S^2(H^0(X,\Omega))$ is the space of those quadratic differentials which come from products of differentials of the first kind on X. There is a natural homomorphism

$$H^1(X_0, \Theta_X) \to H^0(X, \Omega^{\otimes 2})^* \to S^2(H^0(X_0, \Omega))^*\,. \tag{2.14}$$

According to [WE 1958] this composite map is exactly the derivative of the period map at the point X.

3. Hyperbolicity of the Teichmüller Space

A natural problem was to determine the automorphisms of the Teichmüller space. It was settled by the work of Royden in [RO 1971] using geometric methods for the space of closed Riemann surfaces. The result was extended by Earle and Kra to the Teichmüller space of punctured surfaces in [E-K 1974].

The main idea consisted in a characterization of biholomorphic mappings as isometries with respect to suitable metrics. It led to a comparison of the Kobayashi and Teichmüller metrics.

In Sect. 1.1 the Teichmüller metric was defined as logarithm of the maximal dilatation of quasiconformal maps between the resp. marked Riemann surfaces. Since by Teichmüller's theorems extremal quasiconformal maps from the reference structure X_0 say to any other surface X are induced by a Teichmüller differential $\mu = k\overline{\eta}/|\eta|$ for a holomorphic quadratic differential η on X_0 and some $0 < k < 1$, the maximal dilatation is just $(1+k)/(1-k)$ and

$$\tau(X_0, X) = (1/2) \log \frac{1+k}{1-k} \,. \tag{3.1}$$

This is for small k up to second order just k. At least at the reference point the (real) tangent space of the Teichmüller space can be determined in terms of differentials $k'\overline{\eta}/|\eta|$, where k' is some non-negative real number. A norm of such a Beltrami differential is just the number k'.

What is the induced norm on the (complex) cotangent space of $\mathcal{T}_p$, if one assumes the natural pairing $(\mu, \phi) = \int_{X_0} \mu\phi dz\, \overline{dz}$ for Beltrami differentials μ and holomorphic quadratic differentials ϕ on X_0 ?

The result is

$$\|\phi\|_T = \int_{X_0} |\phi| dz\, \overline{dz} \tag{3.2}$$

(cf. [RO 1971]).

Lemma. *The above norm $\| \;\; \|_T$ on the holomorphic cotangent space of the Teichmüller space is the infinitesimal form of the Teichmüller metric which is thus a Finsler metric.*

It turns out that there is more than a formal analogy of the Teichmüller distance function in terms of the excentricity k to the formula for the distance of two points in the unit disk with respect to the Poincaré metric.

Given a differential $\overline{\eta}/|\eta|$, where η is a quadratic differential as above, one may consider the differentiable map $F : D \to \mathcal{T}_g$ from the unit disk to the Teichmüller space, defined by $F(\zeta) := [X_{\zeta\cdot\mu}]$. Obviously the pull-back of the Teichmüller metric under F equals the Poincaré metric ρ, and F is a geodesic embedding.

This is in fact an extreme situation.

Theorem (Royden [RO 1971]). *The Teichmüller metric τ is characterized by*

$$\tau(P, Q) = \inf \rho(a, b)$$

where the infimum is taken over all holomorphic maps $\phi : D \to \mathcal{T}_g$ with $\phi(a) = P, \phi(b) = Q$. In particular it coincides with the Kobayashi metric.

Because of the preceding discussion it has to be shown that for any holomorphic map $\tilde{F}$ from the unit disk to $\mathcal{T}_g$ the pull-back $\tilde{F}^*(d)$ is dominated by ρ.

In order to show this by Ahlfors' version of the Schwarz-Pick lemma one has to construct supporting metrics of curvature at most -4 in a neighborhood of any point $p \in D$ with $\tilde{F}'(p) \neq 0$.

One may assume $p = 0$ and set $\overline{\eta}/|\eta| := \tilde{F}\prime(p)$. To the Beltrami differential $\overline{\eta}/|\eta|$ a map F can be associated according to the above construction. Then the metric $F^*\tau$ is up to a term $o(|\zeta|^2)$ supporting.

We have seen that the Teichmüller modular group Γ_g acts on $\mathcal{T}_g$ as a group of isometries. Conversely one has the following result:

Theorem (Royden [RO 1971]). *Any biholomorphic map between domains in $\mathcal{T}_g$ which is an isometry with respect to τ is induced by an element of Γ_g.*

Since any biholomorphic map of $\mathcal{T}_g$ is certainly an isometry with respect to the Kobayashi metric, this implies:

Corollary. *Any biholomorphic map of the Teichmüller space to itself is induced by an element of the modular group Γ_g. For $g > 2$ the groups* Aut $(\mathcal{T}_g)$ *and Γ_g can be identified, and* Aut $(\mathcal{T}_2)$ *is the quotient of Γ_2 by the group $\mathbb{Z}_2$.*

The theorem is of purely local nature: For any two Riemann surfaces, and any $\mathbb{C}$-linear isometry of the spaces of quadratic differentials with respect to the above norm, the Riemann surfaces are isomorphic. The proof depends on a characterization of quadratic differentials η with a zero of order $3g - 4$ by an estimate of the function $t \mapsto \|\eta + t\eta_1\|_T$; $t \in \mathbb{R}$. Such differentials (if $g > 2$) exist for any point and are unique up to a constant factor. Now, under the the 2-canonical embedding, these define the osculating hyperplanes. The given isometry of the spaces of quadratic differentials preserves this situation, and these geometric data are sufficient to show that the corresponding Riemann surfaces are isomorphic.

We add a further consequence which seems not to be contained in the literature:

Proposition [SCH 1990]. *Let $p > 2$. Then the automorphism group of the moduli space $\mathcal{M}_p$ consists only of the identity.*

For $p = 2$ the statement does not hold, since $\mathcal{M}_2 = \mathbb{C}^3/\mathbb{Z}_5$ has many automorphisms.

In the view of Royden's theorem it is sufficient to show that any automorphism of $\mathcal{M}_p$ lifts to the Teichmüller space, if $p > 2$. This fact follows directly from the following lemma, since the Teichmüller space is a cell and the set of singular points of $\mathcal{M}_p$ is of codimension greater or equal to two.

Lemma. *Let $p > 2$. Then the branch locus of the map $\mathcal{T}_p \to \mathcal{M}_p$ equals the set of singularities of $\mathcal{M}_p$.*

According to classical theorems of Chevalley, Gottschling and Prill we have to show that the modular group contains no generalized reflections. An automorphism of a complex manifold is called a *generalized reflection* at one of its fixed points, if its linearization has exactly one root of unity as eigenvalue with all other eigenvalues being equal to 1.

The existence of generalized reflections in the modular group is exluded roughly as follows.

Assume that $\psi \in \Gamma_p$ is such an automorphism with respect to a fixed point given by some marked Riemann surface X. The linearized action of ψ at this point is given by the action of the corresponding automorphism ϕ of X on the space of quadratic holomorphic differentials. Consider the 2-canonical embedding of X into some projective space $\mathbb{P}_{2p-3}$. The map ϕ extends to some automorphism Φ of the projective space whose fixed point set consist of an isolated point x and a hyperplane H where Φ is a generalized reflection. The point x may or may not be contained in X, and the intersection number of X and H is the degree of thc 2-canonical divisor. The hyperplane H intersects X transversally since Φ is a reflection at these points. Now one considers the map from X to its quotient by the group generated by ϕ. A contradiction follows from the Riemann-Hurwitz formula.

4. The Petersson-Weil Metric

4.1 Basic Properties

The Petersson Scalar Product. In [PE 1949] H. Petersson had defined a scalar product for automorphic forms. For holomorphic quadratic differentials on a compact Riemann surface X with hyperbolic metric $g\,dz\,\overline{dz}$ this is

$$(\phi dz^2, \psi dz^2)_P = \int_X \frac{\phi\overline{\psi}}{g^2} g dz\,\overline{dz}\,. \tag{4.1}$$

The induced norm of an element of the tangent space $H^1(X, \Theta_X)$ of the Teichmüller space with respect to the duality (2.12) can be easily computed: Let $\mu\frac{\partial}{\partial z}\overline{dz}$ be a Beltrami differential on X. Then the norm of its cohomology class is the supremum of $|(\mu, \psi)|/\|\psi\|_P$, where ψ runs through all non-zero holomorphic quadratic differentials. It is convenient to evaluate (2.11) using the representative $\mu = \overline{\phi}/g$, where ϕ is a quadratic holomorphic differential in order to arrive at:

$$\|\mu\|^2 = \int_X \frac{|\phi|^2}{g} dz\,\overline{dz}\,. \tag{4.2}$$

This equation should be interpreted as follows: The Petersson inner product induces on $H^1(X, \Theta_X)$ the L^2-norm of harmonic representatives. This is the infimum of the L^2-norms over all representatives. The corresponding hermitian inner product $(\ ,\ )_{PW}$ on the tangent space of $\mathcal{T}_p$ is called the *Petersson-Weil inner product.* It depends differentiably on the paramenter.

Kähler Property of the Petersson-Weil Metric. In 1958 André Weil suggested the investigation of this hermitian metric on the Teichmüller space and posed the problem of its Kähler property. It was established by Ahlfors [AH 1961] using Bers coordinates (cf. Sect. 2).

Theorem. *On the Teichmüller space $\mathcal{T}_p$, $p > 1$ Bers' coordinates are geodesic for the Petersson-Weil metric at the reference point. In particular this metric is Kähler.*

We denote by $\omega_{PW} = g_{i\bar{j}} ds^i ds^{\bar{j}}$ the induced Kähler form on $\mathcal{T}_p$ which is called Petersson-Weil form.

Royden's infinitesimal description (cf. Sect. 4) implies according to [RO 1974] a comparison of the Teichmüller metric with the Petersson-Weil metric: For any quadratic holomorphic differential ϕdz^2 the following inequality holds

$$\int_X |\phi| dz\,\overline{dz} \le (\operatorname{vol} X)^{1/2} \cdot \|\phi\|_P \tag{4.3}$$

This means for the corresponding Finsler metrics on the Teichmüller space that

$$ds_{PW} \le [4\pi(p-1)]^{1/2} d\tau\,, \tag{4.4}$$

where $d\tau$ is the line element of the Teichmüller metric. (It is remarkable that an analogous inequality to (4.4) also holds on a bounded symmetric domain for the invariant Kähler metric and the infinitesimal Kobayashi metric on it.) The period map $\mathcal{T}_p \to \mathcal{H}_p$ is certainly distance decreasing with respect to the Kobayashi metric, a fact which can also be established on the infinitesimal level starting from (2.13). Moreover the period map is distance decreasing (up to a numerical factor) with respect to the distinguished Kähler metrics (i.e. the Petersson-Weil metric on $\mathcal{T}_p$ and the Bergman metric on $\mathcal{H}_p$.

The Petersson Weil Metric on the Moduli Space. We have seen that the Petersson-Weil metric is defined by intrinsic properties. Consider the universal family $\mathcal{X}_p \to \mathcal{T}_p$ and an arbitrary family $\mathcal{Y} \to S$ of Riemann surfaces of genus $p > 1$ (provided with a marking such that it is induced by a unique holomorphic map $\phi : S \to \mathcal{T}_p$). Denote by $\Phi : \mathcal{Y} \to \mathcal{T}_p$ the lift of ϕ – its restrictions to fibers are isomorphisms: On the level of the first cohomology groups with values in the tangent bundle the pull-backs of harmonic representatives are harmonic. Thus there is a natural hermitian form (not necessarily positive definite) on S which is just the pull-back of the Petersson-Weil metric. In the situation, where $\mathcal{Y} \to S$ is the Teichmüller family and ϕ comes from an element of the Teichmüller modular group Γ_p, this means that the Petersson-Weil form ω_{PW} is invariant under ϕ. The Petersson-Weil hermitian inner product descends to the moduli space $\mathcal{M}_p = \mathcal{T}_p/\Gamma_p$. On a neighborhood of any point $x \in \mathcal{T}_p$ a $\partial\bar{\partial}$-potential of ω_{PW} can be made invariant under the isotropy group of x. However, the local potential descends to a function which may only be continuous. In this sense the Petersson-Weil form is defined on $\mathcal{M}_p$.

4.2 The Petersson-Weil Metric for Families of Singular Riemann Surfaces

In [BE 1974] Bers considered families of compact hyperbolic Riemann surfaces which degenerate to curves with ordinary double points. The connected components of the singular fibers, with singularities removed are called *parts*, and one assumes that these are all hyperbolic. Bers observed that the Poincaré metrics on the singular fibers converge in any C^k-topology to the Poincaré metrics on the parts. However, these families of metrics do not depend differentiably on the parameter. The question of the degeneracy of the Petersson-Weil metric was settled by H. Mazur in [MA 1976].

Singular Families. From the topological viewpoint a degeneration of a Riemann surface of genus p consists of a contraction of closed geodesics. There are two types of such contractions – one increases the number of parts by one and leaves the sum of the genera over all parts fixed – the other decreases the genus of the corresponding part but leaves the number of parts fixed. (The genus of a part is by definition the genus of its closure).

In both cases the contraction of a cycle can be performed with a complex parameter and is compatible with the holomorphic structure. It yields a compact Riemann surface with an ordinary double point. On its (smooth) normalization one can distinguish the inverse image of the singularity. A compact Riemann surface together with a set of distinguished points is called a *punctured Riemann surface.*

The complex structure of an ordinary double point is as follows: The singularity is in terms of local coordinates (z, w) with $|z| < 1$, $|w| < 1$ given by the equation $z \cdot w = 0$, and a universal deformation of this singularity with a complex parameter $t, |t| < 1$ is just $z \cdot w = t$. For any compact Riemann surface with an ordinary double point, this gives rise to a one-dimensional holomorphic family of compact Riemann surfaces such that the fiber over the origin is the only singular one: It is sufficient to remove a neighborhood of the double point and insert the above one-dimensional family (with biholomorphic transition functions on two pairs of annuli). On the other hand one can use Teichmüller theory of punctured Riemann surfaces. From the deformation theoretic viewpoint local families of punctured Riemann surfaces, called "stacks" in algebraic geometry, correspond to families of mappings of a finite set of say n points (all equipped with the sheaf $\mathbb{C}$) to compact Riemann surfaces with variable complex structure. The dimension of a universal family equals $3p - 3 + n$, also for genus zero or one, where n must be larger than 3 or 0 resp.

A compact Riemann surface X with at most ordinary double points such that its parts are hyperbolic is also called a *stable curve*. Its canonical sheaf can be described as follows: A section of this sheaf, pulled back to the normalization can be identified with a holomorphic 1-form on the normalization which has at most simple poles at the distinguished points and opposite residues at points which are identified under the normalization map.

A local family which is universal can be constructed as follows: The parts induce Riemann surfaces X_i of genus p_i with a number of n_i punctures. These possess in particular local universal families over smooth base spaces S_i of dimension $d_i = 3p_i - 3 + n_i$: families $\mathscr{X}_i \to S_i$ together with sections $\sigma_i : S_i \to \mathscr{X}_i$. Set $S = \prod_i S_i$ and extend all families over S. The surface X is the union of all X_i with transversal intersections. The identification of punctures is then extended along the sections induced by σ_i. Finally at all singularities one applies the above insertion procedure ("opening up of nodes") relative S which yields another set of q local parameters, where q is the number of double points. The dimension of the base is

$$\sum_{i=1}^{r}(3p_i - 3 + n_i) + q \ .$$

It follows from general theory that such a family is universal for all neighboring fibers, in particular for the smooth general fiber. In fact, let a number of q contractions be carried out resulting in r parts X_i. Then the genus of X is

$$p = \sum_{i=1}^{r} p_i + q + 1 - r$$

so that the dimension of the base of this family $f : \mathscr{X} \to S$ is exactly $3p - 3$.

One may denote the coordinates which describe the deformations of singularities by t_j, $j = 1, \ldots, q$ and those which come from deformations of punctured surfaces by $\tau_k, k = q + 1, \ldots, 3p - 3$.

Theorem (H. Mazur [MA 1976]). *Let* $S = \{(t, \tau)\}$ *and* $\mathscr{X} \to S$ *be a family of Riemann surfaces with ordinary double points for* $t_j = 0$ *for at least one* j *and hyperbolic parts as above. Then the following estimates for the components* $g_{i\bar{j}}$ *of the Petersson-Weil metric tensor hold:*

(i) There exist constants $0 < a < A$ *and a neighborhood of* $(0, 0)$, *where*

$$\frac{-a}{|t_i|^2(\log|t_i|)^3} \le g_{i\bar{i}} \le \frac{-A}{|t_i|^2(\log|t_i|)^3} \qquad \text{for} \quad i \le q$$

(ii)

$$|g_{i\bar{j}}(t,\tau)| = O\left(\frac{1}{|t_i||t_j|(\log|t_i|)^3(\log|t_j|)^3}\right)$$
$$\text{as} \quad (t,\tau) \to (0,0) \qquad \text{for} \ \ i,\ j \le q,\ i \ne j$$

(iii)

$$\lim_{(t,\tau)\to(0,0)} g_{i\bar{j}}(t,\tau) = g_{i\bar{j}}(0,0) \qquad \text{if} \quad i,\ j > q,$$

(iv)

$$|g_{i\bar{j}}(t,\tau)| = O\left(\frac{-1}{|t_i|(\log|t_i|)^3}\right) \qquad \text{as} \quad (t,\tau) \to (0,0) \qquad \text{if} \ \ i \le q,\ j > q.$$

These estimates have several consequences. The Petersson-Weil metric is *not complete.* In transversal direction to the divisor $t_1 \cdot \ldots \cdot t_q = 0$ it degenerates, whereas restricted to the "parallel" direction it converges to the Petersson-Weil metric on families of punctured Riemann surfaces. Moreover, one can read off the equations that the *volume of the moduli space* $\mathscr{M}_p$ *is finite.*

In fact Mazur treats components

$$g^{i\bar{j}} = \int \phi_i(t,\tau)\overline{\phi_j(t,\tau)}g(t,\tau)dz\,d\,\bar{z} \tag{4.5}$$

first, where the ϕ_i are holomorphic sections of the relative 2-canonical bundles $\Omega^{\otimes 2}_{\mathscr{X}/S}$ which are dual to the holomorphic vector fields $\partial/\partial t_i$ and $\partial/\partial \tau_j$ on S. The sections ϕ_i originate from a basis of $\Omega_{\mathscr{X}/S}(S)$, where the first q sections related to $\partial/\partial t_i$ contain the factor t_i.

The specific estimates of (4.5) are due to the degeneration of both the family of hyperbolic metrics and of these sections. These relative quadratic differentials are meromorphic with at most double poles in the z_i-direction (in terms of local coordinates (z_i, w_i, t_i) near the singularities). These coordinates are sometimes called "pinching coordinates". The relative metric tensor $g(t,\tau)$ can be estimated using the embedding of the annuli $\{(z_i, t_i); |t_i| < |z_i| < 1\}$ equipped with hyperbolic metrics as follows: For small $|t_i|$, the annuli $\{(z_i, t_i); |t_i| < |z_i| < 1\}$ are holomorphically embedded into the fibers of the family $\mathscr{X} \to S$. This means that the hyperbolic metrics $h(z_i, t_i)$ dominate the restrictions of the Poincaré metrics $g(z,t,\tau)$ of the fibers. Furthermore the maximum principle for elliptic differential equations yields that the supremum of h/g for fixed parameters is attained on the boundary of the respective disks. For $|z_i| = \delta < 1$ or $|w_i| = \delta < 1$ by Bers' result one has uniform convergence of the metric tensors g to the hyperbolic metric on the respective parts in all C^k-topologies.

An alternative approach can be based solely on estimates of the metric tensor, since by (2.13) the tensor of the Petersson-Weil metric can be computed in terms of the variation of the hyperbolic metric on the family.

5. The Curvature of the Petersson-Weil Metric

5.1 Ahlfors' Results

Shortly after the Kähler property of the Petersson-Weil metric was established, Ahlfors showed in [AH 1961b]

Theorem. *The Ricci curvature, the holomorphic sectional curvature and the scalar curvature of the Petersson-Weil metric are negative.*

His proof is a continuation of the computations leading to the Kähler property involving an integral formula for the curvature tensor.

5.2 Bounds of the Curvature

Two questions remained open for more than ten years: The negativity of the sectional curvature and the existence of a negative upper bound for the holomorphic sectional curvature.

In [RO 1974] Royden conjectured a precise such upper bound. It was established by Wolpert in [WO 1986] and Tromba [TR 1986]:

Theorem. *(i) The holomorphic sectional and Ricci curvatures of ω_{PW} on $\mathcal{T}_p$ for $p > 1$ are bounded from above by $\frac{-1}{2\pi(p-1)}$.*

(ii) The sectional curvature of ω_{PW} is negative.

In [WO 1986] such results are also attributed independently to Royden. This theorem is a direct consequence of an explicit formula for the curvature tensor. We refer here to Wolpert's approach. Let X be a compact Riemann surface of genus $p > 1$. Denote by $\{\mu_\alpha; \alpha = 1, \ldots, 3p-3\}$ a basis of the vector space of harmonic Beltrami differentials on X corresponding to a set of tangent vectors $\{\partial/\partial s_\alpha; \alpha = 1, \ldots, 3p-3\}$. The product of such a differential with a conjugate one is a differentiable function, and we have seen that the components of the Petersson-Weil form

$$\omega_{PW} = \sqrt{-1} g_{\alpha,\overline{\beta}} dz^\alpha \wedge dz^{\overline{\beta}}$$

are in terms of the basis

$$g_{\alpha\overline{\beta}} = \int_X \mu_\alpha \overline{\mu_\beta} g \, dz \, d\overline{z} \tag{5.1}$$

Denote by D the real laplacian on L^2-functions on X (with non-positive spectrum). Then the inverse $(D-2)^{-1}$ exists and is a compact integral operator.

Theorem (Wolpert [WO 1986], Tromba [TR 1986]). *The curvature tensor of the Petersson-Weil metric equals*

$$\begin{aligned} R_{\alpha\overline{\beta}\gamma\overline{\delta}} = &-2 \int_X (D-2)^{-1}(\mu_\alpha \overline{\mu_\beta})(\mu_\gamma \overline{\mu_\delta}) dz \, d\overline{z} \\ &-2 \int_X (D-2)^{-1}(\mu_\alpha \overline{\mu_\delta})(\mu_\gamma \overline{\mu_\beta}) g dz \, d\overline{z} \, . \end{aligned} \tag{5.2}$$

We indicate, how to derive the various estimates of the curvatures of ω_{PW} from (5.2): One sets $\Delta = -2(D-2)^{-1}$. This is an integral operator with a positive kernel, a fact which implies that for all L^2-functions ϕ and ψ the inequality

$$|\Delta(\phi\psi)| \leq |\Delta\phi^2|^{1/2} |\Delta\psi^2|^{1/2} \tag{5.3}$$

holds. The negativity of the sectional curvature follows from the explicit formula (5.2) by means of (5.3) and the Hölder inequality.

Given an L^2-function ϕ on X, an eigenfunction expansion of ϕ with respect to the laplacian D implies:

$$\int_X \Delta(\phi) \overline{\phi} \, g \, dz \, d\overline{z} \geq \left| \int_X \phi \, g \, dz \, d\overline{z} \right|^2 \Big/ \int_X g \, dz \, d\overline{z} \tag{5.4}$$

The estimates of the holomorphic sectional and Ricci curvatures are based on (5.2) and (5.4).

The following condition goes beyond negative sectional curvature.

Definition. *A Kähler manifold is of* strongly negative curvature in the sense of Siu, *if its Riemann tensor satisfies*

$$R_{\alpha\bar{\beta}\gamma\bar{\delta}}(A^\alpha\overline{B^\beta} - C^\alpha\overline{D^\beta})(\overline{A^\delta B^\gamma - C^\delta D^\gamma}) \geq 0 \tag{5.5}$$

for all complex vectors $A^\alpha, B^\beta, C^\gamma, D^\delta$, and if equality holds only for $A^\alpha\overline{B^\beta} = C^\alpha\overline{D^\beta}$ for all α and β.

Proposition [SCH 1986]. *The Petersson-Weil metric on $\mathcal{T}_p$ for $p > 1$ has strongly negative curvature in the sense of Siu.*

We note that this condition is not satisfied for the invariant metric on bounded symmetric domains (cf. [SI 1986]).

5.3 The Curvature of the Petersson-Weil Metric for Singular Families

In Sect. 5.2 we discussed the asymptotics of the Petersson-Weil metric for families with singular fibers. We assume again that the parts of the singular fibers are hyperbolic and use the above notations.

Theorem [SCH 1986]. *Let $S = \{(t, \tau)\}$ and $\mathcal{X} \to S$ be a family of Riemann surfaces with ordinary double points and hyperbolic parts. Then*

(i) The absolute values of the sectional, Ricci- and scalar curvature are of type $O(-\sum_{i=1}^q \log|t_i|)$.

(ii) the Ricci tensor of the Petersson-Weil metric satisfies

$$|R_{i,\bar{j}}| = \begin{cases} O\left(\frac{1}{|t_i|\cdot\log|t_i|\cdot|t_j|\cdot\log|t_j|}\right)\cdot\min\left(\frac{1}{\log^2|t_i|}, \frac{1}{\log^2|t_j)|}\right) & \text{if } i, j \leq q \text{ and } i \neq j \\ O\left(\frac{1}{|t_i|^2\cdot\log^2|t_i|}\right) & \text{if } i = j \leq q \\ O\left(\frac{1}{|t_i|^2\cdot\log^3|t_i|}\right) & \text{if } i \leq q \text{ and } j > q \\ O(1) & \text{if } i, j > q \end{cases}$$

In [SCH 1986] also estimates for the Christoffel symbols and the curvature tensor of the Petersson-Weil metric are given.

In order to prove such estimates, one considers partial derivatives of $g^{i\bar{j}}$ according to (4.5) using the setup of Sect. 4.2. The holomorphic vector fields $\partial/\partial t_i$ and $\partial/\partial\tau_k$ are lifted to C^∞ vector fields v_i on $\mathcal{X}$ which are with respect to pinching coordinates around the singularties of the form $\partial/\partial t_i$ and $\partial/\partial\tau_k$ resp. Thus it is sufficient to estimate the respective Lie derivatives of the tensors, which occur inside the integral (4.5). The holomorphic quadratic differentials ϕ_i, whose poles are exactly known don't cause any difficulty. The main point is to estimate the derivatives of the metric tensor g.

Lemma.

$$|L_{v_i}(g)|/g = O\left(\frac{-1}{|t_i| \log |t_i|}\right) , \qquad i = 1,\dots,q$$

$$|L_{v_{\bar{j}}} L_{v_i}(g)|/g = O\left(\frac{1}{|t_i||t_j| \log |t_i| \log |t_j|}\right) ; \qquad i,j = 1,\dots q$$

Here we need a global argument. One intruduces an auxiliary metric $h(z, t_i \tau_k)$ which is around the singularities related to the hyperbolic metric on annuli and punctured discs resp. The term $\log(g/h)$ and its derivatives with respect to the vector fields v_i satisfy global elliptic equations, from which the lemma follows.

These methods are sufficient for the estimates of the curvature tensor of the Petersson-Weil metric except for $R_{i\bar{i}i\bar{i}}$. One observes that the harmonic Beltrami differentials μ_i corresponding to $\partial/\partial t_i$ and $\partial/\partial \tau_k$ can be expressed in terms of the dual basis $\{\phi_i\}$ and the Petersson-Weil metric by $\mu_i = \sum_j g_{i\bar{j}} \cdot \overline{\phi_j}/g$. Thus (4.2) can be used to show

$$|R_{i\bar{i}i\bar{i}}| = O\left(\frac{-1}{|t_i| \log^5 |t_i|}\right) \qquad \textit{for} \;\; i \leq q \; .$$

This method yields weaker estimates for the other terms.

A consequence of the above estimates is that

$$-\int_{\mathcal{M}_p} \mathrm{Ric}(\omega_{PW}) \wedge \omega_{PW}^{3p-4} < \infty \; .$$

6. Harmonic Maps and Teichmüller Space

Description of $\mathcal{T}_p$ in Terms of Hyperbolic Metrics. Another way how to phrase the classification problem of complex structures on a compact surface X of genus $p > 1$ is to consider all Riemannian metrics of constant negative curvature -1 say up to diffeomorphisms. This is due to the fact that any such complex structure gives rise to a unique hyperbolic metric on one hand and that on the other hand any Riemannian metric determines a conformal i.e. complex structure. If (for fixed genus) $\mathcal{M}_{-1}$ denotes the space of such metrics and $\mathcal{D}_0$ the group of diffeomorphisms homotopic to the identity, then the quotient space $\mathcal{M}_{-1}/\mathcal{D}_0$ can be identified with the Teichmüller space $\mathcal{T}_p$. Here we do not specify the analytic structure on $\mathcal{M}_{-1}$ and $\mathcal{D}_0$ resp. – one may consider C^∞-metrics and -diffeomorphisms or Sobolev spaces of H^k-metrics and H^{k+1}-diffeomorphisms. Such quotients have been considered for Riemannian manifolds of arbitrary dimension by Berger and Ebin, who prove a slice theorem in [B-E 1969], and by Fischer and Tromba in [F-T 1984].

Harmonic Maps of Riemann Surfaces. The idea of using harmonic maps to investigate Teichmüller space originates from Gerstenhaber and Rauch (cf. [G-R 1954] and [RE 1985]). We refer here to the approach by Fischer-Tromba

[F-T 1984a,b], Tromba [1986/7], Wolf [W 1989], and Jost [JO 1989]. According to a theorem of Schoen and Yau [S-Y 1978] for any two metrics g and γ in $\mathscr{M}_{-1}$ there exists a *unique harmonic map* $v : (X, g) \to (X, \gamma)$ which is homotopic to the identity. This fact also allows the following interpretation: For a fixed hyperbolic (reference) metric g on X one can choose in any $\mathscr{D}_0$-orbit a unique hyperbolic metric $\tilde{\gamma}$ such that the identity $(X, g) \to (X, \tilde{\gamma})$ is harmonic.

The link between classical Teichmüller theory and the approach via harmonic maps is that harmonic maps are closely related to quadratic holomorphic differentials. Let g and γ be Riemannian metrics on X of constant negative curvature -1, and z, w holomorphic coordinate functions on X related to the metrics g and γ – by abuse of notation we write $g(z)dz\overline{dz}$ and $\gamma(w)dw\overline{dw}$. The *energy density* of such a mapping $z \mapsto w(z)$ of class C^1 say is

$$e(w) = \mathscr{H} + \mathscr{L} , \tag{6.1}$$

where

$$\mathscr{H} = \gamma(w(z))|w_z|^2/g(z) \text{ and } \mathscr{L} = \gamma(w(z))|w_{\overline{z}}|^2/g(z) .$$

The engergy functional on the space of such mappings w is

$$E(w) = \int_X e(z)g(z)dz\overline{dz} . \tag{6.2}$$

It only depends on the conformal structure of the domain, as well as the Euler-Lagrange equation for its critical points, the harmonic maps:

$$w_{z\overline{z}} + (\frac{\gamma_w}{\gamma} \circ w) \cdot w_z w_{\overline{z}} = 0 \tag{6.3}$$

An immediate consequence of (6.2) is the following characterization of the decomposition of the pull back of the metric tensor γ under w:

$$w^*(\gamma)dw\overline{dw} = \phi dz^2 + e(w) \cdot g\, dz\overline{dz} + \overline{\phi}\, \overline{dz}^2 \tag{6.4}$$

where the quadratic differential

$$\phi\, dz^2 := g w_z \overline{w}_z dz^2$$

is holomorphic.

Fixing the hyperbolic metric g on the domain (and its complex structure), one has altogether a natural map from the Teichmüller space to the space of holomorphic quadratic differentials:

$$\Phi : \mathscr{T}_g \to H(X, \Omega_X^{\otimes 2}) \tag{6.5}$$

In fact the map Φ provides global coordinates on the Teichmüller space.

Theorem.
(i) (Sampson [SA 1978]): *The map Φ is injective.*
(ii) (M. Wolf [W 1989]): *The map Φ is surjective.*

Since $\log \mathscr{H}$ satisfies an elliptic equation, the injectivity of Φ is reduced to the maximum principle, and since Φ is continous due to regularity theorems, the surjectivity follows from the properness of Φ which is ultimately the properness of the energy functional.

What is the relation with classical Teichmüller theory, in particular Bers coordinates? We do not expect to find in this way global holomorphic coordinates on the Teichmüller space, but the statement of the theorem suggests to identify maybe the holomorphic tangent space of $\mathscr{T}_p$.

The theorem implies that for a basis $\phi_1, \ldots, \phi_{3p-3}$, $t = (t_1, \ldots t_{3p-3}) \in \mathbb{C}^{3p-3}$ and $\phi(t) = \sum_j t_j \phi_j$ there exist hyperbolic metrics $\gamma(t)$ and harmonic maps $z \mapsto w(z,t)$, satisfying (6.4). Since $w(z,0) = z$, one can easily compute the derivative with respect a the complex parameter t_j of the (0,2)-component of $w^*(\gamma)$ at $t = 0$. On one hand it is certainly $\overline{\phi}_j$, on the other hand must equal $g \cdot (\partial^2 w / \partial t_j \partial \overline{z})|_{t=0}$: The derivative of the associated Beltrami differential (with respect to the complex parameter t_j) is just $\overline{\phi}_j / g$. This means that the map Φ of the above theorem supplies an identification of the complex tangent space of $\mathscr{T}_p$ at the reference point given by (X, g) with the space of quadratic holomorphic differentials. When restricted to the first infinitesimal neighborhood of the distinguished fiber the induced differentiable family of compact Riemann surfaces over $H^0(X, \Omega_X^{\otimes 2})$ can in fact be identified with to the Teichmüller family equipped with the local holomorphic Bers coordinates.

The Petersson-Weil Metric. In the above approach of Wolf, the hyperbolic metric on the domain is fixed – previously Tromba had used the opposite standpoint to describe the Petersson-Weil metric by means of harmonic maps. In both set-ups the total energy of harmonic maps gives rise to a real function on the Teichmüller space.

Theorem (Tromba [TR 1987]). *The energy function on $\mathscr{T}_p$ is a $\partial\overline{\partial}$-potential for the Petersson-Weil metric at the reference point.*

A further application is the computation of the curvature of the Petersson-Weil metric (cf. Sect. 5).

Rigidity of $\overline{\mathscr{M}}_p$. Based on the classical theory of harmonic maps in Riemannian geometry and a $\partial\overline{\partial}$-Bochner-type formula Siu showed in [SI 1980] a strong rigidity theorem for compact Kähler manifolds M of strongly negative curvature. It states that any compact Kähler manifold of the same homotopy type must be either biholomorphic or conjugate biholomorphic to M. The strong negativity of the Petersson-Weil metric was verified in [SCH 1986] and under restrictions, related to the branching of the action of the Teichmüller modular group any harmonic map from $\overline{\mathscr{M}_p}$ to itself is either holomorphic or antiholomorphic. A strong rigidity theorem for $\overline{\mathscr{M}}_p$ was shown by Jost and Yau in [J-Y 1987].

7. The Compactified Moduli Space

7.1 Properties of $\overline{\mathcal{M}}_p$

Local Description. The quasi-projectivity of $\mathcal{M}_p$ was established by different methods. Baily [BA 1962] used the Jacobi-map, and Θ-series. Baily and Borel constructed a compactification $\widetilde{\mathcal{M}}_p$ by means of the period map related to the compactification of the moduli space of polarized abelian varieties in [B-B 1966]. Here the complement of the moduli space in $\widetilde{\mathcal{M}}_p$ is of codimension two. Another approach in terms of Kleinian groups and horocyclic coordinates is described in Kra's article [KR 1989]).

Although a compactification $\overline{\mathcal{M}}_p$ is a priori not uniquely defined, there is a natural choice, where the compactifying divisor $\overline{\mathcal{M}}_p \setminus \mathcal{M}_p$ has a geometric interpretation. We shall refer here to the method due to Mumford [MU 1977] and Knudsen [KN 1983b], who construct $\overline{\mathcal{M}}_p$ as a compact projective variety with at most quotient singularities. In Sect. 4.2 we essentially discussed $\overline{\mathcal{M}}_p$ from the analytic viewpoint, namely local universal families of compact (connected) Riemann surfaces with at most ordinary double points, whose parts are hyperbolic, i.e. stable curves.

The stability condition guarantees that only a finite number of cycles may be contracted yielding terminal singularities. These correspond to points, where the largest possible number of locally irreducible components of the compactifying divisor intersect. Moreover there exist only finitely many choices of contractions which lead to terminal singularities. In terms of Fenchel-Nielsen coordinates (cf. Sect. 2.1) one can see that for geodesic length functions $\ell(\alpha)$ tending to zero, one arrives at an ordinary double point. This description suggests that the divisor $\mathcal{D} = \overline{\mathcal{M}}_p \setminus \mathcal{M}_p$ is a union $\mathcal{D} = \mathcal{D}_0 \cup \ldots \mathcal{D}_{[p/2]}$, where the components $\mathcal{D}_j$ can be described as follows: The generic points of $\mathcal{D}_0$ correspond to complex curves of genus $p-1$ and 2 punctures, and for $\mathcal{D}_j; j > 0$ these are curves consisting of two components of genus j and $p-j$ resp. with one puncture each.

As for the intersection of components of $\mathcal{D}$ one may consider now a (local, universal) family $\mathcal{C} \to S$ of stable curves, where $S \subset \mathbb{C}^{3p-3}$ like in Sect. 5.2 and 5.3 and $p > 1$ is the genus of the general fiber. Let t_j be coordinate functions on the base such that the divisor of singular fibers equals $V(t_1 \cdot \ldots \cdot t_q)$. We have seen that over the sets $V(t_{j_1}, \ldots, t_{j_r}) \setminus \bigcup_{k \neq j_i} V(t_k)$ the curve $\mathcal{C}$ induces a locally universal family of Riemann surfaces with r punctures. The subset of $\overline{\mathcal{M}}_p$ corresponding to S can be identified with the quotient of S by a finite group of automorphisms.

Ample Line Bundles on $\overline{\mathcal{M}}_p$. In [MU 1977], Mumford gives explicitely an ample line bundle on $\overline{\mathcal{M}}_p$. The sheaf of relative holomorphic 1-forms $\Omega^1_{\mathcal{X}/S}$ on a family of stable curves has been defined in Sect. 5.2. For the local description one may consider a one-dimensional family $f : \mathcal{C} \to T$ given locally by $V(z \cdot w - t) \to T$; $(z, w, t) \mapsto t$. Then the relative dualizing sheaf $\omega_{\mathcal{C}/T}$ is the invertible sheaf generated by a differential ζ which equals dz/z for $z \neq 0$ and $-dw/w$ for $w \neq 0$. Let $\mathbf{m}_0 \subset \mathcal{O}_T$ be the maximal ideal sheaf for $0 \in T$. Then the sheaf $\Omega^1_{\mathcal{C}/T}$ can be identified with the subsheaf $\mathbf{m}_0 \cdot \omega_{\mathcal{C}/T}$. Like in the non-singular case sections of

$\omega_C^{\otimes e}$ for $e \geq 3$ provide an embedding of a stable curve into a projective space of dimension $\nu-1 = (2e-1)(p-1)-1$, where the degree of the image is $d = 2e(p-1)$. The Hilbert polynomial $P(n) = \chi(C, \omega_C^{\otimes n})$ equals $dn - (g-1)$. The space of all such e-canonical curves embedded as above is parametrized by a locally closed, smooth subscheme H of the Hilbert scheme $Hilb^P_{\nu-1}$, i.e we are given a family of embedded curves

$$\begin{array}{ccc} \mathscr{X} \hookrightarrow & H \times \mathbb{P}^{\nu-1} \\ \pi \searrow & \downarrow \\ & H \end{array}$$

Any isomorphism of stable curves is induced by a linear map of the ambient space, and this means a natural action of $G = PGL(\nu-1, \mathbb{C})$ on H. The term "stable" refers to this group action – it means that all orbits are closed and all stabilizer groups G_x are finite. Geometric invariant theory provides the existence of $\overline{\mathscr{M}}_p$ in the sense described so far.

The main point is now to associate to a given relatively ample line bundle $\mathscr{L}$ on $\mathscr{X}$ a line bundle on the base H which is close to ample. At least generically the direct images of $\mathscr{L}^{\otimes n}$ are locally free and the highest exterior powers (on the base H) are natural candidates. Moreover, for a construction based on intrinsic objects everything should descend to $\overline{\mathscr{M}}_p$, yielding ultimately an ample line bundle on the compactified moduli space.

We give a few details, since the construction makes an impact on the the theory of moduli of higher dimensional manifolds. The natural choice for $\mathscr{L}$ is $\omega_{\mathscr{X}/H}$. Its powers give rise to the direct image sheaves $\pi_*(\omega_{\mathscr{X}/H}^{\otimes n})$. The first direct images vanish except for $n = 1$, where the result is the structure sheaf of H. Thus the first images are locally free (by the comparison theorem). One denotes

$$\lambda_n = \Lambda^{\max}(\pi_* \omega_{\mathscr{X}/H}^{\otimes n}) = \det(\mathbf{R}^\bullet \pi_*(\omega_{\mathscr{X}/\mathrm{H}}^{\otimes \mathrm{n}})) \ ; \ \lambda := \lambda_1 \ . \tag{7.1}$$

It turns out that the choice of the power n is immaterial. The line bundles λ_n can be gotten hold of by computing their Chern classes. Let $\pi_! : K(\mathscr{X}) \to K(H)$ denote the morphism of Grothendieck groups of coherent analytic sheaves, then $c_1(\lambda_n) = c_1(\pi_!\, \omega_{\mathscr{X}/H}^{\otimes n})$ more or less by definition. The latter term is accessible from the generalized Riemann-Roch formula – it is just the degree 2 component of the direct image of the product of the character of the relative dualizing sheaf and the Todd character of the sheaf of regular relative 1-forms:

$$c_1(\lambda_n) = \left[\pi_* \left(\mathrm{Ch}(\omega_{\mathscr{X}/H}^{\otimes n}) \cdot \mathrm{Td}(\Omega^1_{\mathscr{X}/H})\right)\right]_2 \tag{7.2}$$

This expression is expanded as usual. At this place the above relationship between $\Omega^1_{\mathscr{X}/H}$ and $\omega_{\mathscr{X}/H}$ implies that the quotient of these is the structure sheaf $\mathscr{O}_{\mathrm{Sing}(\mathscr{C})}$ of the singular locus. Its first Chern class vanishes, and the second is represented by the associated 2-codimansional cycle $[\mathrm{Sing}(\mathscr{C})]$. The evaluation of the right-hand side of (7.2) yields the the following theorem, since the Picard group of $\overline{\mathscr{M}}_p$ is torsion free.

Theorem (Mumford [MU 1977]).
(i)

$$\lambda_n = \lambda^{\otimes(6n^2-6n+1)} \otimes (\delta^{-1})^{\otimes\frac{n(n-1)}{2}} , \tag{7.3}$$

where $\delta = [\mathscr{D}]$.
(ii) The bundles λ_n descend to the compactified moduli space $\overline{\mathscr{M}}_p$.
(iii) The bundle

$$\lambda^{\otimes b} \otimes (\delta^{-1})^{\otimes a}$$

is ample on $\overline{\mathscr{M}}_p$ if $a > 0$ and $b > 11.2a$, and not ample if $a \leq 0$ or $b \leq 11a$.

The last statement contains the fact that $\lambda_2 = \det(\pi_*(\omega^{\otimes 2}_{\mathscr{X}/H})) = \lambda^{\otimes 12} \otimes \delta^{-1}$ is ample.

There is a relation with Baily's result. Let $\mathscr{A}_{p,1}$ the moduli space of principally polarized abelian varieties of dimension p and $J : \mathscr{M}_p \to \mathscr{A}_{p,1}$ the map which sends a compact Riemann surface to its Jacobian. Denote by $\Theta : \mathscr{A}_{p,1} \to \mathbb{P}^N$ the realization via Θ-functions. Then

Theorem (Mumford [MU 1977]). *The map $J \circ \Theta$ extends to $\theta : \overline{\mathscr{M}}_p \to \mathbb{P}^N$ so that for some m,*

$$\theta^*(\mathscr{O}_{\mathbb{P}^N}(1)) = \lambda^m .$$

Again using the generalized Riemann-Roch formula, Harris and Mumford computed in [H-M 1982] the canonical class of the compactified moduli space:

$$K_{\overline{\mathscr{M}}_p} \equiv 13\lambda - 2\delta_0 - 3\delta_1 - 2\delta_2 - \ldots - 2\delta_{[\frac{g}{2}]}$$

where the δ_j correspond to the components of the compactifiying divisor.

Theorem (Harris-Mumford [H-M 1982], Harris [HA 1985]). *The moduli space $\mathscr{M}_p$ is of general type, if the genus is odd and at least 25, or if the genus is even and at least 40.*

7.2 The Petersson-Weil Class on $\overline{\mathscr{M}}_p$ as Chern Class of a Positive Line Bundle

The unification of algebraic geometry and the analytic approach to Teichmüller theory was achieved by Wolpert's work.

In his results the Petersson-Weil form plays a central role. Its class in $H^2(\mathscr{M}_p, \mathbb{R})$ generates this space and $(1/2\pi^2)[\omega_{PW}]$ is rational and extends to an element of $H^2(\overline{\mathscr{M}}_p, \mathbb{Q})$. It corresponds essentially to a line bundle, and the Petersson-Weil *form* extends as a (1,1)-current which is the Chern form of a continuous hermitian metric on this bundle which is known from algebraic geometry. A smoothing of the metric shows the positivity of this line bundle on $\overline{\mathscr{M}}_p$.

A link with Mumford's results is the following fiber-integral formula. One considers the Teichmüller family where the fibers are equipped with the hyperbolic metrics. These turn the relative anti-canonical bundle into a hermitian line bundle. Denote by $c_1(\mathscr{X}_p/\mathscr{T}_p, g)$ its Chern form. Its second power is a (2,2)-form whose fiber integral is a (1,1)-form on the Teichmüller space. (By functoriality the construction is compatible with base change and everything descends to the moduli space).

Theorem (Wolpert [WO 1986]).

$$\frac{1}{2\pi^2}\omega_{PW} = \int_{\mathscr{X}_p/\mathscr{T}_p} c_1(\mathscr{X}_p/\mathscr{T}_p, g)^2$$

This formula can be viewed upon as an extension of (7.2) to Chern forms rather than classes emphasizing the importance of the Petersson-Weil form.

In the sequel we shall describe the results mentioned above in more detail.

Extension of the Petersson-Weil Class to the Compactified Moduli Space. The device, how to construct an extension $[\omega_{PW}] \in H^2(\overline{\mathscr{M}}_p, \mathbb{R})$ is to change the differentiable structure near the compactifying divisor $\mathscr{D}$. The symplectic structure on the Teichmüller space can be expressed in an amazingly simple way in terms of Fenchel-Nielsen coordinates:

Theorem (Wolpert [WO 1985a]). *Let $0 < \ell_j < \infty$ and $-\infty < t_j < \infty$ for $j = 1, \ldots, 3p-3$ be Fenchel-Nielsen coordinates on $\mathscr{T}_p$. The the Petersson-Weil form equals:*

$$\omega_{PW} = -\sum_j d\tau_j \wedge d\ell_j \tag{7.3}$$

The proof of the duality formula $\omega_{PW}(\partial/\partial\tau_\alpha, \ ) = -d\ell_\alpha$ is based on a very detailed investigation of Fenchel-Nielssen vector fields and hyperbolic geometry, in particular relations for the "twist derivatives" $(\partial/\partial\tau_\alpha)(\ell_\beta)$ and $(\partial^2/\partial\tau_\alpha\partial\tau_\beta)(\ell_\gamma)$ using Petersson series.

A rather direct consequence of (7.3) is that the Petersson-Weil form (pushed down to the moduli space) possesses a differentiable extension ω^{FN} (in the sense of V-manifolds) to $\overline{\mathscr{M}}_p$.

On the other hand, the Petersson-Weil form is singular near the compactifying divisor with respect to the complex structure of the moduli space. The exact estimates are quoted in Sect. 5.2. These imply immediately that ω_{PW} is of class L^2 – the extension of ω_{PW} to $\overline{\mathscr{M}}_p$ as a (1,1)-current with respect to the complex structure is denoted by ω^C. Moreover Mazur's estimates imply that the integral of the product of ω_{PW} with the exterior derivative of a differentiable $(6p-9)$-form with compact support in a neighborhood of some point of $\mathscr{D}$ vanishes. In particular (by de Rham cohomology on V-manifolds) it induces a linear functional on $H^{6p-8}(\overline{\mathscr{M}}_p, \mathbb{R})$, by Poincaré duality this gives an element of $H^2(\overline{\mathscr{M}}_p, \mathbb{R})$.

Theorem (Wolpert [WO 1985a]). *The cohomology classes of ω^{FN} and ω^{C} in $H^2(\overline{\mathscr{M}}_p, \mathbb{R})$ are equal.*

The proof of this fact requires certain prerequesits: One is a combination of Čech- and Dolbeault cohomology theory. The other is the fact that the identity $i : \overline{\mathscr{M}}_p^{FN} \to \overline{\mathscr{M}}_p^{C}$ of the compactified moduli space equipped with the Fenchel-Nielsen and complex structures resp. is Lipschitz continuous (as opposed to i^{-1} which is not even Hölder continuous). The Lipschitz condition is finally reduced to the boundedness of the Fenchel-Nielsen coordinate vector fields with respect to L^2- and L^1-norms and estimates of $\|ds_k\|$, where s_k are holomorphic coordinate functions near the compactifying divisor.

Rationality of the Petersson-Weil Class. Now two methods became available for investigating the Petersson-Weil class. The smooth 2-form may be integrated over 2-cycles or paired with $(n-2)$-forms; the Kähler form ω^C reflects the complex geometry of $\overline{\mathscr{M}}_p$, ω^C is positive on all complex tangent planes.

The rationality of $(1/\pi^2)[\omega_{PW}]$ (on the compactified moduli space) is based on a computation of the rational homology of $\overline{\mathscr{M}}_p$. According to a result of Harer [HA 1983] the group $H_2(\mathscr{M}_p, \mathbb{Q})$ is of rank one for $p > 2$. Wolpert considers in [WO 1983] the components $\mathscr{D}_0, \ldots, \mathscr{D}_{[p/2]}$ of the compactifying divisor $\mathscr{D}$. He can show that the real first cohomology classes $[\omega_{PW}] \in H_{6p-8}(\overline{\mathscr{M}}_p, \mathbb{R})$ and those induced by $\mathscr{D}_0, \ldots, \mathscr{D}_{[p/2]}$ constitute a basis by considerations of the intersection pairing – the rank of $H_2(\overline{\mathscr{M}}_p, \mathbb{Q})$ is $2 + [p/2]$. The components of $\mathscr{D}$ together with the line bundle λ are known to form a set of generators of $\operatorname{Pic}(\overline{\mathscr{M}}_p) \otimes \mathbb{Q}$. A computation of the intersection pairing implies that $(1/\pi^2)[\omega_{PW}]$ is in fact rational – it replaces the class of the Hodge line bundle in the above basis of $H^2(\overline{\mathscr{M}}_p, \mathbb{Q})$. Here the Fenchel-Nielsen description (6.1) of the extended Petersson-Weil class is essential. In particular $[\omega_{PW}]$ is the Chern class of a line bundle. The final step of this program is to show that the Petersson-Weil form is induced by a hermitian metric on an ample line bundle which is continuous over $\overline{\mathscr{M}}_p$ and of class C^∞ over $\mathscr{M}_p$.

A Local Continuous Kähler Potential for ω_{PW}. As above we use the setup of Sect. 5.2. A local Kähler potential F is a solution of the equation $\sqrt{-1}\partial\bar{\partial}H = \omega_{PW}$.

The estimates of the Petersson-Weil metric on the cotangent bundle, in particular its boundedness, guarantee that $g_{i\bar{j}}$ possesses a smooth Hermitian metric as minorant. Thus the associated (1,1)-current $[\omega_{PW}]$ on $\overline{\mathscr{M}}_p$ is positive. The $\partial\bar{\partial}$-Poincaré lemma yields locally a strictly plurisubharmonic potential P in the sense of currents.

Any potential H satisfies $\Delta H = \operatorname{Trace}(\sqrt{-1}\partial\bar{\partial}H) = \operatorname{Trace}(\omega_{PW}) =: h$, where h is comparable to $|t_1|^{-2}(\log 1/|t_1|)^{-3} + \ldots + |t_q|^{-2}(\log 1/|t_q|)^{-3}$.

A (weak) solution H of the Laplace equation $\Delta H = h$ is a convolution with the Green's function G. It can be verified, using the asymptotics of h and G that H is continuous, and from the regularity theorem for solutions of the homogeneous equation it follows that the distribution P can be represented by a contionous plurisubharmonic function.

A Continuous Hermitian Metric. So far all computations are related to local families of compact Riemann surfaces (with singularities). The base of such a family yields a coordinate neighborhood for the V-structure of $\overline{\mathcal{M}}_p$. The notions of hermitian metrics and Chern forms are carried over to to V-manifolds. Let λ be a holomorphic line bundle on $\overline{\mathcal{M}}_p$, whose real Chern class is an integer multiple of $(1/\pi^2)[\omega_{PW}]$. The local continuous Kähler potentials give rise to a continuous hermitian metric on λ with the prescribed curvature form.

Smoothing of the Kähler Current and Projective Embedding of $\overline{\mathcal{M}}_p$. Although the Petersson-Weil class on the compactified moduli space is related to a positive current and up to a numerical factor to the Chern class of a line bundle λ, this bundle need not be a priori positive. This is accomplished by Richberg's theorem. His result is applied to show the following general fact: Let L be a line bundle on a complex manifold an h a continuous hermitian metric which is locally given by strongly plurisuperharmonic functions. Then there exists a hermitian metric of class C^∞ and strictly positive Chern form which is cohomologous to the Chern form of f as current. In particular, L is a positive line bundle.

In this way λ becomes a positive holomorphic line bundle on the V-manifold $\overline{\mathcal{M}}_p$, i.e. with respect to local uniformizing systems the hermitian metric is smooth with positive curvature. Since positive line bundles on V-manifolds yield embeddings into projective spaces (cf. [BA 1957]), the compactified moduli space $\overline{\mathcal{M}}_p$ has a projective embedding determined by the positive line bundle associated to the Petersson-Weil current $(1/\pi^2)\omega_{PW}$.

Chapter B

Moduli Spaces of Compact Kähler Manifolds

8. Existence Theorems for Moduli Spaces of Polarized Kähler Manifolds

Notions. What is the proper analogue of the Teichmüller space and the moduli space of compact Riemann surfaces for compact complex manifolds of arbitrary dimension?

A moduli space also called "coarse moduli space" is a complex space which classifies compact complex manifolds up to isomorphism. Its points correspond to such isomorphism classes and its complex structure reflects the variation of the complex structure on a fixed differentiable manifold in a holomorphic family. This idea will be made more precise. (As it stands, it gives the notion of a moduli space in the reduced category, but at the little expense of technical extensions everything can be generalized to the non-reduced case).

Set-theoretically the points of a moduli space $\mathcal{M}$ consist of isomorphism classes $[X]$ of compact complex manifolds X. A *holomorphic family* of compact complex manifolds $\mathscr{X}_s; s \in S$, parametrized by a (reduced) complex space S, is given by a *proper, smooth, holomorphic* map $f : \mathscr{X} \to S$ of complex spaces with connected fibers, such that $\mathscr{X}_s = f^{-1}(s)$ for all s. Thus one can assign to any such family f the natural map

$$\phi_f : S \to \mathscr{M}$$

which relates a point $s \in S$ to the isomorphism class $[\mathscr{X}_s]$ of its fiber.

The above set of isomorphism classes is called a *moduli space*, if it carries the structure of a complex space such that all maps of the type ϕ_f are *holomorphic*. (For non-reduced moduli spaces one has simply to require the existence of such maps, compatible with base change, whose reduction has the above property). This condition determines the holomorphic structure of a moduli space uniquely. The general problem is not well-posed: There exist families f of compact manifolds over a one-dimensional base, such that all fibers but one are isomorphic to each other – the map ϕ_f has to be constant, unless one introduces non-hausdorff moduli spaces. Certain families of Hopf surfaces or Hirzebruch surfaces have this property. The former may stand for non-Kähler manifolds, the latter for ruled manifolds. Furthermore one knows from the moduli theory of abelian varieties that one has to assign a *polarization*.

The notion of a *polarization* deserves a short discussion: In the algebraic situation it is related to a *model* i.e. a realization of variety X in a projective space. The embedding is defined by a couple of sections of a line bundle $\mathscr{L}$ on X, in fact, $\mathscr{L}$ is just the restriction of the hyperplane section bundle of the respective projective space. A polarization in the algebraic sense is induced by a line bundle, a power of which has this property. For Riemann surfaces of genus larger than one or manifolds with negative first Chern class it is given in an implicit way: These are called canonically polarized, since the canonical bundle, i.e. the bundle of holomorphic differential forms of highest degree, yields a polarization.

The interpretation of a polarization as an integer-valued Kähler class supposes the correct notion – a *polarized Kähler manifold* (X, λ_X) is by definition equipped with a Kähler class $\lambda \in H^2(X, \mathbb{R})$, and an isomorphism of polarized manifolds by definition is a biholomorphic map which carries one polarization to the other.

Moduli of abelian varieties, the period map and the Riemann relations initiated the theory of variations of Hodge stucture with implications for the theory of moduli of K3-surfaces, symplectic and unitary manifolds (cf. [TO 1988]).

Our aim here is to develop a *general theory of moduli spaces of Kähler manifolds*, based on deformation theory.

Methods. We shortly review basic notions of deformation theory.

Definition. *Let X be a compact complex manifold and (S, s_0) a complex space with a distinguished point s_0.*

A deformation *of X over (S, s_0) consists of a family $\mathscr{X} \to S$ of compact complex manifolds together with an isomorphism $X \xrightarrow{\sim} \mathscr{X}_{s_0}$.*

A holomorphic map $(R, r_0) \to (S, s_0)$ induces a deformation of X over R which is defined by the family $\mathscr{X}_R := \mathscr{X} \times_S R \to R$ together with the induced isomorphism between X and the fiber of r_0. This is called base-change.

An isomorphism of deformations is an isomorphism of families which induces the identity on X.

A deformation of a manifold X *is called* complete, *if it generates all deformations by means of base-change up to isomorphism after replacing the base spaces by neighborhoods of the distinguished points, if necessary.*

A deformation is called effective *if the derivative of a base change in the sense above at the distinguished point is uniquely determined – complete and effective deformations are called* versal.

A a universal *deformation is a complete deformation, where the above base-change itself is uniquely determined.*

The main theorem of deformation theory states the existence of versal deformations of compact complex spaces.

In the sense of a general theory of moduli spaces based on deformation theory an initial result for polarized Kähler manifolds was:

Theorem [SCH 1983b]. *There exists the moduli space of polarized Kähler manifolds with vanishing first real Chern class.*

In order to construct a moduli space for manifolds of a certain type or with additional structure one has to perform the following steps which shall be explained below more precisely:

- construction of versal (complete and effective) deformations (of polarized manifolds)
- union of quotients with identification of those points which correspond to isomorphic fibers

At this stage the set of isomorphism classes carries a natural topology. For a complex structure the following is essential:

- universality of versal deformations of manifolds of the type under consideration.
- action of the automorphism group of the central fiber on base such that the orbits consist of exactly the equivalent points.

The automorphism group of a polarized manifold (X, λ_X) is a complex Lie group which contains the identity component $\mathrm{Aut}^0(X)$, since the action of this group on $H^2(X, \mathbb{R})$ is trivial.

The notion of a family of polarized Kähler manifolds is very natural. The Kodaira-Spencer stability theorem means that neighboring fibers of a Kähler manifold are again Kähler – however, the Kähler class only depends differentiably on the parameter. Thus it is reasonable to define a polarization of a family $f : \mathscr{X} \to S$ as a (holomorphic) section $\lambda_{\mathscr{X}/S}$ of the direct image $\mathbf{R}^1 f_* \Omega_{\mathscr{X}/S}$, whose restrictions $\lambda_s := \lambda | \mathscr{X}_s$ are polarizations of the fibers. Heuristically such a real-valued holomorphic function ought to be constant: in fact it determines a section of the locally constant sheaf $\mathbf{R}^2 f_* \mathbb{R}$, whose restrictions to arbitrary fibers are Kähler classes. The converse is also true (up to some extra condition for non-reduced bases spaces).

On the other hand there is the notion of a *Kähler morphism* $f : \mathscr{X} \to S$ which is by definition a proper smooth map of complex spaces equipped with a locally $\partial\bar{\partial}$-exact real (1,1)-form, whose restriction to any fiber is a Kähler form. (It comes from a Čech-1-cochain of differentiable functions which induces a 2-cocycle with values in the sheaf of pluri-harmonic functions $\mathscr{H}$ on $\mathscr{X}$.

It may be remarkable that any family f of polarized Kähler manifolds is a Kähler morphism, even if S is singular. This follows from a very simple argument, whereas the Kodaira-Spencer stability theorem involves forth order elliptic equations.[1]

The introduction of polarized families suggests that the corresponding deformation theory makes sense; with respect to any holomorphic map $R \to S$ the pull-back $\mathscr{X}_R = \mathscr{X} \times_S R \to R$ carries a natural polarization. On the other hand, for any proper, smooth morphism $\tilde{f} : \tilde{\mathscr{X}} \to \tilde{S}$ and a polarization on a distinguished fiber $\mathscr{X}_{s_0}$, there is a maximal subgerm of (S, s_0) over which the polarization can be extended (uniquely). In such a way the following existence theorem can be proved (cf. [SCH 1983a,1984]):

Proposition. *Let (X, λ_X) be a polarized Kähler manifold. Then there exists a versal deformation.*

This first step in the necessary program for the construction of a moduli space could be done with no further assumption. It ensures that $\mathscr{M}$ carries a natural topology.

In order to get hold of isomorphic fibers in a family one may look out for a classifying space for isomorphisms between two given families and their pull-back with respect to a base change: Let $(\mathscr{X} \to S, \lambda_{\mathscr{X}/S})$ and $(\mathscr{Y} \to S, \lambda_{\mathscr{Y}/S})$ be such families. The relevant theorem also exists in the polarized case (cf. [SCH 1984] also [SCH 1983b]):

Proposition. *There exists a holomorphic map $I \to S$ and a universal isomorphism $\Phi : \mathscr{X}_I \to \mathscr{Y}_I$ over I of polarized families. One sets $I = \mathrm{Isom}_S^\lambda(\mathscr{X}, \mathscr{Y})$.*

The fibers I_s of $I = \mathrm{Isom}_S^\lambda(\mathscr{X}, \mathscr{Y}) \to S$ can be identified with the sets of isomorphisms between $\mathscr{X}_s$ and $\mathscr{Y}_s$; thus these are, if not empty, isomorphic to the groups of automorphisms of such a fiber – in particular the fibers are smooth.

We have the following general criterion for the existence of a moduli space.

---

[1] Consider on $\mathscr{X}$ the short exact sequence $0 \to \mathbb{R} \to \mathscr{O} \to \mathscr{H} \to 0$, where $\mathscr{H}$ denotes the sheaf of pluri-harmonic functions. Assume S to be Stein and reduced such that $\mathbf{R}^2 f_* \mathbb{R}$ is constant over S. The polarization determines an element of $H^2(\mathscr{X}, \mathbb{R})$, whose image in $H^2(\mathscr{X}, \mathscr{H})$ vanishes, since all restrictions to fibers do, ($\mathbf{R}^2 f_* \mathscr{O}$ is locally free). Thus the polarization comes from a 1-cocycle in $\mathscr{H}$ which can be written as the coboundary of differentiable functions giving rise to an exact (1, 1)-form. This differs from a given Kähler form on a distinguished fiber by $\partial\bar{\partial}$ of a differentiable function which can be extended to all of $\mathscr{X}$.

Theorem [SCH 1983b, 1984]. *Let $\mathcal{K}$ be a collection of compact polarized manifolds which possess versal deformations (in $\mathcal{K}$). Then there exists a moduli space for $\mathcal{K}$, if for any two families $(\mathcal{X} \to S, \lambda_{\mathcal{X}/S})$ and $(\mathcal{Y} \to S, \lambda_{\mathcal{Y}/S})$ of polarized manifolds the morphism $I = \mathrm{Isom}_S^\lambda(\mathcal{X}, \mathcal{Y}) \to S$ is proper.*

The properness of the above holomorphic map just means that for any sequence s_ν in S converging to some $s_0 \in S$ and any isomorphisms ϕ_ν of the polarized fibers $\mathcal{X}_{s_\nu}$ and $\mathcal{Y}_{s_\nu}$ there exists a sub-sequence converging to an isomorphism of the fibers of s_0.

As it stands the criterion is just the condition for the topological space $\mathcal{M}$ to be *hausdorff*. However, it automatically guarantees that the further analytic steps can be performed.

Its proof is based on purely deformation theoretic arguments – the only assumptions are the existence of versal deformations, the existence of a classifying space of isomorphisms of holomorphic families, and the above properness.[2] In this situation we have the following lemma:

Let (X, λ_X) be a polarized Kähler manifold in $\mathcal{K}$ and S the base of its versal deformation. Denote by $\sim$ the equivalence relation on S induced by the isomorphy of fibers.

Lemma [SCH 1983b, 1984]. *Under the assumption of the theorem, any versal deformation of X is universal.*

After replacing S by a neighborhood of the distinguished point the finite group $G = \mathrm{Aut}^\lambda(X)/\mathrm{Aut}^0(X)$ acts on S, and the quotient $S/\sim$ can be naturally identified with S/G.

We indicate very shortly the method of the *proof*: Let the versal deformation of X be noted in terms of a cartesian diagram:

$$\begin{array}{ccc} X & \overset{i}{\hookrightarrow} & \mathcal{X} \\ \downarrow & & \downarrow \\ 0 & \overset{\iota}{\hookrightarrow} & S \end{array}$$

One can check directly that the graph $\Gamma \subset S \times S$ of $\sim$ is the image of the (proper) canonical map $\kappa : I = \mathrm{Isom}_{S\times S}^\lambda(\mathcal{X} \times S, S \times \mathcal{X}) \to S \times S$. Any $\phi \in \mathrm{Aut}^\lambda(X)$ gives rise to another deformation of X. By the versality of the given deformation there exists a cartesian diagram

---

[2] The general version is as follows: Let **As** and **Ag** be the categories of complex spaces and germs of complex spaces resp. Let $p : \mathbf{F} \to \mathbf{An}$ be a homogeneous fibered groupoid and $q : \mathbf{G} \to \mathbf{Ag}$ the induced groupoid. Assume that any object a_0 of $\mathbf{G}$ such that $q(a_0)$ is the reduced point possesses a versal deformation a. Then there exists a moduli space for $p : \mathbf{F} \to \mathbf{Ag}$ under the condition: For any complex space S and any b, c in $\mathbf{F}$ with $p(b) = p(c) = S$ the functor $\mathrm{Isom}(b, c) : \mathbf{As} \to \mathbf{Sets}$ is representable by a space $\iota : I \to S$, where ι is proper.

$$\begin{array}{ccccc} X & \xrightarrow{\phi} & X & \xhookrightarrow{i} \mathscr{X} \xrightarrow{A} & \mathscr{X} \\ \downarrow & & & \downarrow & \downarrow \\ 0 & & \longrightarrow & S \xrightarrow{\alpha} & S \end{array}$$

where $A \circ i \circ \phi = i$. We assign to any ϕ such a base change morphism $\alpha = \alpha(\phi)$ which is *a priori not uniquely determined* by ϕ and the diagram.

The map $(\alpha, id) : S \to S \times S$ has values in Γ, and from the versality of the given deformation one can derive that the projection from $S \times S$ onto the second component, restricted to Γ is finite around (s_0, s_0), where $s_0 = \imath(0)$. In particular, there exist only finitely many choices for α. The assumption implies that $Aut^\lambda(X)$ is compact, in particular abelian. Let $\mathrm{Aut}^\varepsilon(X) \subset \mathrm{Aut}^\lambda(X)$ denote the subgroup of automorphisms which are extendable to the whole family over some neighborhood of s_0. Let $\phi, \psi \in \mathrm{Aut}^\lambda(X)$. Then $\alpha(\phi) = \alpha(\psi)$ implies that the deformations induced by ϕ and ψ are isomorphic, i.e. $\phi \circ \psi^{-1} \in \mathrm{Aut}^\varepsilon(X)$. Thus the quotient set $\mathrm{Aut}^\lambda(X)/\mathrm{Aut}^\varepsilon(X)$ is finite. In particular $\mathrm{Aut}^\varepsilon(X) \subset \mathrm{Aut}^\lambda(X)$ which implies $\mathrm{Aut}^0(X) \subset \mathrm{Aut}^\varepsilon(X)$, i.e. the dimension of the automorphism groups $\mathrm{Aut}(\mathscr{X}_s)$ is constant: we are dealing in fact with a *universal* deformation. Now the assignment $\alpha \mapsto \phi$ is unique and gives rise to a representation $\rho : G = \mathrm{Aut}^\lambda(X)/\mathrm{Aut}^\varepsilon(X) \to \mathrm{Aut}(S, s_0)$. If we let act G on $S \times S$ trivially on the first component and by ρ on the second, then we can see that the orbits of points (s, s) are contained in Γ.

We claim that on the other hand, the irreducible components Γ_j of Γ through (s_0, s_0) are of the form $\{(s, g \cdot s); s \in S\}$ for some $g \in G$: The embedding of a component Γ_j into $S \times S$ followed by either projection yields two families with isomorphic fibers, i.e. the image of $I \to S \times S$ contains Γ_j, and since this map is *proper* with smooth fibers, it possesses a section over Γ_j – both families over Γ_j are isomorphic. The existence of the desired group element $g \in G$ follows from deformation theory.

Altogether, both quotients $S/\sim$ and S/G are homeomorphic and one can see that the complex structure on the quotient, induced by the finite group G, has the necessary properties.

Remark. The universality of the versal deformation follows already from the compactness of $\mathrm{Aut}^0(X)$.

To see this, we need $\mathrm{Aut}^0(X) \subset \mathrm{Aut}^\varepsilon(X)$ in the above argument: The compactness of $\mathrm{Aut}^0(X)$ implies that a neighborhood of the "point" $\phi \in I$ in I is mapped in a proper way to a neighborhood of $(s_0, s_0) \in S \times S$ (cf. [F-S 1988b]).

Results. How could one verify the assumptions of the criterion? The following idea proved crucial: represent the polarization by a distinguished Kähler form such that any *biholomorphic map* between polarized manifolds exactly corresponds to an *isometry*. This is already sufficient – the properness of the classifying space of isomorphisms over the base follows now from the classical theorems of van Dantzig-van der Waerden and Myers-Steenrod.

In order to apply this argument to prove an existence theorem for the moduli space of polarized manifolds with vanishing first Chern class, one can take the unique Ricci flat metrics (representing the polarization) according to Yau's solution of the Calabi problem as such distinguished metrics. This method using Kähler-Einstein metrics also yields an analytic approach to the moduli space of canonically polarized manifolds. Furthermore one can combine it with the proof of the Matsusaka-Mumford theorem to show the existence of the moduli space of those polarized Kähler manifolds, for which some power of the canonincal bundle is generated by global sections (cf. [SCH 1984]).

Based upon the methods in [SCH 1983b], using the Kähler analogue of the Matsusaka-Mumford theorem in [FU 1981] the general theorem was proved:

Theorem [FU 1984, SCH 1984]. *There exists the moduli space of polarized non-uniruled Kähler manifolds.*

(Uniruled manifolds possess many rational curves – by definition these manifolds are the images of $\mathbb{P}_1$-bundle spaces under meromorphic maps which are not constant on the fibers of the bundle).

One may note that the topology of the moduli space is countable, if the underlying differentiable manifold M say is fixed and equipped with a class $\lambda \in H^2(M, \mathbb{R})$ (cf. [F-S 1988a]).

Another situation, where the criterion is applicable, are polarized *Hodge manifolds.* These are compact complex manifolds together with the isomorphism class of an ample line bundle[3] (cf. [F-S 1988b]). The result is the existence theorem of the moduli space of non-uniruled Hodge manifolds.[4]

9. Moduli Spaces of Kähler-Einstein Manifolds

9.1 The Generalized Petersson-Weil Metric

Definitions. So far, we already emphasized the close relationship between complex and metric structure on Kähler manifolds as related to the existence of a moduli space.

Let X be a complex Kähler manifold with Kähler form ω_X and Ricci form $\mathrm{Ric}(\omega_X)$. The Kähler form is called *Kähler-Einstein,* if

$$\mathrm{Ric}(\omega_X) = k \cdot \omega_X \tag{9.1}$$

for some real number k which can be normalized to $-1, 0$ or 1. (The first Chern form equals $(1/2\pi)\,\mathrm{Ric}(\omega_X)$).

[3] It is convenient to use *isomorphism classes* in order to avoid $\mathbb{C}^*$-parts in the isomorphism space.

[4] A further generalization is the existence of the moduli space of non-uniruled manifolds X with *refined* Kähler class (these are elements of $H^1(X, \mathscr{H})$, induced by Kähler forms).

In terms of local holomorphic coordinates $(z^1, \dots, z^n)$ we have

$$\omega_X = \sqrt{-1} g_{\alpha\overline{\beta}} dz^\alpha \wedge dz^{\overline{\beta}} ,$$

where $g_{\alpha\overline{\beta}}$ denotes the metric tensor. One denotes $g = \det g_{\alpha\overline{\beta}}$ so that the induced volume form equals $(\omega_X)^n = 2^n n! g dv$, dv the euklidian volume element. Then $\mathrm{Ric}(\omega_X) = -\sqrt{-1}\partial\overline{\partial}\log g$ which equals $\sqrt{-1} R_{\alpha\overline{\beta}} dz^\alpha \wedge dz^{\overline{\beta}}$ in terms of the Ricci tensor $R_{\alpha\overline{\beta}}$.

In order to describe the generalized Petersson-Weil metric we have to define the corresponding hermitian form on the tangent spaces of universal deformations. This form is compatible with base-change so that it descends to the moduli space. Let $(X, g_{\alpha\overline{\beta}})$ be a compact Kähler-Einstein manifold of constant (non-positive) curvature. Let (S, s_0) be the base of a universal deformation. By means of the Kodaira-Spencer map ρ, the tangent space of S at s_0 is identified with $H^1(X, \Theta_X)$.

This cohomology group is just the space of obstructions against a holomorphic lift of a tangent vector of the base to the total space (to be precise: to a holomorphic vector field on the infinitesimal neighborhood of the fiber which projects to the given tangent vector). Thus in terms of Dolbeault cohomology ρ is defined in the following way: let $(z^1, \dots, z^n)$ be local holomorphic coordinates of the fiber and $(s^1, \dots, s^k)$ holomorphic coordinates of S with respect to a smooth ambient space such that (z, s) can be taken as local coordinates on $\mathscr{X}$. A tangent vector on S at s_0 different from zero is the of the form $\partial/\partial s$ for some coordinate function s. Now take a differentiable lift of the form $b^\alpha(z)\partial/\partial z^\alpha + \partial/\partial s$. Then $B^\alpha_{\overline{\beta}} \frac{\partial}{\partial z^\alpha} dz^{\overline{\beta}}$ represents the Kodaira-Spencer class, where $B^\alpha_{\overline{\beta}} = \partial b^\alpha / \partial z^{\overline{\beta}}$.

Let $u \in T_{s_0}(S)$ be a tangent vector and

$$\eta = A^\alpha_{\overline{\beta}} \frac{\partial}{\partial z^\alpha} dz^{\overline{\beta}}$$

be the *harmonic* representative of $\rho(u)$ with respect to the Kähler metric $g_{\alpha\overline{\beta}}$ in terms of local holomorphic coordinates $(z^1, \dots, z^n)$.

Definition. *The Petersson-Weil inner product on T_{s_0} is given by the following norm*

$$\|u\|^2_{PW} := \int_X |\eta|^2 g dv = \int_X A^\alpha_{\overline{\beta}} \overline{A^{\overline{\delta}}_{\gamma}} g^{\alpha\overline{\delta}} g_{\gamma\overline{\beta}}\, g dv , \tag{9.2}$$

where $g dv$ denotes the volume element with respect to the given metric.

The harmonicity of η is just the following equation:

$$0 = \overline{\partial}^* \eta = g^{\overline{\beta}\gamma} A^\alpha_{\overline{\beta};\gamma} \frac{\partial}{\partial z^\alpha} . \tag{9.3}$$

(The semi-colon denotes covariant derivatives).

First the Kähler property of the generalized Petersson-Weil metric on non-singular base spaces was introduced by Koiso in [KO 1983] starting from the point of Riemannian geometry and Einstein manifolds. A technical assumption for the case of Ricci-flat metrics could be replaced by the assumption of a polarized family (for this observation cf. [SCH 1985]).

Siu computed explicitely the curvature of the generalized Petersson-Weil metric in [SI 1986].

The verification of the Kähler property heavily depended on the fact that the above harmonic representatives of the Kodaira Spencer classes give rise to symmetric 2-tensors:

$$A_{\overline{\beta}\,\overline{\delta}} = A_{\overline{\delta}\,\overline{\beta}} \tag{9.4}$$

where $A_{\overline{\beta}\,\overline{\delta}} = g_{\alpha\overline{\beta}} A^{\alpha}_{\overline{\delta}}$.

The tensor $A_{\overline{\beta}\,\overline{\delta}}$ describes in fact the variation of the Kähler-Einstein metrics on the fibers of our universal family in the direction of the tangent vector u. This shall be made more precise below.

Some results of [F-S 1988b] shall now be displayed.

A Fiber-Integral Formula. Wolpert's formula for the classical Petersson-Weil metric on the Teichmüller space of Riemann surfaces of genus larger than one can be generalized:

Let $f : \mathscr{X} \to S$ be a family of Kähler manifolds with negative first Chern class over a (reduced) space S. According to the Calabi-Yau theorem there exist unique Kähler metrics $\omega_{\mathscr{X}_s}$, $s \in S$ with some fixed $k < 0$, say $k = -1$. The relative volume forms $g(s)$ consitute a hermitian metric on the relative anti-canonical bundle $\mathscr{K}^{-1}_{\mathscr{X}/S}$. The $(2\pi/k)$-fold of the Chern form now is a real, locally $\partial\,\overline{\partial}$-exact $(1,1)$-form $\omega_{\mathscr{X}}$ of class C^∞ on $\mathscr{X}$.[5] The Kähler-Einstein condition implies that all restrictions $\omega_{\mathscr{X}}|\mathscr{X}_s$ equal $\omega_{\mathscr{X}_s}$.

Observe that this construction also works for universal families of Kähler-Einstein manifolds with positive curvature. (For existence theorems cf. [TI 1987] and [T-Y 1987]).

Although the form $\omega_{\mathscr{X}}$ is only positive definite, when restricted to fibers, it can serve to define *horizontal* lifts of tanget vectors of the base. Such horizontal lifts give now rise to distinguished representatives of the Kodaira-Spencer classes.

We have $\omega_{\mathscr{X}} = -(1/k)\sqrt{-1}\partial\,\overline{\partial}\log g(z,s)$.

The horizontal lift of $\partial/\partial s$ is just

$$a^\alpha(z)\partial/\partial z^\alpha + \partial/\partial s\,,$$

where

$$a_{\overline{\beta}} = \frac{1}{k}\frac{\partial^2 \log g}{\partial z^{\overline{\beta}}\partial s}\,.$$

(We use covariant derivatives as well as raising and lowering of indices with respect to our metrics on the fibers, e.g. $a_{\overline{\beta}} = g_{\alpha\overline{\beta}}a^\alpha$). Now (4) is automatic and (3) follows from a direct computation.

---

[5] The strategy, how to get differentiability for singular spaces S is as follows: Given a holomorphic family of compact complex manifolds $\mathscr{X}_s$, fix the underlying differentiable structure and embed S locally in some smooth space. The corresponding family of holomorphic structures can be extended to a differentiable family of almost complex structures. Now all relevant differential operators on the fibers can be extended differentiably to the ambient space. For the differentiability of the Kähler-Einstein metrics with respect to the parameter refer to the next part.

We remark that horizontal lifts in fact coincide with canonical lifts of tangent vectors in the sense of Siu [SI 1986].

Let n be the dimension of the fibers of our family. The fiber-integral of the $(n+1)$-fold power of the Chern form of the relative anti-canonical bundle equipped with our metric g is a real $(1,1)$-form.

Theorem. *The generalized Petersson-Weil form on the base of a universal family of Kähler-Einstein manifolds of Ricci curvature $k \neq 0$ can be represented as a fiber-integral:*

$$\omega_{PW} = \alpha_n \int_{\mathscr{X}/S} c_1^{n+1}(\mathscr{X}/S, g) \tag{9.5}$$

where

$$\alpha_n = -\frac{2\pi^{n+1}}{k^n(n+1)!}$$

In particular ω_{PW} possesses locally (also around singular points) a $\partial\bar{\partial}$-potential of class C^∞.

For $n = 1$ (9.5) is exactly Wolpert's formula.

The existence of a differentiable Kähler potential of ω_{PW} follows from the integral formula by means of a result of Varouchas [VA 1989] since the Chern form has this property.

We indicate, how to prove (9.5): Both sides of the equation define hermitian forms, hence one has to compare the induced (semi-)norms of any tangent vector $\partial/\partial s \in T_{s_0}(S)$. Therefore it is sufficient to restrict the map f to the subspace of the first infinitesimal neighborhood of $s_0 \in S$ which is defined by $\partial/\partial s$ and its preimage under the map f whose reduction equals the fiber X. We can assume that S is already of this kind.

Now the right hand side of (9.5) equals

$$\frac{(-\sqrt{-1})^{n+1}}{(2k)^n}\left(\int_X \phi\, dz^1 \wedge dz^{\bar{1}} \wedge \ldots \wedge dz^{\bar{n}}\right) ds \wedge d\bar{s}$$

where ϕ is the determinant of the complex Hessian of $\log g$ with respect to the coordinates $(z^1, \ldots, z^n, s)$. The Kähler-Einstein condition

$$\frac{\partial^2 \log g}{\partial z^\alpha \partial z^{\bar{\beta}}} = -k\, g_{\alpha\bar{\beta}}$$

yields

$$\phi = (-k)^n \left(\frac{\partial^2 \log g}{\partial s \partial \bar{s}} + k\, \bar{a}_\alpha a_{\bar{\beta}} g^{\alpha\bar{\beta}}\right)$$

The partial derivative with respect to s and $\bar{s}$ can be eliminated using:

$$\frac{1}{(-k)^n g}(\phi + \frac{1}{k}\sqrt{\phi}) = A^\alpha_{\bar{\beta}} \overline{A}^{\bar{\delta}}_{\gamma} g^{\alpha\bar{\delta}} g_{\gamma\bar{\beta}}\,,$$

from which (9.4) follows finally.

9.2 A Positive Line Bundle on the Moduli Space of Canonically Polarized Manifolds and the Petersson-Weil Metric

Determinant Line Bundles. Let $f : \mathcal{X} \to S$ be a universal family of canonically polarized manifolds. As for the construction of a positive line bundle on S, one can certainly not expect an entire analogy to the one-dimensional situation. The determinants of the sheaves $f_*(\mathcal{K}_{\mathcal{X}/S}^{\otimes k})$ may not be positive and proportional for $k \in \mathbb{N}$. The problem is to find a suitable substitute for such a bundle. Donaldson [DO 1987] uses in a somewhat different situation virtual vector bundles – an approach which works also in a more general way.

Let $f : \mathcal{X} \to S$ be for the moment any family of compact complex manifolds and $\mathcal{F}$ a coherent sheaf on $\mathcal{X}$. Then the theory of Knudsen-Mumford [K-M 1976] associates to the direct image $\mathbf{R}^\bullet f_* \mathcal{F}$ an invertible $\mathcal{O}_S$-module called the *determinant* of $\mathbf{R}^\bullet f_* \mathcal{F}$. [6] One uses the notation $\lambda(\mathcal{F}) := (\det \mathbf{R}^\bullet f_* \mathcal{F})^{-1}$. Let $\mathcal{F}$ be locally free and let the corresponding vector bundle be equipped with a hermitian metric $h_{\mathcal{F}}$. Assume that the fibers $\mathcal{X}_s$ carry a differentiable family of Kähler metrics. Then there exists on the determinant bundle a distinguished hermitian metric $k_{\mathcal{F}}$, the *Quillen metric*.

Chern Forms of Determinant Bundles. The theorem of Bismut, Gillet and Soulé [B-G-S 1987] states that the Chern form of the determinant bundle can be evaluated as a fiber integral:

$$c_1(\lambda(\mathcal{F}, k_{\mathcal{F}})) = -\left[\int_{\mathcal{X}/S} \mathrm{td}(\mathcal{X}/S, g)\mathrm{ch}(\mathcal{F}, h_{\mathcal{F}})\right]_2 \tag{9.6}$$

where $[\ \ ]_2$ denotes the component of degree 2, $\mathrm{ch}(\mathcal{F}, h_{\mathcal{F}})$ and $\mathrm{td}(\mathcal{X}/S, g)$ the Chern and Todd character forms resp.

The formula should be considered a generalization of the Grothendieck Riemann-Roch theorem to Chern forms rather than classes.

On may remark that by checking details of the proof and using the methods of [A-G 1962], [F-K 1972] and [S 1972] concerning relative analytic spaces, the theorem of Bismut, Gillet and Soulé can be extended to the case, where S is singular.

We call an expression of the form $\mathcal{E} = \mathcal{F} - \mathcal{G}$, where $\mathcal{F}, \mathcal{G}$ are holomorphic vector bundles on $\mathcal{X}$, a *virtual* vector bundle and define $\lambda(\mathcal{E}) := \lambda(\mathcal{F}) \otimes \lambda(\mathcal{G})^{-1}$. Its Quillen metric is $k_{\mathcal{E}} := k_{\mathcal{F}} \cdot k_{\mathcal{G}}^{-1}$. The construction of both the determinant bundle and the Quillen metric is compatible with base change. If $h_{\mathcal{F}}$ and $h_{\mathcal{G}}$ are hermitian metrics, then the Chern character *form* of the virtual bundle is defined by $\mathrm{ch}(\mathcal{E}, h_{\mathcal{E}}) := \mathrm{ch}(\mathcal{F}, h_{\mathcal{F}}) - \mathrm{ch}(\mathcal{G}, h_{\mathcal{G}})$. Since both sides of (9.6) are additive, the formula holds also for virtual bundles.

An essential device consists in *evaluating the Chern character form for virtual bundles of degree zero*, like

[6] The direct image of $\mathcal{F}$ equals in the derived category locally with respect to S a bounded complex of free sheaves of finite type on S, for which the determinant sheaf is defined in the obvious way as product of determiniants with alternating exponents ± 1.

$$\mathscr{E} = \left(\mathscr{L} - \mathscr{L}^{-1}\right)^{\otimes n} \otimes \left(\mathscr{K}_{\mathscr{X}/S} - \mathscr{K}_{\mathscr{X}/S}^{-1}\right)$$
$$\mathscr{F} = \mathscr{L} \otimes \left(\mathscr{L} - \mathscr{L}^{-1}\right)^{\otimes n},$$

where $\mathscr{L}$ is a hermitian line bundle on $\mathscr{X}$.

For these, the lowest terms of the Chern character forms are in degree $(n+1, n+1)$ and (n, n) resp.

Proposition. *Let $\mathscr{E}$ and $\mathscr{F}$ be the above virtual line bundles. Then*

$$c_1\left(\lambda(\mathscr{E}, k_{\mathscr{E}})\right) = -2^{n+1} \int_{\mathscr{X}/S} c_1(\mathscr{L}, h)^n c_1(\mathscr{X}/S, g) \tag{9.7}$$

and

$$c_1\left(\lambda(\mathscr{F}, k_{\mathscr{F}})\right) = 2^{n-1} \int_{\mathscr{X}/S} c_1(\mathscr{L}, h)^n c_1(\mathscr{X}/S, g) - 2^n \int_{\mathscr{X}/S} c_1(\mathscr{L}, h)^{n+1} \tag{9.8}$$

Construction of the Line Bundle. For $\mathscr{L} = \mathscr{K}_{\mathscr{X}/S}$ our fiber-integral formula (9.5) and formula (9.7) yield:

Theorem. *Let $\mathscr{X} \to S, \omega_{\mathscr{X}/S}$ be a universal family of Kähler-Einstein manifolds of non-zero curvature k. Then there exists a natural hermitian line bundle on S, whose Chern form is up to a numerical constant the Petersson-Weil form. Namely*

$$\omega_{PW} = \beta_n\, c_1\left(\lambda((\mathscr{K}_{\mathscr{X}/S} - \mathscr{K}_{\mathscr{X}/S}^{-1})^{n+1}), \tilde{k}\right), \tag{9.9}$$

where

$$\beta_n = \frac{(-\pi)^n}{2^n k^n (n+1)!},$$

(the Quillen metric on the respective line bundle being denoted by $\tilde{k}$).

This line bundle descends to the moduli space of canonically polarized varieties. Its hermitian metric also descends (to a continous metric).

As mentioned before, the moduli space is locally the quotient of the base spaces of versal deformations by finite groups of automorphisms – it may be called a *V-space* in an analogous way to a V-manifold. The generalized Petersson-Weil metric on S satisfies the strongest possible condition on differentiability on S – its $\partial\bar{\partial}$-potential on the quotient may just be continous though.

10. Moduli Spaces of Extremal Kähler Manifolds

10.1 Construction of the Moduli Space

Extremal Kähler Metrics. Let (X, λ_X) be a polarized Kähler manifold. According to Calabi [CA 1979, 1985] one considers on the space U of all Kähler forms ω on X which represent the polarization, the following functional Φ defined by

$$\Phi(\omega) := \int_X R^2(\omega)\, \omega^n , \qquad \omega \in U, \ n = \dim X , \tag{10.1}$$

where R denotes the *scalar curvature* $R = g^{\alpha\overline{\beta}} R_{\alpha\overline{\beta}}$ of ω.

Definition. *A critical point of the functional Φ is called an* extremal *Kähler form on X.*

The *Euler-Lagrange equation* of ϕ is (cf. [CA 1982])

$$R_{;\overline{\beta}\,\overline{\delta}} = 0 . \tag{10.2}$$

Remark. The Euler-Lagrange equation just means that the scalar curvature as a differentiable function gives rise to a *holmorphic vector field*, since (10.2) is equivalent to

$$\overline{\partial}\left(g^{\alpha\overline{\beta}} R_{;\overline{\beta}} \frac{\partial}{\partial z^\alpha}\right) = 0$$

or in the notation of [CA 1982]

$$\sqrt{-1}\partial \uparrow_g \overline{\partial} R = 0 .$$

One can see that any Kähler manifold of constant scalar curvature, in particular any Kähler-Einstein manifold, is extremal.

Let $J : \mathrm{Aut}^0(X) \to Alb(X)$ be the Jacobi map of an arbitrary polarized manifold. Then the Lie algebra $a(X)$ of its kernel consists of those holomorphic vector fields which are orthogonal to holomorphic one-forms with respect to the natural pairing and thus coincides with the set of holomorhic vector fields originating from covariant derivatives of complex-valued differentiable functions by "pulling up indices". Note that $\ker J$ is a linear algebraic group by a theorem of Fujiki [FU 1978].

Thus the compactness of $\mathrm{Aut}^0(X)$ is equivalent to

$$a(X) = 0$$

We call this *condition (A)*.

Under this condition, any versal deformation is universal (cf. Sect. 8).

By a theorem of Matsushima [MA 1957], Lichnerowitz [LI 1959] and Calabi [CA 1979] on an extremal Kähler manifold the Lie algebra of holomorphic vector fields is the direct sum of the ideal $a(X)$ and the algebra of parallel holomorphic vector fields.

In particular, on an extremal Kähler manifold (X, ω_X) with compact $\mathrm{Aut}^0(X)$ any biholomorphic map in this group is an isometry. Moreover, any extremal Kähler manifold with $a(X) = 0$) is obviously of *constant scalar curvature*; although we will allways have to assume *(A)* from now on, we will keep the notion of extremal manifolds. (If we waive this condition, holomorphic families of extremal Kähler manifolds are hard to define).

Deformations of Extremal Manifolds. In our situation, all versal deformations of a manifold X with condition (A) are universal (cf. Sect. 8), we may expect that the combination of extremality and compactness of $\mathrm{Aut}^0(X)$ yields the assumptions of our criterion for the existence of a moduli space.

What is the correct definition of a family of extremal Kähler manifolds? In any case, it should induce a polarized family.

Definition. *A family $\underline{\mathscr{X}t}_e$ of extremal Kähler manifolds is a family of compact manifolds $f : \mathscr{X} \to S$ together with a C^∞-family of extremal Kähler metrics $\tilde{\omega} = \{\omega_s\}$ on the fibers $\mathscr{X}_s$, $s \in S$ which represent a polarization.*

This becomes a reasonable definition from the analytic viewpoint because of the following

Theorem. *Let $(f : \mathscr{X} \to S, \lambda_{\mathscr{X}/S})$ be a family of polarized Kähler manifolds with condition (A) and (X, λ_X) the fiber of a point $s_0 \in S$. Let ω_X be an extremal Kähler form which represents the polarization. Then ω_X possesses an extension to a C^∞-family of extremal Kähler metrics on the neighboring fibers which represents the polarization,* uniquely.

As usual, the notion of a "holomorphic familiy" generates the notion of a deformation:

Corollary. *Any compact extremal Kähler manifold X with compact* $\mathrm{Aut}^0(X)$ *possesses a universal deformation.*

Such universal deformations are just universal deformations of the underlying polarized families with extremal metrics extended according to the theorem.

We indicate the proof of the theorem: As observed in Sect. 8, we can extend ω_X to a loacally $\partial\bar{\partial}$-exact $(1,1)$-form $\tilde{\omega}_{\mathscr{X}}$ of class C^∞ (even if S is singular). Our approach is to set

$$\omega_{\mathscr{X}} = \tilde{\omega}_{\mathscr{X}} + \sqrt{-1}\partial\bar{\partial}\Phi\,,$$

where Φ is a real-valued function, uniquely determined by $\omega_{\mathscr{X}}$ up to a function on S which can be normalized. Using a differentiable trivialization of the underlying holomorphic family, one considers Φ as a map from the base with values in a Sobolev space $H^k(X)$ of normalized differentiable functions on the fiber X.(Constant functions are zero). Any $s \in S$ and function $\psi \in H^k(X)$ close to s_0 and zero resp. determine a Kähler form on the respective fiber. Denote by $R(\psi, s)$ the scalar curvature of the Kähler form corresponding to ψ and s. Then $R(\psi, s)$ can be considered as a map with values in $H^{k-4}(X)$ up to some error term which is due to the fact that the integral of the scalar curvature over a fiber is no longer a constant, if we use a trivialization and a fixed volume form. We want to solve the equation $R(s, \psi(s)) = \text{const.}$ by means of the generalized implicit function theorem. The partial derivative $L : H^k(X) \to H^{k-4}(X)$ of this function with respect to the second derivative at $(s_0, 0)$ is

$$L(\phi) = g^{\alpha\overline{\beta}} g^{\gamma\overline{\delta}} \left(\phi_{\alpha\overline{\beta}\gamma\overline{\delta}} + R_{\overline{\beta}\gamma}\phi_{\alpha\overline{\delta}}\right) ,$$

in terms of the metric and Ricci tensor on the central fiber X. A direct computation, based upon R =const. on X, shows that

$$L(\phi) = g^{\alpha\overline{\beta}} g^{\gamma\overline{\delta}} \phi_{\overline{\beta}\,\overline{\delta}\alpha\gamma} .$$

In this setting, the assumption $a(X) = 0$ is applicable; one can see that the elliptic operator L has a bounded inverse. The claim follows from classical regularity theorems.

Remark. Although the extremal metrics on the neighboring fibers are uniquely determined, uniqueness does in general not hold for the inducing locally $\partial\overline{\partial}$-exact $(1,1)$-form $\omega_{\mathscr{X}}$ unless the first Betti number of the fibers vanishes.

Another consequence of the theorem concerns isomorphisms of families of extremal manifolds:

Let $\underline{\mathscr{X}_e} = (\mathscr{X} \to S, \tilde{\omega})$ and $\underline{\mathscr{Y}_e} = (\mathscr{Y} \to S, \tilde{\omega}')$ be families of extremal Kähler manifolds with (A), and $\underline{\mathscr{X}}$, $\underline{\mathscr{Y}}$ the underlying families of polarized manifolds.

Proposition. *The isomorphism functor of extremal families is representable by a a complex space $I_e = \mathrm{Isom}_S(\underline{\mathscr{X}_e}, \underline{\mathscr{Y}_e}) \to S$ such that the natural map from I_e to the classifying space I of isomorphisms between $\underline{\mathscr{X}}$ and $\underline{\mathscr{Y}}$ is locally biholomorphic. The map $I_e \to S$ is proper.*

The first claim follows from the theorem. Since we only know that extremal metrics are $\mathrm{Aut}^0(X)$-invariant (under condition (A)), and no general uniqueness or existence theorem is known, one cannot say more. The classifying space of isomorphisms carries the topology of compact convergence. As isomorphisms of extremal Kähler manifolds are *isometries*, the properness follows from classical theorems mentioned in the beginning of our discussion.

All of the preceeding considerations can be realized for *extremal Hodge manifolds*, i.e. for Hodge manifolds (X, L) equipped with an extremal Kähler form which represents the Chern class of L.

We are now able to apply our criterion:

Theorem. *There exist the moduli spaces $\mathscr{M}_e$ and $\mathscr{M}_{H,e}$ resp. of extremal Kähler and extremal Hodge manifolds resp. for which the identity components of the automorphism groups are compact.*

Remark. The above theorem includes also the existence of the moduli space of Kähler-Einstein manifolds of positive curvature with compact automorphism groups.

10.2 The Generalized Petersson-Weil Metric

It was impossible to generalize directly the Petersson-Weil metric to the case of extremal Kähler manifolds. The approach to use harmonic representatives of Kodaira-Spencer classes failed. Any proof of the Kähler property had to depend upon a close relationship with certain symmetric tensors – like the infinitesimal deformations of the metric tensor of a Kähler-Einstein metric. The answer is based upon a formula of Berger and Ebin [B-E 1969] for the derivative of the scalar curvature with respect to some parameter an a decomposition theorem for symmetric tensors in the same article. The proper distinguished representatives satisfy a *forth* order elliptic equation, and from the definition, it is possible to verify the Kähler condition by a lengthy computation – at least when the base is smooth. Here we shall present a fiber integral formula and an approach to the Petersson-Weil metric which proved to be very natural – a posteriori.

Infinitesimal Deformations of Polarized Manifolds. For a family $f : \mathscr{X} \to S$ of complex manifolds (always assume (A)) and a polarization λ_X of a fiber X, the obstructions against its extension consisted in a holomorphic section of $\mathbf{R}^2 f_* \mathcal{O}_{\mathscr{X}}$ (cf. Sect. 8). Infinitesimally the assignment of a tangent vector of S to its obstruction is (via the Kodaira-Spencer map) just the map $\cup_{\lambda_X} : H^1(X, \Theta_X) \to H^2(X, \mathcal{O}_X)$, induced by the cup-product. Its kernel $H^1(X, \Theta_X)_{\lambda_X}$ is the space of infinitesimal deformations of (X, λ_X). With respect to the canonical isomorphism $H^1(X, \Theta_X) \xrightarrow{\sim} \mathrm{Ext}^1(X, \mathcal{O}_X)$ the given polarization λ_X corresponds to the isomorphism class of an extension

$$(10.3) \qquad 0 \to \mathcal{O}_X \to \Sigma_X \to \Theta_X \to 0$$

If the polarization is the Chern class of an element $L \in \mathrm{Pic}(X)$ the sequence (10.3) is just the Atiyah-sequence, and $H^1(X, \Sigma_X)$ is the space of infinitesimal deformations of the Hodge manifold[7] (X, L), and $H^0(X, \Sigma_X)$ is the space of infinitesimal automorphisms of (X, L).

The induced cohomology sequence may give some more insight. We assume $a(X) = 0$, then the following is exact.

$$(10.4) \quad 0 \to H^0(X, \Theta_X) \xrightarrow{\delta_0} H^1(X, \mathcal{O}_X) \to H^1(X, \Sigma_X) \to H^1(X, \Theta_X) \xrightarrow{\delta_1} H^2(X, \mathcal{O}_X) \ldots$$

The edge-homomorphisms δ_0 and δ_1 are just given by the cup product with the polarization. The map δ_0 is induced by the infinitesimal action of $\mathrm{Aut}^0(X)$ on $\mathrm{Pic}(X)$.

The Generalized Petersson-Weil Metric. Let $(\mathscr{X} \to S, \mathscr{L}, \tilde{\omega})$ be a family of polarized, extremal Hodge manifolds – $\mathscr{L}$ is a section of the relative Picard group and $\tilde{\omega}$ a family of extremal Kähler metrics. Let $\mathscr{L}$ be represented by an actual line bundle denoted by the same letter. It is desirable to construct a hermitian metric h

[7] In the general case it corresponds to infinitesimal deformations of the manifold X equipped with a refined Kähler class.

on $\mathscr{L}$ such that $\omega_{\mathscr{X}} := 2\pi c_1(\mathscr{L}, h)$ induces on all fibers the given extremal metrics. Starting from an arbitrary metric on $\mathscr{L}$ the method described in Sect. 10.1 works - and in a sense the ambiguity is taken from h and $\omega_{\mathscr{X}}$. These are now unique up to a factor and summand resp. which depend only on the parameter $s \in S$. [8]

Now as in the Kähler-Einstein case (cf. Sect. 9.1), we take horizontal lifts of tangent vectors $\partial/\partial s$ of the base of the family which give rise to certain representatives $\eta = B^{\alpha}_{\overline{\beta}} \partial/\partial z^{\alpha} dz^{\overline{\beta}}$ of the associated Kodaira-Spencer class on the fiber X. A computation like in the Kähler-Einstein case furnishes the following equations:

$$(10.5)(a) \qquad B_{\overline{\beta}\,\overline{\delta}} = B_{\overline{\delta}\,\overline{\beta}}$$

$$(10.5)(b) \qquad g^{\overline{\beta}\alpha} g^{\overline{\delta}\gamma} B_{\overline{\beta}\,\overline{\delta};\alpha\gamma} = 0$$

$$(10.5)(c) \qquad B_{\overline{\beta}\,\overline{\delta};\overline{\tau}} = B_{\overline{\beta}\,\overline{\tau};\overline{\delta}}$$

Equation (a) is the relationship of η with the first derivative of the metric tensor, and (c) is the $\overline{\partial}$-closedness of η, or the infinitesimal integrability condition of the almost complex structure $B^{\alpha}_{\overline{\beta}}$.

The second formula (b) can be recognized easily as the equation for the constancy of the scalar curvature (cf. [B-E 1969]), where all terms but one vanish on a Kähler manifold. It is related to decomposition theorems of spaces of symmetric tensors with respect to certain elliptic operators in the same article. (A more general construction related to an elliptic complex is due to Calabi [CA 1960]). It just says that $B_{\overline{\beta}\,\overline{\delta}}$ is orthogonal with respect to the L^2-inner product to the space of tensors of the form

$$f_{\overline{\beta}\,\overline{\delta}}$$

where f is a C^{∞}-function. One can check in different ways that $(10.5)(a) - (c)$ define distinguished representatives of classes in $H^1(X, \Sigma_X)$. The image of $B_{\overline{\beta}\,\overline{\delta}}$ in $H^1(X, \Theta_X)$ is the class of η. Given a harmonic $(0,1)$-form $u_{\overline{\beta}} dz^{\overline{\beta}}$ on X, its image in $H^1(X, \Sigma_X)$ equals $u_{\overline{\beta};\overline{\delta}} dz^{\overline{\beta}} \wedge dz^{\overline{\delta}}$ Thus, one can identify the kernel of δ^1, i.e. the space of infinitesimal deformations of the underlying polarized manifold with the space of tensors $A_{\overline{\beta}\,\overline{\delta}}$ which satisfy the additional condition

$$(10.5)(d) \qquad \int_X A_{\overline{\beta}\,\overline{\delta}} v_{\alpha;\gamma} g^{\overline{\beta}\alpha} g^{\overline{\delta}\gamma} g dv = 0\ , \ \text{for all holomorphic 1-forms } v_{\alpha} dz^{\alpha}\ .$$

Definition. *Let a universal family of extremal, polarized Kähler or Hodge manifolds be given. Associate to any Kodaira-Spencer class the distinguished representative satisfying (10.5)(a)–(d) or (a)–(c) resp. Then the* generalized Petersson-Weil

[8] The analogous construction works for manifolds with refined Kähler classes. So it can also be applied in the Kähler case.

metric *on the base of such a family is defined by the* L^2*-inner product (cf.* (9.2)*) of these tensors. We denote the corresponding differential form on* S *by* $\hat{\omega}_{PW}$.

Remark. There is a natural map from the moduli space of (extremal) polarized Hodge manifolds to the moduli space of polarized Kähler manifolds. This map is also defined on the level of universal families. The fibers are smooth related to the quotients of the form $H^1(X, \mathcal{O}_X)/H^0(X, \Theta_X)$. One can read of the formula of the Petersson-Weil metric that it is *flat*, when restricted to the fibers.

A Fiber-Integral Formula. Our general situation can be characterized as follows: We are given a family $f : \mathscr{X} \to S$ of compact complex manifolds over a complex space S (which need not be even reduced at this place), and a (1,1)-form $\omega_{\mathscr{X}}$ on $\mathscr{X}$ which has locally a real $\partial\bar{\partial}$-potential of class C^∞ and induces on all fibers Kähler metrics. Then horizontal lifts of tangent vectors of the base are well-defined, (even if $\omega_{\mathscr{X}}$ is only fixed up to some form $f^*(\omega_S)$, where ω_S is a locally $\partial\bar{\partial}$-exact (1,1)-form on S). Such a differential form may be called *admissible.*

As discussed above the horizontal lift of tangent vectors furnishes representatives of the Kodaira-Spencer classes related to symmetric 2-tensors, and the L^2-inner product of such constitutes a hermitian form on the tangent space of S at the respective point. The form is positive definite, provided the family is versal. Denote by Ω_S the corresponding (1,1)-form on S.[9]

An admissible form $\omega_{\mathscr{X}}$ induces a hermitian metric g on the relative anti-canonical bundle.

Theorem. *Let* $\mathscr{X} \to S, \omega_{\mathscr{X}}$ *be as above. Then*

$$\Omega_S = -\int_{\mathscr{X}/S} \left(2\pi c_1(\mathscr{X}/S, g) \frac{\omega^n}{n!} - R \frac{\omega^{n+1}}{(n+1)!} \right) \tag{10.6}$$

(Here R *is the differentiable function, whose restriction to a fiber is the respective scalar curvature of* $\omega_{\mathscr{X}} | \mathscr{X}_s$*).*

In particular, the real form Ω_S *possesses locally a* $\partial\bar{\partial}$*-potential of class* C^∞.

The *proof* depends on the methods used in the Kähler-Einstein case (cf. Sect. 9.1).

Let $(\mathscr{X} \to S, \mathscr{L}, \tilde{\omega})$ be a family of extremal polarized Hodge manifolds with a hermitian metric h and global (1,1)-form $\omega_{\mathscr{X}} = 2\pi c_1(\mathscr{L}, h)$ arranged like in the preceding section. The general theorem above now yields a fiber-integral formula for the Petersson-Weil form:

Corollary. *The Petersson-Weil form on the base of a universal family of extremal Hodge manifolds equals*

---

[9] Although the dimension of the tangent space need not be constant, differentiability with respect to the parameter is not a problem (cf. footnote 5).

$$(10.7) \qquad \hat{\omega}_{PW} = -(2\pi)^{n+1}\left(\int_{\mathscr{X}/S} c_1(\mathscr{X}/S, g)\, \frac{c_1(\mathscr{L}, h)^n}{n!} - R\int_{\mathscr{X}/S} \frac{c_1(\mathscr{L}, h)^{n+1}}{(n+1)!}\right)$$

in particular, $\hat{\omega}_{PW}$ is a Kähler form which possesses locally a Kähler potential.[10]

The value of the scalar curvature is in terms of the cohomology classes

$$R = c_1(\mathscr{X}_s)\cdot c_1(\mathscr{L}_s)^{n-1}/c_1^n(\mathscr{L})^n$$

which does not depend on the parameter s.

The contribution from the second term in (10.7) can be cancelled in the following way: If $\int \omega_{\mathscr{X}}^{n+1} = \eta$, replace $\omega_{\mathscr{X}}$ by $\omega_{\mathscr{X}} - \alpha^{-1} f^*\eta$, where $\alpha = (n+1)\int_{\mathscr{X}/S} \omega_{\mathscr{X}}^n$ is up to a factor just the (constant) volume of the fibers. This does not affect horizontal lifts neither the metrics on the fibers. This adjustment can be raised locally with respect to S to the level of hermitian metrics on $\mathscr{L}$, since η has a $\partial\bar{\partial}$-potential. One may call such admissible forms and hermitian metrics normalized and make up a deformation theory in this way. However, the hermitian metric on $\mathscr{L}$ is only unique up to a constant, a fact which caused a problem, when putting a positive bundle on the moduli space.

Corollary. *The Petersson-Weil form on the base of a universal family of extremal Kähler manifolds possesses locally a Kähler potential of class C^∞.*

If the first Betti number of the fiber vanishes, or more generally, if b_1 equals the dimension of Aut(X), the corollary follows immediately from (10.7). Otherwise one has to use the remark in Sect. 10.2. For details cf. [F-S 1988b]).

10.3 Positive Line Bundles

We indicate how to construct a positve line bundle on the moduli spaces of extremal Hodge manifolds and extremal Kähler manifolds (with integer-valued or rational) polarization, whose curvature is up to some numerical constant the Petersson-Weil form $\hat{\omega}_{PW}$ or ω_{PW}.

Local Construction. Let $(f : \mathscr{X} \to S, \mathscr{L}, \tilde{\omega})$ be a family of extremal Hodge manifolds. The idea is to combine the formulas (9.7) and (9.8) for the Chern forms of the determinant bundles arising from the virtual bundles $\mathscr{E} = (\mathscr{L} - \mathscr{L}^{-1})^{\otimes n} \otimes (\mathscr{K}_{\mathscr{X}/S} - \mathscr{K}_{\mathscr{X}/S}^{-1})$ and $\mathscr{F} = \mathscr{L} \otimes (\mathscr{L} - \mathscr{L}^{-1})^{\otimes n}$ resp. with the fiber-integral expression (10.7) for $\hat{\omega}_{PW}$. This can be realized by a suitable linear combination of these bundles:

Denote by $a = c_1(\mathscr{X}_s)\cdot c_1(\mathscr{L}_s)^{n-1}$ and $b = c_1(\mathscr{L}_s)^n$ natural numbers (not depending on s) with $R = na/b$ and set $\mathscr{G} = \mathscr{E}^{\oplus(na-2(n+1)b)} \oplus \mathscr{F}^{\oplus 4na}$.

Theorem. *The Petersson-Weil form on the base of a universal deformation of extremal Hodge manifolds equals up to a numerical constant the Chern form of the*

[10] Again there is an analogous theorem for families of extremal Kähler manifolds with refined Kähler classes.

determinant bundle of a virtual vector bundle on the total space, equipped with the Quillen metric:

Namely

$$\hat{\omega}_{PW} = \frac{-(\pi/2)^{n+1}}{n!} c_1(\lambda(\mathcal{G}, k_{\mathcal{G}})) \,. \tag{10.8}$$

For any universal deformation of an extremal Hodge manifold (X, L, ω_X) the action of the automorphism group of the fiber on the base (S, s_0) descends to the determinant bundle $\lambda(\mathcal{G}, k)$ by functoriality of the construction. Denote by $G \subset \mathrm{Aut}(S, s_0)$ the associated finite group. Then G acts trivially on some power $\lambda(\mathcal{G}, k)^m$ (e.g. take as m the order of G). In such a situation the power of determinant bundle descends to a determinant bundle on S/G and the Quillen metric descends to a continuous metric on the push-down. One may call this a hermitian line bundle with respect to the V-structure or V-hermitian line bundle.

Global Construction on the Moduli Space of Extremal Hodge Manifolds. First, we can decompose $\mathcal{M}_{H,e}$ into a disjoint union (of open and closed) subspaces $\mathcal{M}^P_{H,e}$ corresponding to isomorphism classes of (X, L, ω_X), where P is the Hilbert polynomial of L. We consider such a component. We fix a uniform exponent such that all powers of the line bundles L in question are very ample. Then one can show that such a $\mathcal{M}^P_{H,e}$ is essentially a quotient of an open subset U in the corresponding Hilbert scheme. The main point is that one can construct a global family of extremal Hodge manifolds with a global G-invariant hermitian metric on the line bundle. Now the method explained in 10.1 yields globally a hermitian metric, whose Chern form is admissible for the family of extremal manifolds, i.e. its restriction to the fibers yields the extremal metrics. The construction of the determinant bundle as well as (10.8) can be globalized: A power of the determinant bundle, equipped with the Quillen metric exists globally on $\mathcal{M}^P_e$ and yields the Petersson-Weil metric. This is the content of the following

Theorem. *The moduli space $\mathcal{M}_{H,e}$ of extremal polarized Hodge manifolds possesses a V-hermitian line bundle whose Chern form equals up to a numerical constant the Petersson-Weil form $\hat{\omega}_{PW}$.*

Denote by $\mathcal{M}'_e \subset \mathcal{M}_e$ the space of extremal Kähler manifolds with integer-valued polarization and by $p : \mathcal{M}_{H,e} \to \mathcal{M}'_e$ the natural projection which assigns to an extremal Hodge manifold the underlying polarized Kähler manifold. Then p is proper, its fibers are essentially the quotients of the Picard groups by the automorphism groups of the corresponding manifolds. We have seen in the remark following (10.5) that $\hat{\omega}_{PW}$ is flat, when restricted to the fibers of p. In particular on respective connected components of the moduli spaces (which we denote by the same letter)

$$\omega_{PW} = \text{const.} \int_{\mathcal{M}_{H,e}/\mathcal{M}'_e} \hat{\omega}^{m+1}_{PW} \,, \tag{10.8}$$

where m is the dimension of the fibers of p.

On the other hand, let $(\mathcal{G}, k)$ be the determinant bundle on $\mathcal{M}_{H,e}$ which induces $\hat{\omega}_{PW}$. Then like in Sect. 9.2 we get

$$c_1((\mathcal{G} - \mathcal{G}^{-1})^{\otimes m+1}) = 2^{m+1} \int_{\mathcal{M}_{H,e}/\mathcal{M}'_e} c_1(\mathcal{G}, k)^{m+1} . \tag{10.9}$$

Thus

Theorem. *The moduli space of $\mathcal{M}'_e$ of extremal Kähler manifolds with integer-valued polarization carries a hermitian line bundle, whose Chern form is up to a constant the Petersson-Weil form.*

Corollary. *All compact complex subspaces of the moduli spaces $\mathcal{M}'_e$ and $\mathcal{M}_{H,e}$ resp. including the moduli spaces of Kähler-Einstein manifolds are* projective.

Note added in proof: In the meantime A. M. Nadel has introduced multiplier ideal sheaves in order to prove the existence of Kähler-Einstein metrics. Examples include Del Pezzo surfaces and complete intersections of low degree, the blow-up of $\mathbb{P}_4$ along the intersection of two quartic hypersurfaces, and the cubic threefold along an elliptic curve.

E. Viehweg has proved the quasi-projectivity of the moduli scheme of polarized compact manifolds (with fixed Hilbert polynomial) and semi-ample canonical bundle.

References

[AB 1980] Abikoff, W.: The real analytic theory of Teichmüller space. Lecture Notes in Mathematics, vol. 820) Springer, Berlin Heidelberg 1980

[AH 1938a] Ahlfors, L.V.: On quasiconformal mappings. Journal d'Analyse Math. **3** (1938) 359–364

[AH 1938b] Ahlfors, L.V.: An extension of Schwarz's lemma. Transactions of the AMS **43** (1938) 359–364

[AH 1953] Ahlfors, L.V.: On quasiconformal mappings. J. d'Analyse Math. **3** (1953) 1–58

[AH 1960] Ahlfors, L.V.: The complex analytic structure of the space of closed Riemann surfaces. In: Analytic Functions. Princeton University Press 1960

[AH 1961a] Ahlfors, L.V.: Some remarks on Teichmüller's space of Riemann surfaces. Ann. Math. **74** (1961) 171–191

[AH 1961b] Ahlfors, L.V.: Curvature properties of Teichmüller's space. J. d'Analyse Math. **9** (1961) 161–176

[A-B 1960] Ahlfors, L., Bers, L.: Riemann mapping theorem for variable metrics. Ann. Math. **72** (1960) 385–404

[A-G 1962] Andreotti, A., Grauert, H.: Théorème de finitude pour la cohomologie des espace complexes. Bull. Soc. Math. France **90** (1962) 193–259

[BA 1957] Baily, W.L.: On the imbedding of V-manifolds in projective space. Am. J. Math. **79** (1957) 403–430

[BA 1962] Baily, W.L.: On the theory of Θ-functions, the moduli of abelian varieties and the moduli space of curves. Ann. Math. **75** (1962) 342–381

[B-B 1966] Baily, W.L., Borel, A.: Compactification of arithmetic quotients of bounded symmetric domains. Ann. Math **84** (1966) 442–528

[B-E 1969] Berger, M., Ebin, D.G.: Some decompositions on the spaces of symmetric tensors on a Riemannian manifold. J. Diff. Geom. **3** (1969) 379–392

[BE 1960] Bers, L.: Spaces of Riemann surfaces. Proc. Int. Cong. 1958, Cambridge 1960

[BE 1970] Bers, L.: On boundaries of Teichmüller spaces and Kleinian groups I. Ann. Math. **91** (1970) 570–600

[BE 1974] Bers, L.: Spaces of degenerating Riemann surfaces, discontinous groups and Riemann surfaces. Princeton University Press, Princeton 1974

[B-G-S 1987] Bismut, J.M., Gillet, H., Soulé, Ch.: Analytic torsion and holomorphic determinant bundles, I, II, III. Comm. Math. Phys. **115** (1987) 49–87, 79–126, 301–351

[CA 1960] Calabi, E.: On compact Riemann manifolds with constant curvature, I. AMS Proc. Symp. Pure Math. **III** (1960) 155–180

[CA 1979] Calabi, E.: Extremal Kähler metrics. In: Yau, S.T. (ed.) Seminars on differential geometry. Princeton 1979

[CA 1985] Calabi, E.: Extremal Kähler metrics II. In: Cheval, I., Farkas, H.M. (eds.) Differential geometry and complex analysis, dedicated to E. Rauch. Springer, Berlin Heidelberg 1985, pp. 259–290

[C-S 1989] Campana, F., Schumacher, G.: A geometric algebraicity property for moduli spaces of compact Kähler manifolds with $h^{2,0} = 1$. To appear in Math. Z.

[DO 1987] Donaldson, S.K.: Infinite determinants, stable bundles and curvature. Duke Math. J. **54** (1987) 231–247

[E-K 1974] Earle, C.J., Kra, I.: On holomorphic mappings between Teichmüller spaces. In: Ahlfors, L., Kra, I., Maskit, B., Nirenberg, L., (eds.). Contributions to Analysis. New York London 1974

[F-N] Fenchel, W., Nielsen, J.: J. Discontinous groups of non-Euklidean motions. Unpublished manuscript

[F-T 1984a] Fischer, A.E., Tromba A.J.: On a purely Riemannian proof of the structure and dimension of the unramified moduli space of a compact Riemann surface. Math. Ann. **267** (1984) 311–345

[F-T 1984b] Fischer, A.E., Tromba A.J.: On the Weil-Petersson metric on Teichmüller space. Trans. AMS **284** (1984) 311–345

[F-T 1987] Fischer, A.E., Tromba, A.J.: A new proof that Teichmüller's space is a cell. Trans. AMS **303** (1987) 257–262

[F-K 1972] Forster. O., Knorr, K.: Relativ-analytische Räume und die Kohärenz von Bildgarben. Invent. math. **16** (1972) 113–160

[F-K 1926] Fricke, R., Klein, F.: Vorlesungen über die Theorie der automorphen Funktionen. Leipzig 1926

[FU 1978] Fujiki, A.: On automorphism groups of compact Kähler manifolds. Invent. math **44** (1978) 226–258

[FU 1981] Fujiki, A.: A theorem on bimeromorphic maps of Kähler manifolds and its applications. Publ. RIMS Kyoto **17** (1981) 735–754

[FU 1984] Fujiki, A.: Coarse moduli spaces for polarized Kähler manifolds. Publ. RIMS, Kyoto **20** (1984) 977–1005

[F-S 1988a] Fujiki, A., Schumacher, G.: The moduli space of Kähler structures on a real symplectic manifold. Publ. RIMS, Kyoto **24** (1988) 141–168

[F-S 1988b] Fujiki, A., Schumacher, G.: The moduli space of extremal, compact Kähler manifolds and generalized Weil-Petersson metrics. Preprint 1988, 157 p., to appear in Publ. RIMS, Kyoto

[G-H 1988] Gerritsen, L., Herrlich, F.: The extended Schottky space. J. reine angew. Math. **389** (1988) 190–208

[G-R 1954] Gerstenhaber, M., Rauch, H.E.: On extremal quasi-conformal mappings, I, II. Proc. Nat. Acad. Sci. **40** (1954) 808–812, 991–994

[GR 1928] Grötzsch, H.: Über die Verzerrung bei schlichten nichtkonformen Abbildungen und über eine damit zusammenhängende Erweiterung des Picardschen Satzes. Leipz. Ber. **80** (1928)

[GR 1961] Grothendieck, A.: Technique de construction en géométrie analytique. Sém. Cartan no. 7–17 (1960/61)

[HA 1983] Harer, J.: The second homology group of the mapping class group of an orientable surface. Invent. math. **72** (1983) 221–231

[H-M 1982] Harris, J., Mumford, D.: On the Kodaira dimension of the moduli space of curves. Invent. math. **67** (1982) 23–86

[HA 1984] Harris, J.: On the Kodaira dimension of the moduli space of curves, II. The even-genus case. Invent. math. **75** (1984) 437–466

[HE 1989] Herrlich, F.: The extended Teichmüller space. To appear in Math. Z.

[J-Y 1987] Jost, J., Yau, S.T.: On the rigidity of certain discrete groups and algebraic varieties. Math. Ann. **278** (1987) 481–496

[JO 1989] Jost, J.: Harmonic maps and curvature computations in Teichmüller theory. Preprint 1989

[KE 1971] Keen, L.: On Fricke moduli. Ann. Math. Studies **66** (1971) 205–224

[K-M 1976] Knudsen, F., Mumford, D.: The projectivity of the moduli space of stable curves, I: Preliminaries on "det" and "div". Math. Scand. **39** (1976) 19–55

[KN 1983a] Knudsen, F.: The projectivity of the moduli space of stable curves, II: The stacks $M_{g,n}$. Math. Scand. **52** (1983) 161–199

[KN 1983b] Knudsen, F.: The projectivity of the moduli space of stable curves, III: The line bundles on $M_{g,n}$ and a proof of the projectivity of $\overline{M_{g,n}}$ in characteristic 0. Math. Scand. **52** (1983) 200–212

[KO 1983] Koiso, N.: Einstein metrics and complex structure. Invent. math. **73** (1983) 71–106

[KR 1989] Kra, I.: Horocyclic coordinates for Riemann surfaces and moduli spaces, I: Teichmüller and Riemann spaces of Kleinian groups. Preprint 1989

[LI 1959] Lichnerowicz, A.: Isométrie et transformations analytique d'une variété Kählerienne compacte. Bull. Soc. Math. France **87** (1959) 427–437

[LI 1978] Liebermann, P.: Compactness of the Chow Scheme: application to automorphisms and deformations of Kähler manifolds. Sém Norguet. (Lecture Notes in Mathematics, vol. 670.) Springer, Berlin Heidelberg 1978

[MT 1977] Maskit, B.: Decomposition of certain Kleinian groups. Acta math. **130** (1977) 63–82

[M 1957] Matsushima, Y.: Sur la structure du groupe d'homomorphismes analytique d'une certaine variété Kählerienne. Nagoya Math. J. **11** (1957) 145–150

[MA 1976] Mazur, H.: The extension of the Weil-Petersson metric to the boundary of Teichmüller space. Duke Math. J. **43** (1976) 623–635

[MU 1977] Mumford, D.: Stability of projective varieties. L'enseign. math. **23** (1977) 39–11

[PE 1949] Petersson, H.: Über die Berechnung der Skalarprodukte ganzer Modulformen. Comment. Math. Helv. **22** (1949) 168–199

[PO 1977] Popp, H.: Moduli theory and classification theory of algebraic varieties. (Lecture Notes in Mathematics, vol. 620). Springer, Berlin Heidelberg 1977

[RB 1968] Richberg, R.: Stetig, streng pseudokonvexe Funktionen. Math. Ann. **175** (1968) 257–286

[RE 1985] Reich, E.: On the variational principle of Gerstenhaber and Rauch. Ann. Acad. Sci. Fenn. Ser. A I Math. **10** (1985) 469–475

[RI 1857] Riemann, B.: Theorie der Abel'schen Functionen. Borchardt's Journal für reine und angewandte Mathematik, Bd. 54 (1857)

[RO 1971] Royden, H.L.: Automorphisms and isometries of Teichmüller space. Advances in the theory of Riemann surfaces, Stony Brook, 1969. Ann. Math. Studies **66** (1971)

[RO 1974a] Royden, H.L.: Invariant metrics on Teichmüller space. In: Ahlfors, L., Kra, I., Maskit, B., Nirenberg, L. (eds.) Contributions to Analysis. New York London, 1974

[RO 1974b] Royden; H.L.: Intrinsic metrics on Teichmüller space. Proc. Int. Cong. Math. **2** (1974) 217–221

[SAI 1988] Saito, Kyoji: Moduli space for Fuchsian groups. Alg. Analysis II (1988) 735–787

[SA 1978] Sampson, J.H.: Some properties and applications of harmonic mappings. Ann. Sci. École Norm. Sup. **4** (1978) 211–228

[S 1972] Schneider, M.: Halbstetigkeitssätze für relativ analytische Räume. Invent. math. **16** (1972) 161–176

[S-Y 1978] Schoen, R., Yau, S.T.: On univalent harmonic maps between surfaces. Invent. math **44** (1978) 265–278

[SCH 1983] Schumacher, G.: Construction of the coarse moduli space of compact polarized Kähler manifolds with $c_1 = 0$. Math. Ann. **264** (1983) 81–90

[SCH 1984] Schumacher, G.: Moduli of polarized Kähler manifolds. Math. Ann. **269** (1984) 137–144

[SCH 1986] Schumacher, G.: Harmonic maps of the moduli space of compact Riemann surfaces. Math. Ann. **275** (1986) 466–466

[SCH 1990] Schumacher, G.: A remark on the automorphisms of the moduli space $\mathscr{M}_p$ of compact Riemann surfaces. Preprint 1990

[SI 1980] Siu, Y.T.: The complex analyticity of harmonic maps and the strong rigidity of compact Kähler manifolds. Ann. Math. **112** (1980) 73–111

[SI 1986] Siu, Y.T.: Curvature of the Weil-Petersson metric in the moduli space of Kähler-Einstein space of negative first Chern class. Aspect of Math. **9**. Vieweg, Braunschweig Wiesbaden 1986, pp. 261–298

[SI 1987] Siu, Y.T.: Lectures on Hermitian-Einstein metrics for stable bundles and Kähler-Einstein metrics. Birkhäuser, Basel Boston 1987

[TE 1939] Teichmüller, O.: Extremale quasikonforme Abbildungen und quadratische Differentiale. Preuß. Akad. math. Wiss., nat. Kl. **22** (1939) 1–197

[TE 1943] Teichmüller, O.: Bestimmung der extremalen quasikonformen Abbildungen bei geschlossenen Riemannschen Flächen. Preuß. Akad. math. Wiss., nat. Kl. **4** (1943) 1–42

[TE 1944] Teichmüller, O.: Veränderliche Riemannsche Flächen. Deutsche Math. **7** (1944) 344–359

[TE 1982] Teichmüller, O.: Gesammelte Abhandlungen. Collected papers. (Ahlfors, L.V. and Gehring, F.W., eds.). Springer, Berlin Heidelberg 1982

[TI 1987] Tian, G.: On Kähler-Einstein metrics on certain Kähler manifolds with $C_1(M) < 0$. Invent. math. **89** (1987) 225–246

[T-Y 1987] Tian, G., Yau, S.T.: Kähler-Einstein metrics on complex surfaces with $C_1 > 0$. Comm. Math. Phys. **112** (1987) 175–203

[TO 1988a] Todorov, A.: The Weil-Petersson geometry of the moduli space of $SU(n \geq 3)$ (Calabi-Yau) manifolds I. (Preprint)

[TO 1988b] Todorov, A.: Weil-Petersson geometry of Teichmüller space of Calabi-Yau manifolds II. (Preprint)

[TR 1986] Tromba, A.J.: On a natural affine connection on the space of almost complex structures and the curvature of the Teichmüller space with respect to its Weil-Petersson metric. Man. math. **56** (1986) 475–497

[TR 1987] Tromba, A.J.: On an energy function for the Weil-Petersson metric on Teichmüller space. Man. math. **59** (1987) 249–260

[VA 1984] Varouchas, J.: Stabilité de la class des variétés Kählériennes par certaines morphismes propres. Invent. math. **77** (1984) 117–127

[VA 1989] Varouchas, J.: Kähler spaces and proper open morphisms. Math. Ann. **283** (1989) 13–52

[W 1958] Weil, A.: On the moduli of Riemann surfaces. Coll. Works [1958b] Final report on contract AF 18(603)-57; Coll. Works [1958c] Module des surfaces de Riemann, Séminaire Bourbaki, no. 168 (1958)

[W 1989] Wolf, M.: The Teichmüller theory of harmonic maps. J. Diff. Geom. **29** (1989) 449–479

[WO 1983] Wolpert, S.: On the homology of the moduli space of stable curves. Ann. Math. **118** (1983) 491–523

[WO 1985a] Wolpert, S.: On the Weil-Petersson geometry of the moduli space of curves. Am. J. Math. **107** (1985) 969–997

[WO 1985b] Wolpert, S.: On obtaining a positive line bundle from the Weil-Petersson class. Am. J. Math. **107** (1985) 1485–1507

[WO 1986] Wolpert, S.: Chern forms and the Riemann Tensor for the moduli space of curves. Invent. math. **85** (1986) 119–145